ORCHID BIOTECHNOLOGY II

ORCHID BIOTECHNOLOGY II

Edited by

Wen-Huei Chen
National Cheng Kung University, Taiwan

Hong-Hwa Chen
National Cheng Kung University, Taiwan

World Scientific

NEW JERSEY • LONDON • SINGAPORE • BEIJING • SHANGHAI • HONG KONG • TAIPEI • CHENNAI

Published by

World Scientific Publishing Co. Pte. Ltd.
5 Toh Tuck Link, Singapore 596224
USA office: 27 Warren Street, Suite 401-402, Hackensack, NJ 07601
UK office: 57 Shelton Street, Covent Garden, London WC2H 9HE

Library of Congress Cataloging-in-Publication Data
Orchid biotechnology II / edited by Wen-Huei Chen, Hong-Hwa Chen.
p. cm.
Includes bibliographical references and index.
ISBN-13: 978-981-4327-92-3 (hardcover : alk. paper)
ISBN-10: 981-4327-92-1 (hardcover : alk. paper)
1. Orchids--Biotechnology. I. Chen, Wen-Huei. II. Chen, Hong-Hwa.
QK495.O64O553 2011
635.9'344--dc22

2011003152

British Library Cataloguing-in-Publication Data
A catalogue record for this book is available from the British Library.

Typeset by Stallion Press
Email: enquiries@stallionpress.com

Printed in Singapore by Mainland Press Pte Ltd.

Foreword

Taiwan is one of the world-renowned exporting countries of orchid, not only because the climate congenial to the growth of orchid, but also its rich resource of native varieties and the advances in breeding techniques to breed more beautiful and elegant new varieties. While Taiwan's orchid industry faces competition from all over the world, modern biotechnologies are certainly the key to the success in maintaining a leading role in the global orchid market. Along with the sound development of agricultural business in general, numerous scientists in Taiwan have devoted themselves to orchid research for years. How to consolidate the academic strength in orchid basic and applied research is one of the main concerns of national agricultural biotechnology.

National Science and Technology Program for Agricultural Biotechnology (NSTP.AB), a joint program for the National Science Council (NSC), Council of Agriculture (COA) and Academia Sinica (AS), has supported seven items of agricultural researches for ten years from 1998 to 2008. These included floral and ornamental plants, plant protection, vaccines for animals and fishes, plant useful genes, post-harvest, marine aquaculture, medicinal plants and health food. To promote the output and application of the research, the Development Program of Industrialization for Agricultural Biotechnology (DPIAB) was lunched in 2009, and I am honored to have served as the Director for DPIAB since then. The main focus of DPIAB includes aquaculture germplasm, floral germplasm, live stock germplasm, safe agriculture, animal vaccine, medicinal herbs and health food, microbial products, molecular farming and molecular animal farming. The financial support for DPIAB extend from the original sponsors of NSC, COA, AS, to the Departments of Education, Economy, and Health. All these efforts will bring better joint cooperative programs between academia and industrial sectors and to build a solid biotechnology-based agriculture in Taiwan. Since the third stage of NSTP.AB from 2004, orchid has been selected as one of the main target crops to develop floral and ornamental plants. Orchid research has remained high priority in the DPIAB to produce better quality orchid germplasm as well as to have systemic integration of key technology for orchid cultivation and exportation. It is now the right time to review past achievements of the orchid researchers and to build the foundation for future development of Taiwan's orchid industry.

In 2007, Drs. Wen-Huei Chen and Hong-Hwa Chen published the first volume of "Orchid Biotechology", which comprised fundamental knowledge about breeding new varieties, *in vitro* morphogenesis, somaclonal variation, and analysis of orchid genomes. In response to rapid development of agricultural biotechnology, especially transgenic strategies in these years, they further compiled papers from seventeen orchid research frontiers in Taiwan to publish Orchid Biotechnology II.

Hopefully for Taiwan, as well as the global orchid industry and researchers, all can benefit enormously from this book and be brought to a bright new era for the floral advancement internationaly.

Jen-Leih Wu
Distinguished Research Fellow
Institute of Cellular and Organismic Biology
Academic Sinica, Taiwan

Director
Development Program of Industrialization for
Agricultural Biotechnology, Taiwan

Preface

The orchid production in the world has increased 12%–18% every year since 2000. The demand of the orchid pot-flower is enormous in Japan, Europe and the United States of America. It is the top seller among all pot-flowers in the Netherlands and Japan. The *Phalaenopsis* is the main orchid in the world market. It had 66% share of the orchid market in FlorHolland in 2006. Taiwan's orchid industry has been recognized by the international community as a prosperous one in terms of sales volume, average of orchid plantation, orchid growing technologies, and financial resources. *Phalaenopsis* seedling in Taiwan has maintained to be the main product for domestic and export markets since 1995. It has been selected as one of the four flaghsip export products for Taiwan by the Agriculture Product Competition Module in Council of Agriculture since 2004.

From the technology side, orchid research has long been supported by the National Science and Technology Program for Agriculture Biotechnology for ten years. Recently, the National Development Program of Industrialization for Agricultural Biotechnology has been initiated in 2009 to bring Taiwan orchid industry to a new level of more profitability. The authors of this book have collaborated and dynamically interacted with local orchid growers to develop orchid biotechnology for the application of orchid industry in Taiwan.

To compile the results from Taiwan's orchid research teams, the first Orchid Biotechnology was published by World Scientific Publishing Co. in 2007. This effort not only benefits researchers but also is useful for education of young students as well as orchid growers. In addition, it enhances the visibility and the soft strength of Taiwan's orchid research capability. The current book is the second volume dedicated enormously to orchid biotechnology, mainly for *Phalaenopsis* and *Oncidium*. It includes eighteen chapters from seventeen orchid research groups. The contents cover molecular phylogeny and biogeography, endopolyploidy and its application in breeding, growth responses to orchid mycorrhizae, a practical method for *Phalaenopsis* cultivation, inflorescence development, flower color, floral scent, flower development, ovule developmental, chloroplast genome and chloroplast RNA editing, early response of *Erwinia* infection, and orchid viruses infection. Its publication would be tremendously helpful for readers such as scientists, graduate students, orchid growers interested in orchid biology, and orchid biotechnology.

The Editors
Wen-Huei Chen
Orchid Research Center
National Cheng Kung University
Taiwan

Hong-Hwa Chen
Department of Life Sciences
Institute of Tropical Plant Sciences
Orchid Research Center
National Cheng Kung University,
Taiwan

Contents

CHAPTER 1

Molecular Phylogeny and Biogeography of *Phalaenopsis* Species

Chi-Chu Tsai*,†

Christenson (2001) treated *Kingidium* and *Doritis* as synonyms of *Phalaenopsis*, and divided the latter into five subgenera, *Phalaenopsis, Polychilos, Parishianae, Proboscidioides* and *Aphyllae*. Molecular techniques are used to clarify the phylogeny of *Phalaenopsis*. Based on the internal transcribed spacer (ITS) of ribosomal DNA, *Kingidium* and *Doritis*, as well as *Polychilos* could be treated as synonyms of *Phalaenopsis*, as suggested by Christenson (2001). However, the phylogeny of the subgenera *Phalaenopsis, Polychilos* and *Aphyllae* was not supported by evidence in nature. According to the geographical distribution of the genus *Phalaenopsis*, the phylogenetic relationship of the genus *Phalaenopsis* is basically congruent with its biogeographies. In addition, there were two evolutionary trends of *Phalaenopsis* emanating from its center of origin to Southeast Asia. In one trend, the *Phalaenopsis* species dispersed and evolved from southern China to Southeast Asia, while in the other, the *Phalaenopsis* species dispersed and evolved from southern China to Southeast Asia. Furthermore, the molecular phylogenetics and biogeography of four species complexes, including *P. amabilis* complex, *P. sumatrana, P. violacea* complex, and *P. lueddemanniana* complex were also described based on molecular evidence.

1.1. Introduction

Genus *Phalaenopsis* Blume (Orchidaceae), a beautiful and popular orchid, comprises approximately 66 species, according to the latest classification (Table 1.1). Species of the genus *Phalaenopsis* are found

*Kaohsiung District Agricultural Improvement Station, Pingtung 900, Taiwan.
†National Pingtung University of Science and Technology, Pingtung 912, Taiwan.
E-mail: tsaicc@mail.kdais.gov.tw

Table 1.1 Taxon Names, Systematics and Geographical Distribution

Taxa and Systematic Classification[a]	Geographical Distribution
Subgenus *Proboscidioides* (Rolfe) E. A. Christ.	
Phalaenopsis lowii Rchb.f.	Myanmar, adjacent to western Thailand
Subgenus *Aphyllae* (Sweet) E. A. Christ.	
Phalaenopsis braceana (J. D. Hook.) E. A. Christ.	Bhutan and China at 1,100–1,700 m elevation
Phalaenopsis hainanensis Tang and Wang	China (Hainan, Yunnan)
Phalaenopsis honghenensis F. Y. Liu	China (Yunnan)
Phalaenopsis minus (Seidenf.) E. A. Christ.	Endemic to Thailand
Phalaenopsis taenialis (Lindl.) E. A. Christ. and U. C. Pradhan	Nepal, northeast India, Bhutan, Myanmar, and Thailand at 1,000–2,500 m elevation
Phalaenopsis wilsonii Rolfe	China (Szechuan, Yunnan) and eastern Tibet at 800–2,200 m elevation
Subgenus *Parishianae* (Sweet) E. A. Christ.	
Phalaenopsis appendiculata C. E. Carr	Endemic to Malaysia (Malay Peninsula)
Phalaenopsis gibbosa Sweet	Vietnam and Laos
Phalaenopsis lobbii (Rchb.f.) Sweet	India, Bhutan, Myanmar and Vietnam
Phalaenopsis parishii Rchb.f.	Eastern Himalayas, India, Myanmar and Thailand
Subgenus *Polychilos* (Breda) E. A. Christ.	
Section *Polychilos* (Breda) Rchb.f.	
Phalaenopsis borneensis Garay	Endemic to Borneo
Phalaenopsis cornu-cervi (Breda) Bl. and Rchb.f.	Northeast India and the Nicobar Islands to Java and Borneo from sea level to 800 m elevation
Phalaenopsis lamelligera Sweet	
Phalaenopsis mannii Rchb.f.	Northeast India, Nepal and China to Vietnam at 500–1,400 m elevation
Phalaenopsis pantherina Rchb.f.	Endemic to Borneo from sea level to 800 m elevation
Section *Fuscatae* Sweet	
Phalaenopsis cochlearis Holtt.	Malaysia (Malay Peninsula) and Indonesia (Sarawak) at 500–700 m elevation
Phalaenopsis fuscata Rchb.f.	Malaysia (Malay Peninsula), Borneo (West Koetai)

Table 1.1 (*Continued*)

Taxa and Systematic Classification[a]	Geographical Distribution
Phalaenopsis kunstleri J. D. Hook.	Myanmar and Malay Peninsula
Phalaenopsis viridis J. J. Sm.	Endemic to Indonesia (Sumatra) at 1,000 m elevation
Section *Amboinenses* Sweet	
Phalaenopsis ×*gersenii* (Teijsm. and Binn.) Rolfe	Indonesia (Sumatra)
Phalaenopsis ×*singuliflora* J. J. Sm.	Western Borneo
Phalaenopsis amboinensis J. J. Sm.	Indonesia (Molucca Islands and Sulawesi)
Phalaenopsis bastianii Gruss and Rollke	Endemic to the Philippines
Phalaenopsis bellina (Rchb.f.) E. A. Christ.	Malaysia (Malay Peninsula) and East Malaysia (Sarawak)
Phalaenopsis doweryensis Garay and E. A. Christ.	East Malaysia, Sabah, without precise locality
Phalaenopsis fasciata Rchb.f.	Endemic to the Philippines (Luzon, Bohol and Mindanao)
Phalaenopsis fimbriata J. J. Sm.	Indonesia (Java, Sarawak and Sumatra)
Phalaenopsis floresensis Fowlie	Endemic to the island of Flores at 300–500 m
Phalaenopsis gigantea J. J. Sm.	Endemic to Sabah in East Malaysia and adjacent to Kalimantan Timur, from sea level to 400 m elevation
Phalaenopsis hieroglyphica (Rchb.f.) Sweet	Endemic to the Philippines
Phalaenopsis javanica J. J. Sm.	Endemic to Indonesia (Java)
Phalaenopsis lueddemanniana Rchb.f.	Endemic to the Philippines
Phalaenopsis maculata Rchb.f.	Malaysia (Pahang), East Malaysia (Sabah, Sarawak), Indonesia (Kalimantan Timur)
Phalaenopsis mariae Burb.ex Warn. and B. S. Wms.	Endemic to the Philippines and Indonesia (Kalimantan, Borneo)
Phalaenopsis micholitzii Rolfe	The Philippines (Mindanao)
Phalaenopsis modesta J. J. Sm.	Endemic to the island of Borneo in East Malaysia (Sabah) and Indonesia (Kalimantan)
Phalaenopsis pallens (Lindl.) Rchb.f.	Endemic to the Philippines
Phalaenopsis pulchra (Rchb.f.) Sweet	Endemic to the Philippines (Luzon, Leyte) at 100–650 m elevation
Phalaenopsis reichenbachiana Rchb.f. and Sander	Endemic to the Philippines
Phalaenopsis robinsonii J. J. Sm.	Endemic to Indonesia (Ambon)

Table 1.1 (*Continued*)

Taxa and Systematic Classification[a]	Geographical Distribution
Phalaenopsis venosa Shim and Fowlie	Endemic to Indonesia (Sulawesi)
Phalaenopsis violacea Witte	Indonesia (Sumatra) and Malaysia (Malay Peninsula)
Section *Zebrinae* Pfitz.	
Phalaenopsis corningiana Rchb.f.	Borneo (Sarawak and elsewhere on the island)
Phalaenopsis inscriptiosinensis Fowlie	Endemic to Indonesia (Sumatra)
Phalaenopsis speciosa Rchb.f.	Endemic to India (Andaman and Nicobar Islands)
Phalaenopsis sumatrana Korth. and Rchb.f.	Widespread from Myanmar, Thailand, Vietnam, to Indonesia (Java, Sumatra), Malaysia (Perak, Johore), East Malaysia (Sabah), and the Philippines (Palawan)
Phalaenopsis tetraspis Rchb.f.	India (Andaman and Nicobar Islands) and Indonesia (Sumatra)
Phalaenopsis zebrina Witte	Borneo, the Philippines (Palawan)
Subgenus *Phalaenopsis* E. A. Christ.	
Section *Phalaenopsis* Benth.	
Phalaenopsis amabilis	Widespread from Sumatra and Java to the southern Philippines, and east to New Guinea and Queensland, Australia
Phalaenopsis aphrodite Rchb.f.	The Northern Philippines and southeastern Taiwan
Phalaenopsis intermedia Lindl.	The Philippines
Phalaenopsis philippinensis Golamco ex Fowlie and Tang	Endemic to the Philippines
Phalaenopsis sanderiana Rchb.f.	Endemic to the Philippines
Phalaenopsis schilleriana Rchb.f.	Endemic to the Philippines
Phalaenopsis stuartiana Rchb.f.	Endemic to the island of Mindanao in the southern Philippines
Phalaenopsis ×*amphitrite* Kraenzl.	Endemic to the Philippines
Phalaenopsis ×*leucorrhoda* Rchb.f.	Endemic to the southern Philippines
Phalaenopsis ×*veitchiana* Rchb.f.	Endemic to the Philippines
Section *Deliciosae* E. A. Christ.	
Phalaenopsis chibae Yukawa	Endemic to Vietnam at 400–600 m elevation
Phalaenopsis deliciosa Rchb.f.	Widespread from Sri Lanka and India to the Philippines and Sulawesi
Phalaenopsis mysorensis Saldanha	Endemic to southern India

Table 1.1 (*Continued*)

Taxa and Systematic Classification[a]	Geographical Distribution
Section *Esmeralda* Rchb.f.	
Phalaenopsis buyssoniana Rchb.f.	Thailand and Indochina
Phalaenopsis pulcherrima (Lindl.) J. J. Sm.	Widespread from northeast India and southern China throughout Indochina to Malaysia (Malay Peninsula), Indonesia (Sumatra), and East Malaysia (Sabah)
Phalaenopsis regnieriana Rchb.f.	Endemic to Thailand
Section *Stauroglottis* (Schauer) Benth.	
Phalaenopsis celebensis Sweet	Endemic to Indonesia (Sulawesi)
Phalaenopsis equestris (Schauer) Rchb.f.	The Philippines and Taiwan
Phalaenopsis lindenii Loher	Endemic to the Philippines

[a]The classification of *Phalaenopsis* based on Christenson (2001).

throughout tropical Asia, namely southern China, Indochina, India and Southeast Asia, and in Australia. The western distribution of *Phalaenopsis* is in Sri Lanka and southern India. The eastern limit of the range is in Papua New Guinea. To the north, they are distributed in Yunnan Province (southern China) and Taiwan. The southern limit is in northern Australia (Christenson, 2001). Traditonally, the systematics of *Phalaenopsis* is confused with those of both genera of *Kingidium* and *Doritis*. A systematics of genus *Phalaenopsis* introduced by Sweet (1980) still treated *Doritis* and *Kingidium* as separate genera from *Phalaenopsis*. According to the latest systematics of *Phalaenopsis* suggested by Christenson (2001), *Kingidium* and *Doritis* are treated as synonyms of *Phalaenopsis*, which is divided into five subgenera, namely *Aphyllae, Parishianae, Proboscidioides, Phalaenopsis* and *Polychilos*.

Different subgenera of *Phalaenopsis* have distinct geographic distributions. The subgenera *Aphyllae, Parishianae* and *Proboscidioides* are distributed in southern China and India, extending to northern Vietnam, Myanmar and Thailand, respectively. The subgenus *Polychilos* has a few species distributed as far west as northeastern India, but it is primarily centered in Indonesia and the Philippines (Christenson, 2001). Furthermore, the subgenus *Phalaenopsis* is centered in the Philippines with two species extending to Taiwan (*P. aphrodite* subsp.

formosana and *P. equestris*) and one wide-ranging species (*P. amabilis*) found from the Philippines and Indonesia to northern Australia (Christenson, 2001). Furthermore, as we know, *Phalaenopsis* is only distributed in tropical Asia, namely southern China, Indochina, India, and Southeast Asia (including Malaysia, Indonesia, the Philippines, New Guinea, etc.) (Christenson, 2001). In these regions, Southeast Asia in particular, there have been numerous biogeographical studies on flora or fauna. Since Southeast Asia was created from the collisions between several oceanic plates (Pacific, Indian, Philippine) and land plates (Eurasian, Indian, Australian) (Hall, 1996), the flora and fauna of these regions are highly complicated and have characteristics shared between Asia and Australia. Several biogeographical boundaries have been introduced in these regions, such as the Wallace Line and Weber's Line, since deep straits limit the dispersal of flora and fauna (Van Oosterzee, 1997; Pianka, 1994).

1.2. Biogeographical Pattern of Genus *Phalaenopsis*

In the late 19th century, six major biogeographic regions in the world were recognized, namely Palaearctic, Nearctic, Neotropical, Ethiopian, Oriental, and Australian regions (Pianka, 1994) (Fig. 1.1). Lyell suggested that these six regions were separated based on dispersal barriers such as oceans and high mountain ranges (Van Oosterzee, 1997). In the Oriental region, there are several deep straits in Southeast Asia that separate the Oriental from the Australian region. On the other hand, the Himalayas divides the Oriental and the Palaearctic region (Pianka, 1994). All of the distribution regions of the genus *Phalaenopsis*, namely southern China, India, Indochina, Southeast Asia (Malaysia, Indonesia and the Philippines) (Sweet, 1980; Christenson, 2001) belong to the Oriental region. Southeast Asia was created from the collisions between several oceanic plates (Pacific, Indian, Philippine) and land plates (Eurasian, Indian, Australian). The process may already have started in the early Palaeozoic (about 400 Mya) (Hall, 1996). Furthermore, one widespread species of *Phalaenopsis*, *P. amabilis*, extended to the Australian region, namely New Guinea and northern Australia.

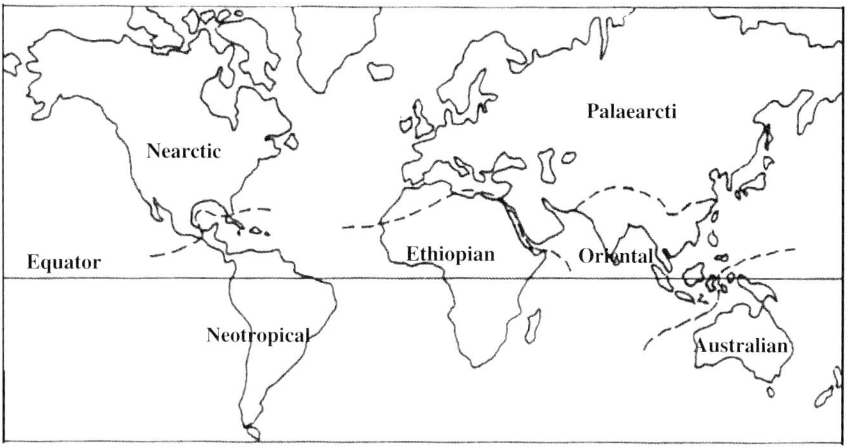

Fig. 1.1. The world's six major biogeographical regions. [Redrawn from Pianka (1994).]

Within the distributions of genus *Phalaenopsis*, there are ten different geographical regions, namely India, Sri Lanka, southern China, Indochina, the Malay Peninsula, Borneo, Sumatra, Java, Flores, Sulawesi, the Moluccan Islands, New Guinea, the Philippines, Taiwan, and Australia. According to the distribution matrix of the *Phalaenopsis* species (Tsai, 2003a), a biogeographical tree has been constructed based on the neighbor-joining method (Fig. 1.2). In that tree, New Guinea and Australia form a group. The second group includes Sulawesi and the Moluccas. The third includes Taiwan and the Philippines; the fourth, the Malay Peninsula, Borneo, Sumatra, Java and Flores; and the fifth, India and Indochina. Furthermore, southern china and Sri Lanka are in independent groups (Fig. 1.2). This scheme is in agreement with biogeographical regions of the world (Pianka, 1994), since the first group belonging to the Australian region is not nested within the groups of the Oriental region (Figs. 1.1 and 1.2). Furthermore, the group comprising the Malay Peninsula, Borneo, Sumatra and Java belonging to Sunda Shelf show the close relationship between one another based on the biogeographical tree. This is in agreement with the historical geology of these regions, which were interconnected during Pleistocene times (about 0.01~1.8 Mya)

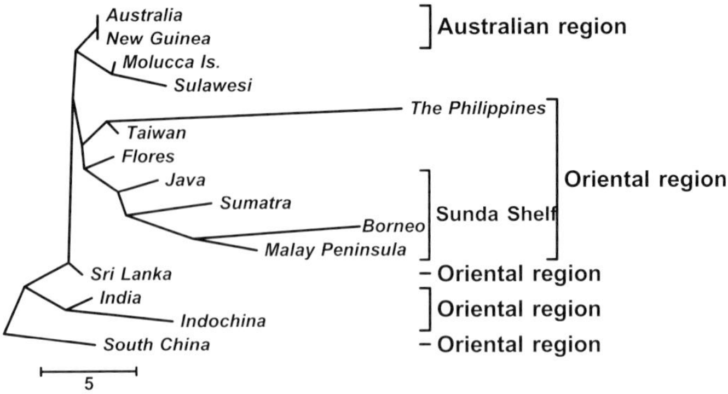

Fig. 1.2. Biogeographical tree of the genus *Phalaenopsis* constructed using the Neighbor-joining method.

(Van Oosterzee, 1997). This allowed easy species migration among these regions. In addition, the high divergence of species between Sulawesi and Borneo is also in agreement with the historical geology of both regions, as they were isolated from each other at least 14 Mya or 50 Mya by the Makassar Strait (Morley, 1998). Studies of the flora and fauna in both regions led to the introduction of the Wallace Line in the late 19th century (Van Oosterzee, 1997). However, the biogeography of the genus *Phalaenopsis* was not in agreement with the natural barriers of Weber's Line between Sulawesi and the Moluccan Islands, since the biogeographical relationship of the *Phalaenopsis* species distributed in both those regions were close (Fig. 1.2).

1.3. Molecular Phylogeny and Biogeography of the *Phalaenopsis* Species

1.3.1. *Phylogenetics and biogeography of the genus Phalaenopsis*

Based on the number of pollinia, traditional species of *Kingidium* (which have four pollinia) have been distinguished from the genus *Phalaenopsis* (which has two pollinia) (Sweet, 1980; Seidenfaden, 1988b). Christenson (2001) treated the traditional genus *Kingidium* as

Phalaenopsis and split it into different parts of *Phalaenopsis*, placing some species into the subgenus *Aphyllae* (*P. braceana, P. minus* and *P. taenialis*) and others into the section *Deliciosae* (*P. chibae* and *P. deliciosa*) of the subgenus *Phalaenopsis*. Another group, the *P. parishii* complex, having four pollinia, was proposed as a separate genus, *Grafia*, by Hawkes [see Christenson (2001)]. This complex was first treated as the section *Parishianae* of the genus *Phalaenopsis* by Sweet (1968). Christenson (2001) agreed with Sweet's treatment and placed this complex in the subgenus *Parishianae* of *Phalaenopsis*. Furthermore, the other four-pollinia species, *P. lowii*, was also placed in the section *Proboscidioides* of the genus *Phalaenopsis* (Sweet, 1980) and in the subgenus *Proboscidioides* of the genus *Phalaenopsis* by Christenson (2001). Shim (1982), however, disagreed with Sweet's Concept (1980) and separated sections *Proboscidioides, Aphyllae, Parishianae, Polychilos, Zebrinae, Fuscatae* and *Amboinenses* as the genus *Polychilos* from a narrowly defined *Phalaenopsis*. In addition, the genus *Doritis* was traditionally separated from *Phalaenopsis* because of its pollinium number, lip structure, and adaptations to a terrestrial habitat (Sweet, 1980; Seidenfaden, 1988a). This group was also treated as the genus *Phalaenopsis* and placed in the section *Esmeralda* of the subgenus *Phalaenopsis* (Christenson, 2001). A molecular phylogenetic tree of genus *Phalaenopsis* was reconstructed based on the nucleotide sequences of the ITS region of the nrDNA by Tsai *et al.* (2006a) (Fig. 1.3).

1.3.1.1. *Molecular phylogenetics of Phalaenopsis*

1.3.1.1.1. Monophyly of the genus *Phalaenopsis*

Christenson (2001) presented a generic revision of *Phalaenopsis* and treated *Doritis* as a synonym of the genus *Phalaenopsis* based on the high hybrid fertility and similar morphology of microspores between *P. pulcherrima* (syn. *Doritis pulcherrima*) and parts (*P. lobbii* or *P. parishii*) of *Phalaenopsis* (Aoyama *et al.*, 1994; Christenson, 2001). In addition, the genus *Kingidium* was also treated as a synonym of the genus *Phalaenopsis* based on the fact that it shares small subsaccate lip bases with parts of *Phalaenopsis*, namely the subgenus

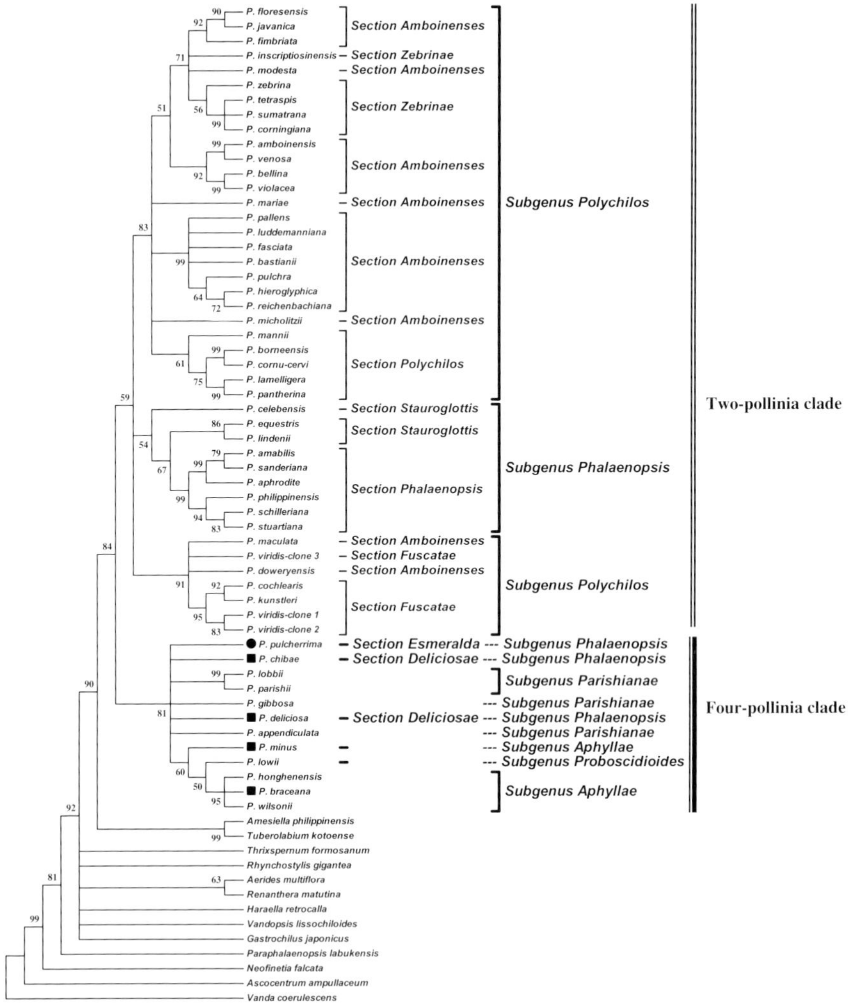

Fig. 1.3. The strict consensus parsimonious tree of 53 *Phalaenopsis* species plus 13 outgroups obtained from sequence comparisons of the ITS region of rDNA. Bootstrap values greater than 50% are shown on each branch. A solid circle (•) on the tree indicates that this species was traditionally treated as the genus *Doritis*. Solid squares (■) on the tree indicate that these species were traditionally treated as the genus *Kingidium*. [Redrawn from Tsai *et al.* (2006).]

Aphyllae, as well as having four pollinia as have parts of *Phalaenopsis*, namely the subgenera *Proboscidioides*, *Aphyllae* and *Parishianae* (Christenson, 2001). Molecular data from the ITSs of nrDNA showed that outgroup species were not nested within the genus *Phalaenopsis*. In addition, traditional *Kingidium* and *Doritis* species were not separated from *Phalaenopsis* species. Therefore, the monophyly of the genus *Phalaenopsis* as described by Christenson (2001) was supported.

1.3.1.1.2. Infrageneric relationships within *Phalaenopsis*

(i) Subgenera *Proboscidioides, Aphyllae* and *Parishianae*

Molecular data from ITS sequences showed that the monotypic subgenus *Proboscidioides*, namely *P. lowii*, formed a clade with the subgenera *Aphyllae* and *Parishianae* plus the sections *Esmeralda* and *Deliciosae* of the subgenus *Phalaenopsis*. The result is in agreement with the morphological characteristics of this clade, bearing four separate pollinia, as well as with the fact that they are geographically distributed in close proximity to one another (Table 1.1) (Sweet, 1980; Christenson, 2001). This four-pollinia clade, namely subgenera *Aphyllae* and *Parishianae*, plus the sections *Esmeralda* and *Deliciosae* of the subgenus *Phalaenopsis*, is separate from the two-pollinia clade, namely the subgenus *Polychilos* and sections *Phalaenopsis* and *Stauroglottis* of the subgenus *Phalaenopsis*, based on phylogenetic tree derived from ITS data.

Within the four-pollinia clade, neither subgenus *Aphyllae* nor *Parishianae* were a monophyletic group based on the ITS sequences. Furthermore, the monotypic subgenus *Proboscidioides*, namely *P. lowii*, had a close relationship to the subgenus *Aphyllae*. Therefore, the unique characteristics in *P. lowii*, namely the long beak-like rostellum and lateral lobes of the lip in the form of recurved hooks, might have been overemphasized in the systematics of *Phalaenopsis* proposed by Christenson (2001). In addition, section *Deliciosae* of the subgenus *Phalaenopsis* was not monophyletic either.

(ii) Subgenus *Phalaenopsis*

The monophyly of the subgenus *Phalaenopsis* was not supported based on ITS sequences. Excluding sections *Deliciosae* and *Emeralda* of the subgenus *Phalaenopsis*, the rest of the subgenus *Phalaenopsis*, namely sections *Phalaenopsis* and *Stauroglottis*, formed a monophyletic group supported by moderate bootstrap values based on the ITS sequences. The close relationship between sections *Phalaenopsis* and *Stauroglottis* was also supported by morphological characteristics sharing similar geographic ranges, small chromosome sizes, and flowers lacking transversely barred patterns (Christenson, 2001), RAPD analyses (Chen *et al.*, 1995), and the intergenic spaces (IGS) sequence of 5S rDNA (Kao, 2001).

Within subgenus *Phalaenopsis*, the monophyly of the section *Phalaenopsis* was supported in this study. The section *Phalaenopsis* bears flowers with broad petals, which are much broader than the sepals. Additionally, they bear prominent, erect, somewhat glossy calli (Christenson, 2001). Furthermore, based on geographical distributions (Sweet, 1980; Christenson, 2001), and molecular data of the IGS of 5S nrDNA (Kao, 2001), the monophyly of the section *Phalaenopsis* was supported as well. The monophyly of the section *Stauroglottis* was not supported in this study, since *P. celebensis* was unique from other members of the section *Stauroglottis*. According to the geographical distribution of this section, *P. celebensis* (distributed in Sulawesi, Indonesia) was also separated from other species of the section *Stauroglottis* (distributed in the Philippines) (Christenson, 2001). Furthermore, section *Deliciosae* was also not shown to be monophyletic based on ITS sequences.

(iii) Subgenus *Polychilos*

This large subgenus is subdivided into the four sections *Polychilos*, *Fuscatae*, *Amboinenses* and *Zebrinae* based on the systematics of Christenson (2001). The monophyly of the subgenus *Polychilos* was not supported, since parts of the subgenus *Phalaenopsis*, namely sections *Phalaenopsis* and *Stauroglottis*, were nested within subgenus *Polychilos*. Within subgenus *Polychilos*, section *Polychilos* was monophyletic. This finding is in agreement with the morphological characteristics of this section, which features a fleshy, flattened rachis (with the exception of *P. mannii*); non-fragrant flowers; petals narrower than sepals; a triseriate callus; a slightly saccate lip base; a transversely lunate mid-lobe of the lip; a lip base continuous with the column foot; and a pair of knee-like projections at the base of the column (Christenson, 2001). However, other sections of the subgenus *Polychilos*, namely *Amboinenses*, *Zebrinae* and *Fuscatae*, were not monophyletic. Based on ITS data, the section *Fuscatae* and two species of the section *Amboinenses*, namely *P. doweryensis* and *P. maculata*, formed a unique clade and were separated from other members of the subgenus *Polychilos*.

1.3.1.2. *Biogeography*

According to the evolutionary trend of the pollinia number (Holttum, 1959; Dressler, 1993), the four-pollinia clade was suggested as the basal clade of the genus *Phalaenopsis*. The four-pollinium clade of *Phalaenopsis* developed in southern China and Indochina and then dispersed into Indonesia, Malaysia and the Philippines. Thereafter, the two-pollinia clade developed. Within the two-pollinia clade, two major groups were shown, sections *Phalaenopsis* and *Stauroglottis* of subgenus *Phalaenopsis* and subgenus *Polycholis*. According to the present geographical distribution of *Phalaenopsis*, sections *Phalaenopsis* and *Stauroglottis* are only distributed in the Philippines with the exception of the widespread species, *P. amabilis* and *P. celebensis*, which are distributed in Sulawesi, and *P. aphrodite* subsp. *formosana* and *P. equestris*, which are distributed in Taiwan. In contrast, subgenus *Polychilos* is distributed in Indonesia and Malaysia with the exception of the *P. luddemanniana* complex and *P. mitcholitzii*, which are distributed in the Philippines. Therefore, the evolutionary trends of subgenus *Phalaenopsis* sections *Phalaenopsis* and *Stauroglottis* and subgenus *Polychilos* might be different.

During Pleistocene times (about 0.01∼1.8 Mya), and when sea levels were low, the Malay Peninsula, Borneo, Sumatra, Java, Bali, and various parts of the Philippines would have been interconnected. For example, the sea level was approximately 120 meters below the current level in the Last Glacial Maximum (Hanebuth *et al.*, 2000). The

Sunda Shelf connected the Thai-Malay Peninsula and Borneo, forming Sundaland (0.02 Mya) (Sathiamurthy and Voris, 2006). This would have made crossings relatively easy among these regions (Van Oosterzee, 1997). Therefore, the *Phalaenopsis* species might have dispersed from southern China and Indochina to Indonesia and Malaysia, using the Malay Peninsula as a stepping stone, from which the subgenus *Polychilos* developed. In addition, historical geology suggests that most of the Philippine islands are young (<5 Mya) with the exception of Palawan, Mindoro, Zamboanga and parts of the western Philippines (Aurelio *et al.*, 1991; Quebral *et al.*, 1994). The older islands of the Philippines, including Palawan and Mindoro, are on the margin of the Eurasian Plate and may have begun to slide away from the main mass in the middle Oligocene (~30 Mya). Therefore, the *Phalaenopsis* species might have dispersed from southern China and Indochina to the Philippines, using some older lands of the Philippines (e.g., Mindoro and Palawan) as stepping stones, from which the sections *Phalaenopsis* and *Stauroglottis* developed (Fig. 1.4).

1.3.2. *Phylogenetics and biogeography of the Phalaenopsis amabilis complex*

Tsai (2003b) examined the *Phalaenopsis amabilis* species complex, including *P. amabilis, P. amabilis* subsp. *moluccana, P. amabilis* subsp. *rosenstromii, P. aphrodite, P. aphrodite* subsp. *formosana* and *P. sanderiana*. The internal transcribed spacers 1 and 2 (ITS1 and ITS2) region of nuclear ribosomal DNA (nrDNA) was applied to reconstruct the phylogeny of this complex. Rooted at outgroups, the monophyly of the species complex was significantly supported in the neighbor-joining tree. Within accessions of *P. amabilis* and its subspecies, different locations of *P. amabilis* and its subspecies formed different separated clades with the exceptions of Palawan and Borneo populations plus the Timor population and *P. amabilis* subsp. *rosenstromii*. Furthermore, *P. aphrodite* from different locations and its subspecies could not be separated from each other, but all of them were separable from others of the *P. amabilis* complex. In addition, accessions of *P. sanderiana* were nested within accessions of both *P. amabilis* and its subspecies.

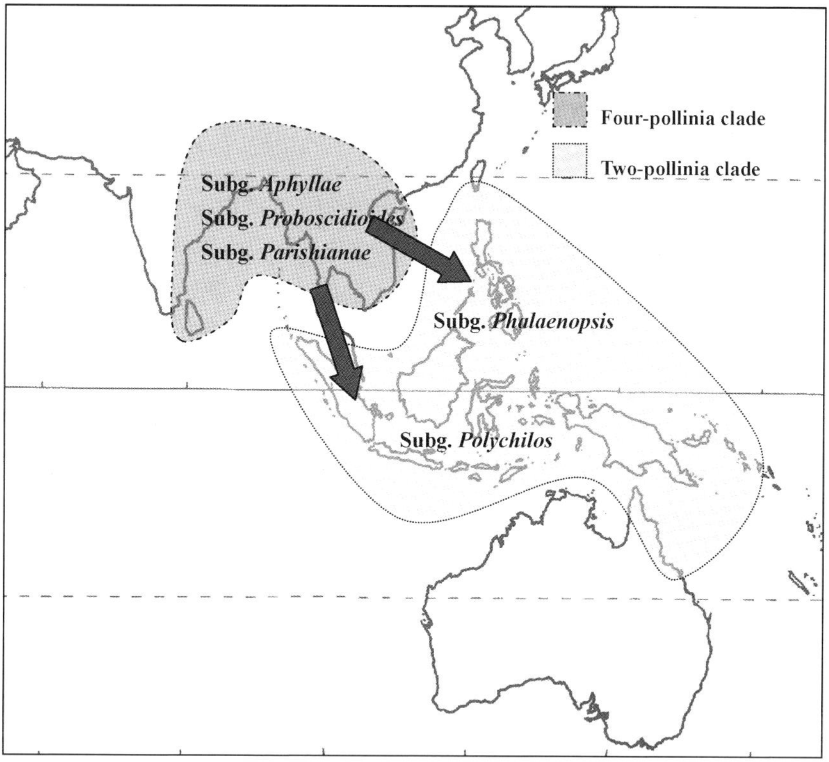

Fig. 1.4. The evolutionary trends of the genus *Phalaenopsis* obtained from this study plotted on a map of the geographical distribution of this genus.

This result does not support *P. sanderiana* being treated as a separate species from *P. amabilis*. According to the phylogenetic tree derived from the ITSs of nrDNA, the Palawan population of *P. amabilis* was suggested as being the origin group of the *P. amabilis* complex. *P. aphrodite* and *P. sanderiana* descended from *P. amabilis* (or their most recent common ancestor) were suggested. In addition, the evolutionary trend of the *P. amabilis* complex included three different lineages corresponding to three different dispersal pathways. First, *P. amabilis* distributed in Palawan dispersed into southern Mindanao and evolved into *P. sanderiana*, thereafter further dispersing into Sulawesi and New Guinea, from which *P. amabilis* subsp. *moluccana* and *P. amabilis* subsp. *rosenstromii* respectively developed. *P. amabilis* subsp. *rosenstromii* further dispersed into northern Australia and Timor. Second, *P. amabilis* distributed in Palawan dispersed into Borneo, thereafter further dispersing into the Mentawai Islands. Third, the Palawan population of *P. amabilis* dispersed into other islands of the Philippines and evolved into *P. aphrodite*. *P. aphrodite* is distributed throughout the Philippines with the exceptions of Palawan and southern Mindanao.

1.3.3. *Phylogenetics and biogeography of the Phalaenopsis sumatrana complex*

The *Phalaenopsis sumatrana* complex includes *P. sumatrana* and *P. corningiana* plus one questionable species, *P. zebrina*. Of these, *P. sumatrana* is widely distributed in Sumatra, Java, Borneo, the Mentawei Islands, Malaysia, Perak, Johor and Thailand. *P. corningiana* and *P. zebrina* are restricted to Borneo and Sarawak and to Borneo and Palawan, respectively (Sweet, 1980; Masaaki, 2002). Based on their morphology, the characteristics of these three species of this complex are not easily differentiated, even recently. *P. zebrina* was treated as a synonym of *P. sumatrana* by Sweet (1968). Fowlie (1982) attempted to raise plants of the *P. sumatrana* complex with a white ground color into a separate species, *P. zebrina*, from plants of *P. sumatrana* with a yellowish ground color. Christenson (2001), however, disagreed with Fowlie's separation based on the treatment being contrary to the type specimen of *P. zebrina* with a yellowish ground color. Therefore, *P. zebrina* is still retained as a synonym of *P. sumatrana*. Nevertheless, *P. zebrina* is still used in horticulture to represent plants with narrower brown markings on a white or yellowish ground color (Masaaki, 2002). Furthermore, Sweet (1968), Fowlie (1969), and Christenson (2001) are in agreement that *P. sumatrana* and *P. corningiana* both have species-level status, are separate from each other, and are treated as two separate species.

The molecular phylogenetics of this complex was examined based on the ITS of nrDNA and the IGS of *atp*B-*rbc*L of chloroplast DNA by Tsai *et al.* (2009) as described in the following section.

1.3.3.1. *Molecular phylogenetics*

P. sumatrana and *P. corningiana* cannot be separated from each other from the molecular phylogenetic tree derived from ITS sequences (Fig. 1.5). These results do not support *P. corningiana* and *P. sumatrana* being treated as two separate species, as described by Sweet (1968), Fowlie (1969) and Christenson (2001). However, the characteristics of the callus morphology and marking pattern on the petals could distinguish between *P. sumatrana* and *P. corningiana*, as described by Sweet (1980) and Christenson (2001). Accessions of *P. zebrina* in this complex can be separated from both *P. sumatrana* and *P. corningiana* based on the ITS sequences of nrDNA. In analyses of sequences of IGS of *atp*B-*rbc*L, neither species of this complex could be separated from the other. Therefore, the molecular data of this study only support *P. zebrina* being separated from both *P. corningiana* and *P. sumatrana*. This result supports the treatment of *P. zebrina* described by Fowlie

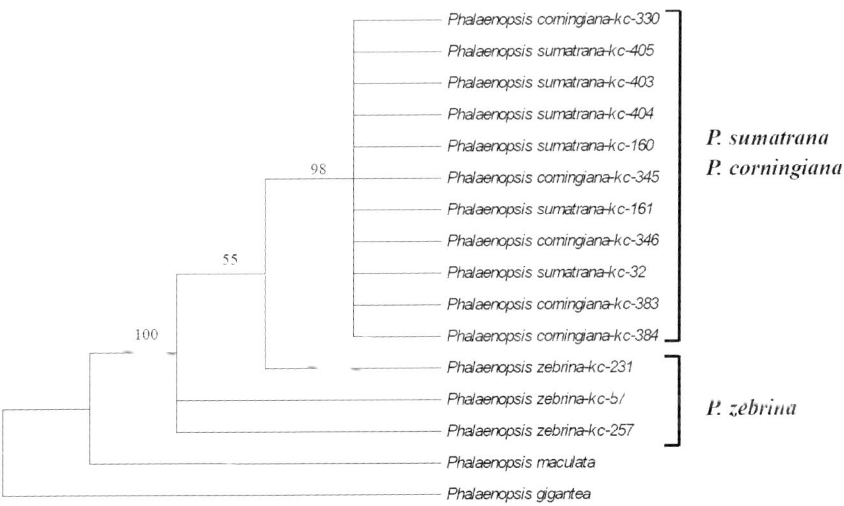

Fig. 1.5. The MP tree of the 14 accessions from the *Phalaenopsis sumatrana* complex plus the outgroups *P. maculata* and *P. gigantea*, obtained from sequence comparisons of the ITS region of nrDNA. The numbers above the internodes indicate the values of the bootstrap test from 1,000 replicates. Bootstrap values greater than 50% are shown. [Redrawn from Tsai *et al.* (2009).]

(1982). However, the floral ground color of *P. zebrina* is not a characteristic of this species, as described by Fowlie (1982). Inspection of the plants of *P. zebrina* revealed that both white and yellowish ground colors in the flower of this species were apparent, as described by Masaaki (2002). Furthermore, all sepals and petals of these three species are retained after pollination (Christenson, 2001). We found that the marking pattern of sepals and petals in both *P. zebrina* and *P. sumatrana* was partly retained after pollination, but not in that of *P. corningiana*. In short, plants of *P. zebrina* could be separated from both *P. sumatrana* and *P. corningiana* based on floral fragrances and molecular data. Therefore, it is not appropriate to treat *P. zebrina* as a synonym of *P. sumatrana*, as described by Sweet (1968, 1980) and Christenson (2001).

1.3.3.2. *Biogeography and evolutionary trends*

The distributions of these three species of the species complex overlap in Borneo. This creates a much greater chance for the three species to hybridize in their natural environment; which makes it difficult to discriminate among the three species. In fact, some naturally hybridized plants between species of *P. sumatrana* and *P. corningiana* have been collected (Christenson, 2001). Apparently, the three species share a common ancestor. Using the phylogenetic tree inferred from the ITS sequences of nrDNA, the evolutionary trend of the *P. sumatrana* complex was deduced. Since accessions of *P. zebrina* were located as the basal group within the *P. sumatrana* complex, *P. zebrina* was suggested to be the relative origin group of the *P. sumatrana* complex. Based on the evolutionary trend derived from molecular data, the dispersal pathway of the *P. sumatrana* complex was deduced and is shown in Fig. 1.6.

P. zebrina developed in Borneo and dispersed into Palawan, the Philippines. *P. corningiana* and *P. sumatrana* might have evolved from *P. zebrina* in Borneo. Thereafter, *P. sumatrana* dispersed into Sumatra, the Malay Peninsula and the Andaman Islands via land bridges in glacial times and it has been shown to be a widespread species. Land bridges between Sumatra, the Malay Peninsula and Borneo could have been formed, since Borneo, Sumatra, the Malay Peninsula and Java

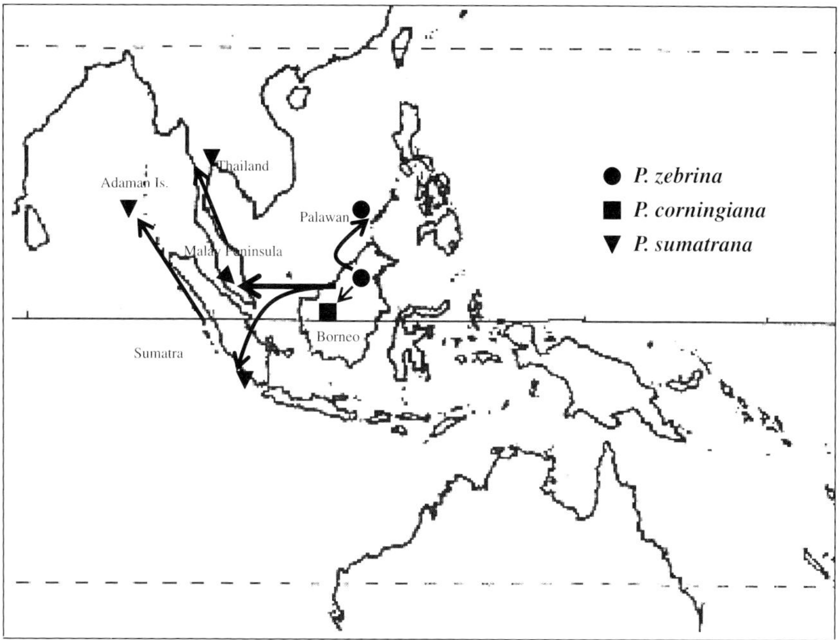

Fig. 1.6. The evolutionary trend of the *P. sumatrana* complex based on the phylogenetic tree. [Redrawn from Tsai *et al.* (2009).]

comprised the Sunda Shelf (Van Oosterzee, 1997). However, to date *P. sumatrana* has not been found in Java. This result is in agreement with the biogeography of the genus *Phalaenopsis*, with *Phalaenopsis* species found in Java differing from those of other lands of the Sunda Shelf (Christenson, 2001; Tsai, 2003a). Furthermore, evidence suggests that the Andaman Islands might have been connected to Sumatra in ancient times.

1.3.4. *Phylogenetics and biogeography of the Phalaenopsis violacea complex*

The *P. violacea* complex includes two species, namely *P. violacea* Witte and *P. bellina* (Rchb.f.) E. A. Christ. In addition, three forms of *P. violacea* have been found according to different distributions, including Sumatra, the Malay Peninsula and the Mentawai Islands.

The *P. violacea* species complex was examined by Tsai (2003c) using molecular data. The phylogenetic trees inferred from the ITS region of nuclear ribosomal DNA (nrDNA), the *trn*L intron, and IGS of *atp*B-*rbc*L of plastid DNA, were used to clarify the phylogenetics and biogeography of the *P. violacea* complex. The sequences of the IGS of *atp*B-*rbc*L of plastid DNA among the accessions of the *P. violacea* complex are identical. No substitution of the *trn*L intron was found from this complex. In contrast, a hot spot region of insertion/deletion was found within the *trn*L introns of plastid DNA among those accessions. However, this hot spot region cannot offer valuable information to discriminate these two species of the *P. violacea* complex. Two valuable polymorphic sites were found within the ITS1 region of nrDNA. Based on the phylogenetic tree inferred from ITS sequences, *P. bellina* cannot be separated from accessions of *P. violacea* with the exception of the population distributed on the Mentawai Islands, Indonesia. Furthermore, in terms of morphological characteristics, the *P. violacea* distributed on the Mentawai Islands has a long and roundish rachis and has been separated from the other groups of the *P. violacea* complex described by Christenson (2001). Hence, the results of this study tend to support the population on the Mentawai Islands of the *P. violacea* complex being a separate species from *P. violacea*. The Mentawai Islands are located near Sumatra, so Mentawai plants of this complex might be descended from those of Sumatra and the Malay Peninsula.

1.3.5. *Phylogenetics, biogeography and evolutionary trends of the Phalaenopsis lueddemanniana complex*

The *Phalaenopsis lueddemanniana* species complex includes *P. bastianii, P. pallens, P. hieroglyphica, P. reichenbachiana, P. lueddemanniana, P. fasciata, P. pulchra* and *P. mariae*. This complex as well as others of the section *Amboinenses* was examined using ITS sequences and chloroplast DNA by Tsai (2003a). Those findings are in agreement with the treatment of Sweet (1980), which differentiated this complex into seven species. It also showed that *P. mariae* is a

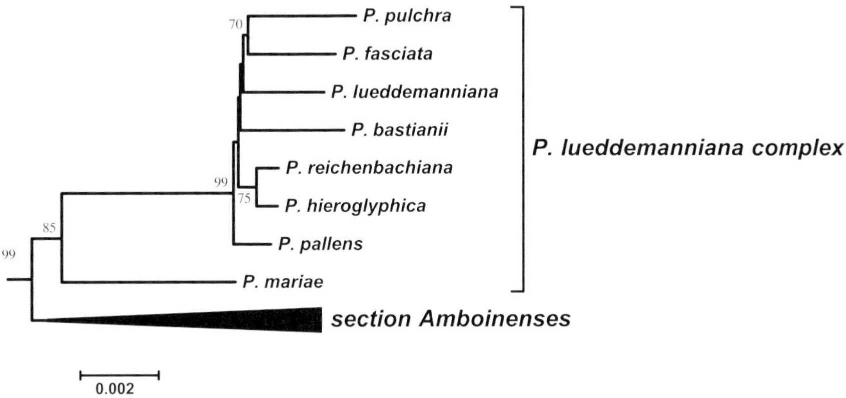

Fig. 1.7. The evolutionary phylogenetic subtree of both the section *Amboinenses* and the *P. lueddemanniana* complex inferred from combined data of the ITS sequences of nrDNA and chloroplast DNA data constructed by the minimum-evolution method [Redrawn from Tsai (2003a)]. Species of the section *Amboinenses* are compressed and shown in a bold branch.

basal species of the *P. lueddemanniana* complex (Fig. 1.7). With the exceptions of the *P. lueddemanniana* complex and *P. micholitzii* (distributed in Mindanao), no species of the section *Amboinenses* has been found in the Philippines. This suggests that the species of the *P. lueddemanniana* complex in the Philippines descended from species of the section *Amboinenses* distributed in Borneo and developed into a unique lineage (Tsai, 2003a) (Fig. 1.8).

Until 5 ~ 10 Mya, the crust of an older plate, Palawan, was combined with Borneo (Karig *et al.*, 1986; Stephan *et al.*, 1986; Hall, 1996). This provided an opportunity for interchange between the Philippines species and Borneo species, and thereafter, *P. amabilis* dispersed into Borneo, Java, New Guinea and Australia, and species of subgenus *Polychilos* dispersed into the Philippines during the glacial period. Because *P. mariae* is distributed in both Borneo and Palawan (Christenson, 2001) and belongs to the basal species of the *P. lueddemanniana* complex according to the combined data of the nrITS and plastid DNA, *P. mariae* has been suggested as a relative origin species of the *P. lueddemanniana* complex. This species dispersed from Borneo into the Philippines, and thereafter, the *P. lueddemanniana* complex

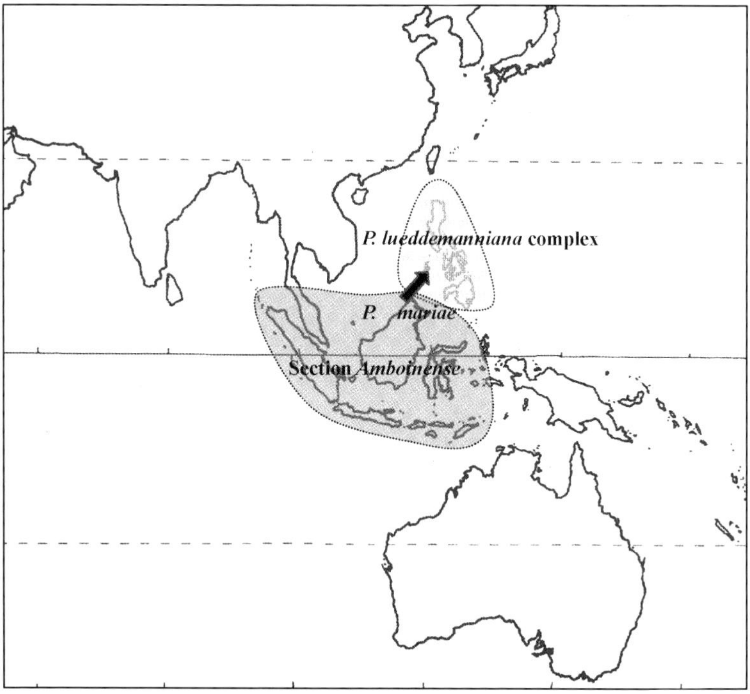

Fig. 1.8. The evolutionary trend of the *P. lueddemanniana* complex as suggested by historical geology and molecular DNA.

developed. Two *Phalaenopsis* taxa, *P. aphrodite* subsp. *formosana* and *P. equestris,* respectively belonging to sections *Phalaenopsis* and *Stauroglottis,* are found in Taiwan. Therefore, these two Taiwanese *Phalaenopsis* species must have separately come from the Philippines.

References

Aoyama M, Kojima K, Kobayashi M. (1994) Morphology of microspores in *Phalaenopsis* hybrids. *Kinki Chugoku Agr Res* **88**: 49–53.

Aurelio MA, Barrier E, Rangin C, Muller C. (1991) The Philippine Fault in the late Cenozoic tectonic evolution of the Bondoc-Masbate-N. Leyte area, central Philippines. *J SE Asian Earth Sci.* **6**: 221–238.

Chen WH, Fu YM, Hsieh RM *et al.* (1995) Application of DNA amplification fingerprinting in the breeding of *Phalaenopsis* orchid. In: Terzi M, Cella R, Falavigna A (eds.), *Current Issues in Plant Molecular and Cellular Biology,* pp. 341–346. Kluwer Academic Publishers, Dordrecht, Boston, and London.

Christenson EA. (2001) *Phalaenopsis.* Timber Press, Portland, Oregon. 330 pp.

Dressler RL. (1993) *Phylogeny and Classification of the Orchid Family*. Dioscorides Press, Portland, OR.

Fowlie JA. (1969) A note on *Phalaenopsis sumatrana* and *Phalaenopsis corningiana*. *Orchid Dig* **33**: 301.

Fowlie JA. (1982) Malaya revised, pt. 21: *Phalaenopsis corningiana* on saplings and streamside trees near waterfalls. *Orchid Dig* **46**: 138–142.

Hall R. (1996) Reconstructing Cenozoic SE Asia. In: Hall R and Blundell DJ (eds.), *Tectonic Evolution of SE Asia*. Geological Society of London Special Publication, No. 106, pp. 153–184.

Hanebuth T, Stattegger K, Grootes PM. (2000) Rapid flooding of the Sunda Shelf: A late-glacial sea-level record. *Science* **288**: 1033–1035.

Holttum RE. (1959) Evolutionary trends in the Sarcanthiine orchids. *Amer Orchid Soc Bull* **5**: 399–423.

Kao YH. (2001) Phylogeny of *Phalaenopsis* species based on 5S rDNA intergenic sequences. Master's Thesis. Institute of Botany National Taiwan University, Taipei, Taiwan. In Chinese with English abstract.

Karig DE, Sarewitz DR, Iiaeck GD. (1986) Role of strike-slip faulting in the evolution of allochthonous terranes in the Philippines. *Geology* **14**: 852–855.

Masaaki C. (2002) *Phalaenopsis* species. Chiba Masaaki *Phalaenopsis* species Publishing Society, Saitama-kenn, Japan.

Quebral RD, Pubellier M, Rangin C. (1994) The Mindanao: A transition from collision to strike-slip environment. *Tectonics* **15**: 713–726.

Sathiamurthy E, Voris HK. (2006) Maps of Holocene sea transgression and submerged lakes on the Sunda Shelf. *Nat Hist J Chulalongkorn Univ Suppl* **2**: 1–43.

Seidenfaden G. (1988a) *Doritis*. *Opera Bot* **95**: 31–34.

Seidenfaden G. (1988b) *Kingidium*. *Opera Bot* **95**: 182–189.

Shim PS. (1982) A new generic classification in the *Phalaenopsis* complex (Orchidaceae). *Malayan Nat J* **36**: 1–28.

Stephan JF, Blanchet R, Rangin C *et al.* (1986) Geodynamic evolution of the Taiwan–Luzon–Mindoro belt since the Late Eocene. *Tectonophysics* **125**: 245–268.

Sweet HR. (1968) Revision of the genus *Phalaenopsis*, Part 2. *Amer Orchid Soc Bull* **37**: 1089–1104.

Sweet HR. (1980) The genus *Phalaenopsis*. The Orchid Digest, Pomona, CA.

Tsai CC. (2003a) Molecular phylogeny, biogeography, and evolutionary trends of the genus *Phalaenopsis* (Orchidaceae), Chapter 1. PhD Thesis. Division of Biological Science, National Sun Yat-sen University, Kaohsiung, Taiwan.

Tsai CC. (2003b) Phylogenetics, biogeography, and the evolutionary trend of the *P. amabilis* complex, Chapter 2, Section 1. PhD Thesis. Division of Biological Science, National Sun Yat-sen University, Kaohsiung, Taiwan.

Tsai CC. (2003c) Phylogenetics, biogeography, and the evolutionary trend of the *P. violacea* complex, Chapter 2, Section 3. PhD Thesis. Division of Biological Science, National Sun Yat-sen University, Kaohsiung, Taiwan.

Tsai CC, Huang SC, Chou CH. (2006a) Molecular phylogeny of *Phalaenopsis* Blume (Orchidaceae) based on the internal transcribed spacers 1 and 2 of nuclear ribosomal DNA. *Plant Syst Evol* **256**: 1–16.

Tsai CC, Huang SC, Huang PL *et al.* (2006b) Molecular evidences of a natural hybrid origin of *Phalaenopsis* × *intermedia* Lindl. *J Hort Sci Biotech* **81**: 691–699.
Tsai CC, Huang SC, Chou CH. (2009) Phylogenetics, biogeography, and evolutionary trends of the *Phalaenopsis sumatrana* complex inferred from nuclear DNA and chloroplast DNA. *Biochem Syst Ecol* **37**: 633–639.
Van Oosterzee P. (1997) *Where Worlds Collide: The Wallace Line*. Cornell University Press, Ithaca.
Van Steenis CGGJ. (1950) The delimitation of Malaysia and its main plant geographical divisions. Flora Malesiana. *Noordhoff-Kolff* **1**: 70–75.

CHAPTER 2

Endopolyploidy in *Phalaenopsis* Orchids and Its Application in Polyploid Breeding

Wen-Huei Chen[†,‡], Yu-Lin Kao[*], Ching-Yan Tang[*] and Goamg-Tyng Jean[*]

Polyploidy plays an important role in the development of different varieties of *Phalaenopsis* orchids and other orchids. However, evidence indicates that the genetic base of commercial orchid hybrids is quite narrow. On the other hand, sterility in progenies derived from hybridization between species with different chromosome sizes or ploidy levels is one of the problems encountered by *Phalaenopsis* breeders. For these reasons, a research project was initiated to study the occurrence of endopolyploidy in *Phalaenopsis* species and to develop simple and effective techniques to determine nuclear DNA content and to double chromosome numbers. By employing flow cytometry, a database was set up for the nuclear DNA content of the wild species and some related hybrids of *Phalaenopsis* collected from several germplasm banks and orchid nurseries in Taiwan. Information in the database will be helpful to orchid breeders to select parental varieties for hybridization to increase their breeding efficiency. In addition, endopolyploidy in different tissues of *Phalaenopsis* species was thoroughly studied using flow cytometry. It was found that different patterns of endopolyploidy occurred in different tissues of *Phalaenopsis* species at various stages of development. Based on these findings, a simple and effective protocol was developed for the production of polyploid plants by sectioning of protocorms or protocorm-like bodies (PLBs) without using anti-microtubule agents. Through this technique, a conversion program to develop a series of tetraploid species of *Phalaenopsis* is in progress. This project will provide a wide range of tetraploid germplasm which will match the ploidy level of the commercial hybrids to accelerate the breeding of novel varieties of *Phalaenopsis* for the market. With the full-scale upgrading of the breeding program to the polyploid level, the production of a

[*]Department of Life Sciences and Institute of Biotechnology,
National University of Kaohsiung, Kaohsiung, Taiwan.
[†]Orchid Research Center, National Cheng Kung University, Tainan 701, Taiwan.
[‡]Corresponding author. E-mail: a08539@gmail.com

wide range of new hybrids of *Phalaenopsis* is expected to create a significant impact on the orchid industry in the coming years.

2.1. Introduction

The *Phalaenopsis* orchid (the moth orchid) is one of the most popular orchids on the market because of the great diversity in its plant types as well as flower colors, shapes and arrangements. Continual development of new and distinct varieties is one of the vital forces that sustain the industry in Taiwan and in other parts of the world. The basic chromosome number of this group of orchids is $2n = 2x = 38$. The most popular commercial varieties/hybrids are usually tetraploid while the wild species are diploid. Therefore, it is difficult to transfer favorable genes from wild species to the most advanced varieties or hybrids. In order to solve this problem, a research program was established in 2004 at the National University of Kaohsiung in Taiwan to study the occurrence of endopolyploidy and, to develop simple and effective technologies for the determination of nuclear DNA content and for chromosome doubling of *Phalaenopsis* species. The application of these methods to develop a comprehensive polyploid breeding program for *Phalaenopsis* orchids is presented in this chapter.

2.1.1. *Endopolyploidy and its occurrence in plants*

Endopolyploidy, the occurrence of different ploidy levels among cells of an organism, is commonly observed in angiosperms and other plant species (Barow and Meister, 2003; Barow, 2006). It is frequently observed to be organ-specific and related to plant development. Mature organs with a high degree of endopolyploidy have been detected in *Arabidopsis* (Galbraith *et al.*, 1991) as well as other plants (Barow, 2006). Endopolyploidy is generated by endoreduplication, a process in which mitosis does not occur completely after DNA replication (Larkins *et al.*, 2001; Barow, 2006; Inze and De Veylder, 2006). Due to the increase in gene redundancy, endopolyploidy can boost metabolism, improve crop yield and quality (Comai, 2005), and also accelerate growth and enhance physiological functions (Barow, 2006).

2.1.2. *Endopolyploidy in orchid species*

Endopolyploidy has been reported in several species and hybrids of orchids, such as in *Dendrobium* spp. (Jones and Kuehnle, 1998), *Phalaenopsis* spp., *Oncidium varicosum* (Lin *et al.*, 2001; Lee *et al.*, 2004), *Vanda* (Alvarez, 1968; Lim and Loh, 2003) and *Spathoglottis plicata* (Yang and Loh, 2004). Diverse endopolyploidy patterns are observed in leaves, roots and flowers of different species, where endopolyploidy is concurrent with maturation or aging. Moreover, these studies also showed that the degree of endopolyploidy was organ specific. In a study of the distribution of endopolyploid nuclei in different tissues of *Phalaenopsis equestris*, including the flower, leaf and root, it was reported that endopolyploidy was related to organ development (Lin *et al.*, 2001). During flower development, endoreduplication is a contributing factor to cell growth because the cell fresh weight is positively correlated to ploidy level (Lee *et al.*, 2004). Since the endoreduplication transition rate is reduced due to low temperature (Lee *et al.*, 2007), it is proposed that a compensatory balance exists between the endoreduplication process and the growth rate.

2.1.3. *Polyploidy in orchids and its significance in variety improvement*

Since the discovery of tetraploidy in *Phalaenopsis amabilis* in 1930 (Vaughn and Vaughn, 1973), efforts were made to use polyploids in the breeding of *Phalaenopsis* orchids (Nakasone and Kamemoto, 1961; Griesbach, 1981, 1985). One form of the famous hybrid *P.* Be Tris was developed by using the colchicine-induced tetraploid *P. equestris* as one of the parents. In addition, one form of *P. rimestandiana* was reported to be tetraploid and was probably used in the development of the tetraploid white *Phalaenopsis* (Griesbach, 2002) in the USA. In Taiwan, one of the two native *Phalaenopsis* species, *P. amabilis* (now renamed *P. aphrodite* subsp. *formosana*) was found to have a tetraploid form which had better shape and substance compared with the diploid. Varieties derived from this form have become some of the most important medium-sized white flower orchids in the Japanese market since 1980s. The series of "Big White Flower" hybrids of *Phalaenopsis*, which are popular in the market, are tetraploids. Development of these hybrids was related to the polyploidization of one of the ancestors, *P.* Doris, which is the most important parent in this group of hybrids (Tang and Chen, 2007). Besides the white flower hybrids, polyploids were reported in many other hybrids with yellow or red flowers, such as *P.* Sunrise Goldamour, *P.* Taipei Gold, etc. (Chuang *et al.*, 2008), and *Dtps.* Sinica Sunday, *Dtps.* Sweet Strawberry "Wei", etc. (Bolanos-Villegas and Chen, 2007). Due to the larger size of the flowers, heavier substance and better shape, polyploidy plants are used extensively for variety improvement of *Phalaenopsis* by breeders with or without the realization of the ploidy nature of their breeding stocks. Thus, many advanced polyploid hybrids have been developed by selecting parents with large, quality flowers for hybridization. Systematic research in the understanding of the ploidy levels of these breeding stocks and to develop polyploid germplasm of *Phalaenopsis* species will accelerate the development of new varieties in this group of orchids.

2.2. Endopolyploidy in *Phalaenopsis*

2.2.1. *Endopolyploidy in ovule and seed development*

Endopolyploidy was studied at different developmental stages in the embryonic tissues of the ovary and protocorms of *Phalaenopsis aphrodite* subsp. *formosana* using a combination of flow cytometry, light microscopy and fluorescence microscopy equipped with ApoTome slider (Jean, 2008). It was found that rapid growth of the ovular tissue after pollination was contributed by nuclear and cell division as in the normal cycle. After meiosis and fertilization (about 43–50 days after pollination), embryo sacs elongated and became spindle-shaped. A high level of endopolyploidy was observed during this stage

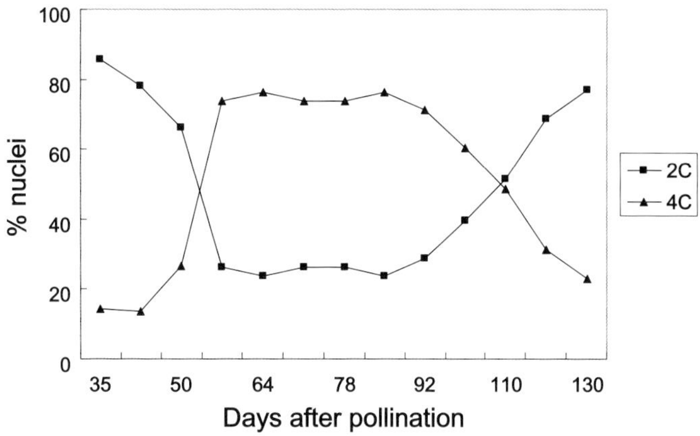

Fig. 2.1. Changes in ploidy pattern in ovules and developing seeds on different days after pollination in *P. aphrodite*.

of seed development. By 90 days after pollination, the seeds became mature and endopolyploidy decreased to a low level (Fig. 2.1). Large and prominent nuclei were found at the basal portion of the mature embryo sac by 140 days after pollination.

Endopolyploidy was observed in the rapidly developing protocorms as early as 4 days after sowing. Large and prominent nuclei were also found at the basal portion of the developing protocorms as revealed by DAPI staining at 20 days after sowing (Jean, 2008). A model of the distribution of nuclei with various ploidy levels in the seeds and protocorms was proposed according to these observations (Fig. 2.2). Different levels of DNA content were observed in the nuclei extracted from the tissue of 30-day-old protocorms. The mean percentages of nuclei having 2C, 4C, 8C and 16C were 47.5, 35.7, 15.6 and 1.2, respectively (Fig. 2.3; Chen *et al.*, 2009). The upper primordium-containing part of the protocorm showed little endopolyploidy while the lower part of protocorm, which contains more mature and differentiated cells, exhibited a higher level of endopolyploidy. These observations indicate the existence of different levels of endopolyploid cells in the protocorms, which are considered to be mixoploids.

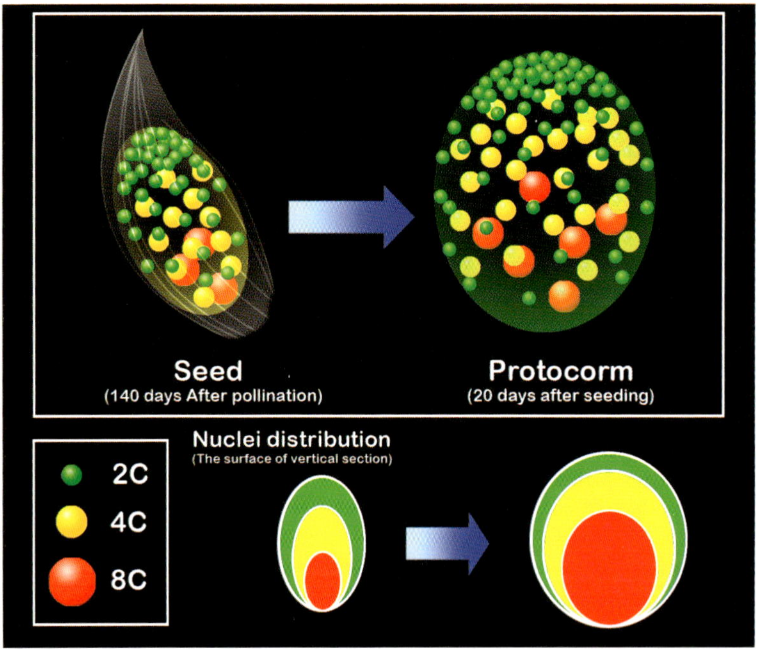

Fig. 2.2. A model of the distribution of nuclei with various ploidy levels in the seeds and protocorms.

2.2.2. *Endopolyploidy in different tissues at different stages of plant development*

Studies on the ploidy pattern were carried out in our laboratory for different tissues at various developmental stages in *P. aphrodite* and *P. bellina*. The level of endopolyploidy increased as the *in vitro* leaves developed. The young leaves, which were actively growing, had the lowest level of endopolyploidy compared to the mature leaves. Tissue in the root tips, composed of apical meristem with high mitotic capacity, predominantly contained nuclei without endoreduplication. However, the cells in the maturation region of the roots had almost entirely endoreduplicated nuclei up to the third cycle. Moreover, endopolyploidy was detected at a high level as early as in the newly emerged leaves in greenhouse-grown plants. Endopolyploidy patterns

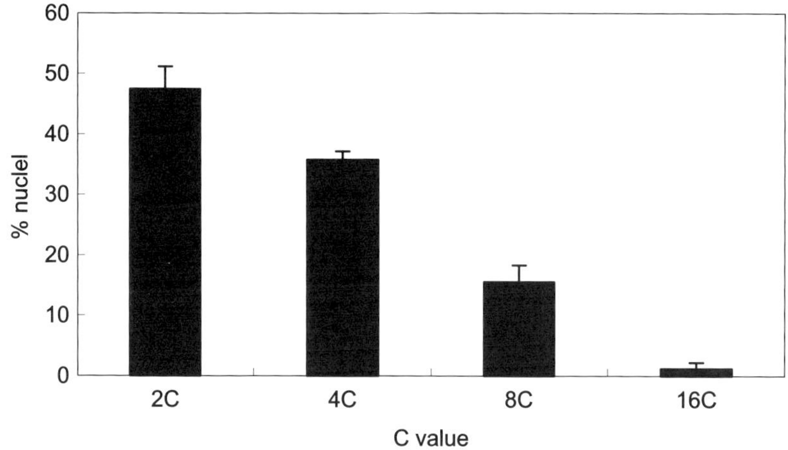

Fig. 2.3. Ploidy patterns in protocorms at 30 days after sowing of seeds in *P. aphrodite.* Data were means and standard deviations (vertical bars) of 20 observations. [From Chen *et al.* (2009).]

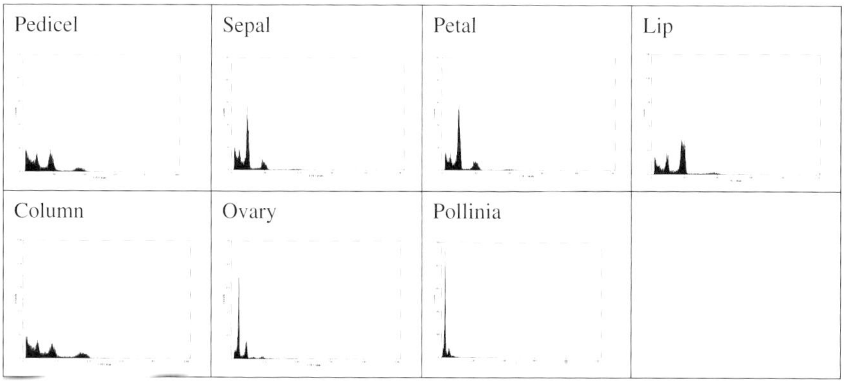

Fig. 2.4. Flow cytometric histograms showing ploidy patterns of different tissues in the flower of *P. aphrodite.*

were diverse and changed progressively in various parts of the flower, including the pedicel, sepal, petal, lip and column, but not in the ovary and pollinia, where endoploypoidy occurred at a very low level (Fig. 2.4).

2.3. Production of Polyploid Plants through Endopolyploidy in *Phalaenopsis*

2.3.1. *Previous research on spontaneous chromosome doubling through tissue culture*

Spontaneous chromosome doubling through tissue culture has been claimed as an alternative in the production of polyploid plants in tobacco (Murashige and Nakano, 1966). Double haploids were reconstituted from excised mature leaf tissue of anther-derived haploids of tobacco using tissue culture (Kasperbauer and Collins, 1972). Besides tobacco, this approach was also applied extensively in potato breeding as a means to manipulate the ploidy levels among monoploids, diploids and tetraploids (Hermsen *et al.*, 1981; Jacobsen, 1981; Karp *et al.*, 1984; Cardi *et al.*, 1992; Chauvin *et al.*, 2003). Compared to chemical induction, it was generally agreed that production of polyploid plants via tissue culture was simpler and more efficient (Hermsen *et al.*, 1981; Cardi *et al.*, 1992). Tetraploid plants were also produced by *in vitro* regeneration of petal marginal tissue from a diploid Belgian azalea (De Schepper *et al.*, 2004) and from callus tissue of a diploid *Dianthus* "Telstar Scarlet" (Nontaswatsri and Fukai, 2005). Although this method of induction of polyploid plants proved to be useful, its application in other plant species was uncommon. However, these findings indicate the possibility of producing polyploid plants through tissue culture without using anti-microtubule agents. Although the mechanism by which these polyploid plants produced is still not well understood, it is logical to propose that regeneration of endopolyploid cells to polyploidy plants by tissue culture is one of the possible pathways.

2.3.2. *Production of polyploid plants by excision of protocorms or protocorm-like bodies (PLBs)*

Micropropagation using tissue culture is very important in the orchid industry. Different *in vitro* methods for orchid multiplication according to the initial explants used (such as leaf, shoot, flowering stalk, root and PLBs) and the developmental pathways (such as direct embryogenesis, callus formation, PLB or adventitious bud formation) have been reported (Tokuhara and Mii, 1993; Park *et al.*, 1996; Ishii *et al.*, 1998; Chen *et al.*, 2000; Chen and Chang, 2004, 2006). Among these *in vitro* methods, production of PLBs from protocorm or from PLBs was shown to be more efficient (Chen *et al.*, 2000; Chen and Chang, 2004). In addition, the horizontal sectioning of PLBs to produce the next cycle of PLBs was demonstrated in *Phalaenopsis* in 1984 by Tanaka and associates (Tanaka *et al.*, 1984; Tanaka, 1987). A high frequency of multiplication using similar techniques was also shown in *Phalaenopsis gigantea* (Murdad *et al.*, 2006). Furthermore, development of PLBs through reactivation of the epidermal and cortical cells on the sectioning surface of the excised protocorms was observed in a study of the cytology of somatic embryogenesis in *Phalaenopsis* species (Huang, 2001).

Based on these research results, an investigation was initiated in our laboratory to explore the feasibility of using horizontal sectioning of protocorms or PLBs as a means of micropropagation to regenerate polyploid plants in eight *Phalaenopsis* species including *P. aphrodite* and *P. bellina*. In this study, 10–40 percent of the PLBs regenerated from the first cycle of sectioned protocorms was found to be polyploids with ploidy doubled once or twice as determined by flow cytometry. The frequency of ploidy doubling increased as the sectioning cycles increased. By the third cycle of sectioning, more than 50 percent of plants regenerated from PLBs were polyploids in all species studied (Table 2.1, Fig. 2.5). In other studies, protocorms of diploid, triploid and tetraploid *P. aphrodite* were used as explants to regenerate PLBs by the sectioning method. It was found that production of polyploidy was highest in diploid followed by triploid and tetraploid plants. Hence, a large number of stable polyploid plants could be produced through repeated sectioning and regeneration of PLBs from the advanced sectioning cycles (Fig. 2.6).

2.3.3. *Performance of polyploidy plants developed from protocorm-like bodies*

After the regeneration of polyploidy plants according to the method described above, the morphology of five-month potted plants of

Table 2.1 Number and Percentage of Polyploidy Plants Obtained by *In Vitro* Culture of Excised Protocorm-like Bodies at 3rd Cycle of Sectioning in Eight Species of *Phalaenopsis*

Species	Total Number of Plants Examined	No. of Polyploid Plants	% of Polyploidy Plants
P. amabilis	142	93	65.4
P. aphrodite	100	66	66.2
P. gigantea	280	190	67.8
P. hieroglyphica	153	85	55.6
P. philippinensis	58	50	85.5
P. shcilleriana	61	50	81.7
P. stuartiana	89	50	56.0
P. bellina	163	95	58.3

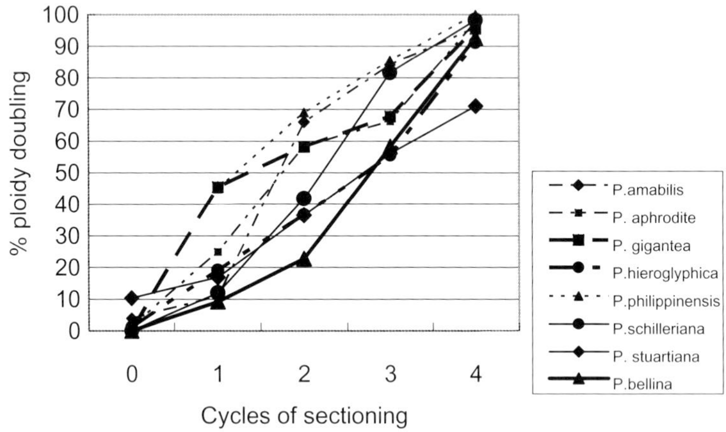

Fig. 2.5. Frequencies of ploidy doubling in plants derived from different cycles of sectioning of protocorms or PLBs in eight *Phalaenopsis* species. [Modified from Chen *et al.* (2009).]

different ploidy levels of two species was studied. No significant difference in the width of the leaves of diploid and tetraploid plants was found (Table 2.2). However, the length and the leaf index of the leaves in the tetraploid *P. hieroglyphica* was significantly shorter and lower than that of the diploid, indicating that differences in leaf size between diploid and tetraploid of *Phalaenopsis* is genotype specific. However,

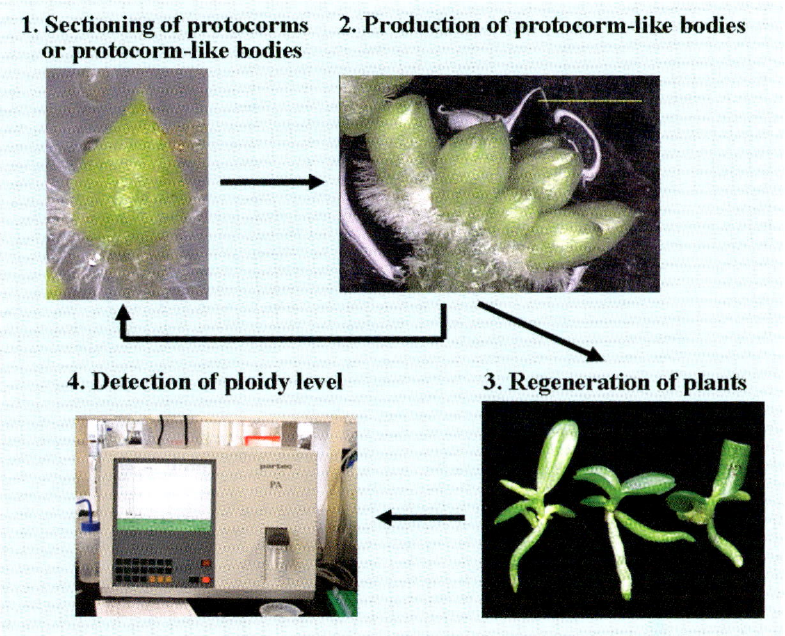

Fig. 2.6. A simple protocol for production and detection of polyploid plants in *Phalaenopsis* orchids.

among five leaf parameters measured in this study, only the stomata density was found to be significantly different between diploids and tetraploids in the two species studied (Table 2.2). Thus, stomata density could be used as a parameter for distinguishing diploid and tetraploid plants in *Phalaenopsis* species. This investigation revealed that some flower characteristics were different between diploid and tetraploid plants. For example, the length of the flower stalk in the tetraploid plant was shorter than that of the diploid plant (Fig. 2.7). Diameter of flower stalk and flower size (Fig. 2.8D) were larger in the tetraploid plants than in the diploid plants. It was also observed that the petals of the tetraploid flowers were heavier and thus had better texture compared to those of the diploid plants.

This study demonstrated that the *in vitro* sectioning of protocorms or PLBs is a simple, effective and reliable technique to produce polyploid

Table 2.2 Means and Standard Deviations of Leaf Characteristics of Diploid and Tetraploid Plants of Two *Phalaenopsis* Species 5 Months after Transplanting to Pots*

Species	Ploidy	No. of Plants	Leaf Length (cm)	Leaf Width (cm)	Leaf Index (Length/Width)	Stomata Density (No./25 mm^2)	Stomata Length (μm)
P. aphrodite	2x	20	25.5 ± 5.8	8.4 ± 1.8	3.2 ± 0.9	9.9 ± 2.2	19.6 ± 2.7
	4x	36	23.6 ± 4.6	8.0 ± 1.6	3.0 ± 0.7	6.1 ± 1.4**	20.5 ± 2.6
P. hieroglyphica	2x	18	36.4 ± 8.5	13.1 ± 4.5	3.1 ± 1.2	9.1 ± 2.5	19.3 ± 2.3
	4x	24	25.1 ± 6.4**	11.2 ± 2.7	2.3 ± 0.3**	6.1 ± 1.1**	22.1 ± 2.7**

*Data from Chen *et al.* (2009). *Plant Cell Tiss Org Cult* **98:** 229–238.

**Significantly different between plants of 2x and 4x of the same species at the $\alpha = 0.05$ level by two-sample *t*-test.

Fig. 2.7. Comparison of diploid (left) and tetraploid (right) plants of *P. stuartiana*.

plants in *Phalaenopsis* orchids. There are advantages in the production of polyploid plants by this method. Firstly, the protocol developed is an efficient method to produce a large number of polyploid plants by repeated sectioning of PLBs. A high frequency of polyploids in the subsequent sectioning cycles of PLBs was observed in eight *Phalaenopsis* species in this study. With this method, polyploid seedlings can be produced in a population which is large enough for selection of superior genotypes at the tetraploid level. Secondly, polyploid plants produced by this method grow normally without adverse effects due to the toxicity of the anti-microtubule agents. Due to inheritance, tetraploid plants usually grow slower than the diploids. From our observations so far, both diploid and tetraploid plants recovered by this method can produce flowers more or less in the same year. Hence, selection of superior polyploid genotypes can be achieved directly in the first generation of the polyploid plants.

Fig. 2.8. Polyploid plants at early (A and B) and flowering (C and D) stages of development. Comparison of the flowers of diploid (right) and tetraploid (left) *P. schilleriana* is shown in D.

2.4. Polyploid Breeding in *Phalaenopsis*

2.4.1. *Major problems in the breeding of Phalaenopsis orchids*

There are two major problems encountered by breeders of *Phalaenopsis* orchids: limitation of tetraploid germplasm and sterility of the progeny derived from interspecific hybridization. In a study of the 12 most popular tetraploid hybrids of white Taisuco *Phalaenopsis* bred by Taiwan Sugar Cooperative, it was found that they were developed from seven wild species of *Phalaenopsis* through more than 10 generations of hybridization. It was also shown that the genetic contribution of three species, namely *P. amabilis, P. aphrodite* and *P. rimestandiana,* was 95.3 percent. One of the parental hybrids, *P.* Doris even contributed 50.5 percent of the genetic materials of these varieties (Chen *et al.*, 1999). This is an indication of the narrow genetic base of this group of hybrids which may become one of the limitations for their further improvement. Thus, there is a need for tetraploid germplasm with a broad genetic base to increase the genetic variability for the selection of better varieties.

Interspecific hybridization is commonly practiced by the breeders of *Phalaenopsis.* One way to broaden the genetic base of the commercial tetraploid hybrids is to cross them with the diploid wild species. The resulting triploid hybrids may perform well and can be selected for commercial production. However, further improvement of these triploid hybrids is very difficult due to the problem of sterility. The famous yellow hybrid *P.* Taipei Gold belongs to this category (Chuang *et al.*, 2008). Another reason for sterility is the crosses involving species having different sizes of chromosomes, which leads to the problem of chromosome pairing during meiosis (Arends, 1970; Kao *et al.*, 2001; Bolanos-Villegas and Chen, 2007). An example of this type is *P.* Deventeriana, a hybrid between *P. amabilis* and *P. amboinensis,* having small and large chromosomes, respectively.

2.4.2. *Chromosome doubling as a means of solution*

One of the strategies to overcome the above problems of sterility caused by nonhomologous pairing of chromosomes is by chromosome doubling of the parental stocks or particular hybrids (Nakasone and Kamemoto, 1961; Griesbach, 1985). Good examples are *P.* Taipei Gold and *P.* Deventeriana, as mentioned in the previous section. *P.* Taipei Gold is a yellow hybrid that was registered in 1984. It is a triploid derived from a cross between *P.* Glads Read, a "Big White Flower" hybrid, and *P. venosa,* a yellow wild species. *P.* Taipei Gold is highly sterile due to the nature of triploidy. However, after the discovery of a new clone named *P.* Taipei Gold "Gold Star", many good yellow hybrids have been developed from it (Fig. 2.9A). We now know that *P.* Taipei Gold "Gold Star" is a tetraploid (Chuang *et al.*, 2008). *P.* Deventeriana was registered as early as in 1927. It is a cross

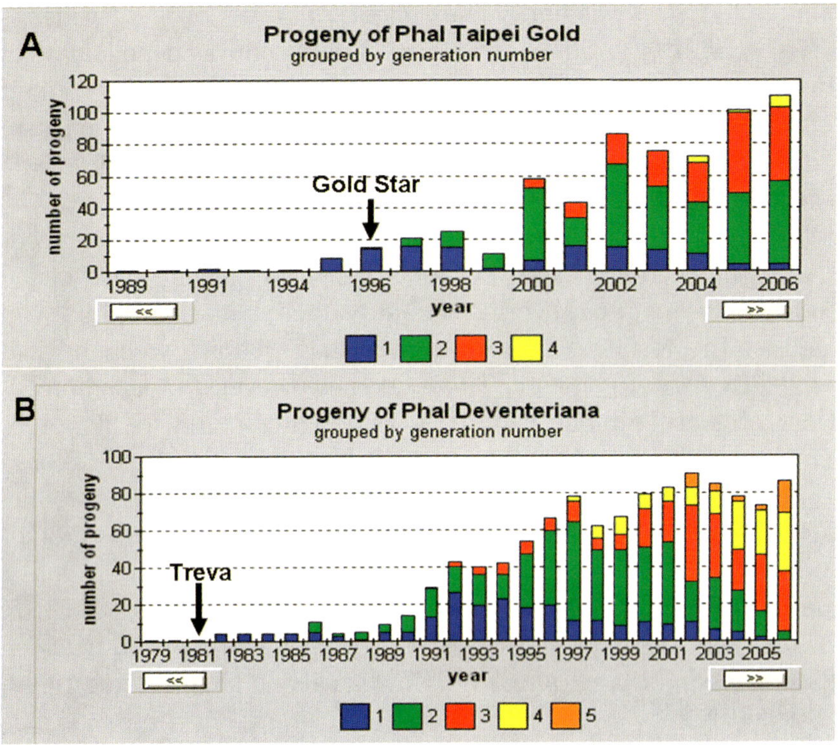

Fig. 2.9. Number of progeny of *P*. Taipei Gold (A) and *P*. Deventeriana (B) registered in Sander's List. [Data from *Wildcatt Orchids*, 2007 by Wildcatt Database Co.]

between *P. amabilis* and *P. amboinensis*, having small and large chromosomes, respectively (Kao *et al.*, 2001) and is therefore sterile due to the nonhomologous paring of chromosomes. There were very few hybrids derived from it before 1960. However, it was a completely different story after the discovery of a tetraploid line *P*. Deventeriana "Treva". Large numbers of hybrids derived from it were registered in Sander's List (Fig. 2.9B). These provide good evidence that polyploidization of the appropriate breeding stocks can overcome the problem of sterility caused by interspecific hybridization. Along this line of thought, the following strategy for the breeding of *Phalaenopsis* orchids is proposed.

2.4.3. *Comprehensive polyploid breeding program for Phalaenopsis orchids*

2.4.3.1. *A database for the nuclear DNA content of Phalaenopsis species*

In order to have a planned breeding program for *Phalaenopsis* orchids, there is a need to understand the genome sizes and ploidy levels of the breeding materials, including the wild species and some important breeding stocks. With the latest technology, detection of the nuclear DNA content by flow cytometry is a useful way to provide the above information needed for the choice of parents for hybridization. The nuclear DNA content of 19 species of *Phalaenopsis* and *Doritis* ranging from 2.7–16.6 pg/2C was reported (Lin *et al.*, 2001). A microscopic study observed a positive correlation between nuclear DNA content and chromosome size for some of these species (Kao *et al.*, 2001). In order to have a complete database of the nuclear DNA content of *Phalaenopsis* species, a simple and efficient flow cytometry protocol was developed by using the tissue of *in vitro* explants or the ovary of the flower of *Phalaenopsis* (Chen *et al.*, 2009). With the above protocol, a survey of the nuclear DNA content of the native species and some related hybrids of *Phalaenopsis* was carried out in several germplasm banks and orchid nurseries in Taiwan. Based on this survey, more than 30 *Phalaenopsis* species were grouped into low, medium and high nuclear DNA content (Table 2.3), which reflects the chromosome size of the orchids. This classification of *Phalaenopsis* species is useful to the breeders in their decision-making on the selection of parental varieties for hybridization since fertility of the hybrid is usually high when crosses are made between parents with chromosomes of similar sizes. The present data revealed that polyploids varying from triploid to hexaploid were found among *Phalaenopsis* species and primary hybrids in different orchid nurseries and markets (Table 2.4). Most of the growers were not aware of the polyploid nature of these varieties. Therefore, identification of the ploidy levels of breeding stocks using this modified protocol to determine the nuclear DNA content in *Phalaenopsis* will increase breeding efficiency through proper choices of the parents in terms of their ploidy levels.

Table 2.3 Classification of Some of the *Phalaenopsis* and *Doritis* Species According to the Nuclear DNA Content

Group	Nuclear DNA Content (pg/2C)	Species
Low	2.5–5.0	*P. amabilis; P. aphrodite; P. equestris; P. gigantea; P. intermedia; P. lindenii; P. philippinensis; P. sanderiana; P. schilleriana; P. stuartiana*
Medium	5.0–10.0	*P. amboinensis; P. bastiana; P. corningiana; P. cornu-cervi; P. fasciata; P. lueddemanniana; P. micholitzii; P. pallens; P. tetraspis; P. wilsonii*
High	>10.0	*P. bellina; P. chibae; P. lobbii; P. mannii; P. parishii; P. violacea; Dor. pulcherima*

Table 2.4 Polyploid Breeding Stocks Found in the Orchid Industry in Taiwan

	Ploidy Level According to Nuclear DNA Content			
Variety	3c	4c	5c	6c
P. aphrodite	x	x		x
P. bellina	x	x		
P. equestris	x	x		
Dor. pulcherrima		x	x	
P. Gelblieber	x			
P. Princess Kaiulani	x			
P. Cassaandra	x			
P. Hannover Passion x *P. bellina*		x		
Dtps. Anna-Larati Soekardi		x		

2.4.3.2. *A conversion program of diploid Phalaenopsis species to tetraploid*

With the development of the new protocol, this project produced a batch of polyploid plants in eight species of *Phalaenopsis* (Chen *et al.*, 2009). In order to meet the needs of the industry, a program has been developed to convert over 20 diploid *Phalaenopsis* species to tetraploid. A polyploid gene bank will be developed to provide various kinds of polyploid germplasm of *Phalaenopsis*. Some

Table 2.5 A List of Tetraploid Germplasms Converted from Diploid *Phalaenopsis* and *Doritis* Species

Serial No.	Species Code	Species Name	Level of Nuclear DNA Content*	Stage of Development
1	K01-02	*P. amabilis* "Philippines"	L	large seedlings
2	K02-02	*P. amboinensis* "Common"	H	small seedlings
3	K03-02	*P. aphrodite*	L	fully grown plants
4	K06-04	*P. equestris* "Rosea"	L	small seedlings
5	K15-02	*P. gigantea*	M	large seedlings
6	K16-02	*P. hieroglyphica*	L	fully grown plants
7	K23-02	*P. lueddemanniana*	M	small seedlings
8	K27-02	*P. micholitzii*	M	small seedlings
9	K37-02	*P. schilleriana*	S	fully grown plants
10	K40-02	*P. stuartiana* AxB	S	fully grown plants
11	K46-04	*Dor. pulcherrima* "coerulea"	L	small seedlings
12	K48-02	*P. intermedia*	S	small seedlings
13	K57-02	*P. philippinensis*	S	large seedlings
14	K62-02	*P. bellina*	L	large seedlings

*Nuclear DNA content. $L = 2.5\text{–}5$ pg/2C; $M = 5\text{–}10$ pg/2C; $H \geq 10$ pg/2C.

of these converted germplasms are listed in Table 2.5. These converted tetraploid germplasms are grown and preserved in the Polyploid *Phalaenopsis* Germplasm Bank located at the Taiwan Sugar Cooperative, which was one of the collaborators in this project. With the participation of the industry, a wide range of superior germplasms that are used as parental stocks to produce various types of hybrids can be converted to polyploids to accelerate the development of new varieties.

2.4.3.3. *Extension and utilization of polyploid germplasm*

Conversion of diploid species of *Phalaenopsis* to polyploid will provide a broad spectrum of genotypes allowing for a full-scale upgrading of the breeding program to the polyploid level. Through this program, the present commercial tetraploid hybrids can be further improved by the introgression of favorable genes from the same or different species. Novel varieties can be created through different combinations of interspecific hybridization including species having

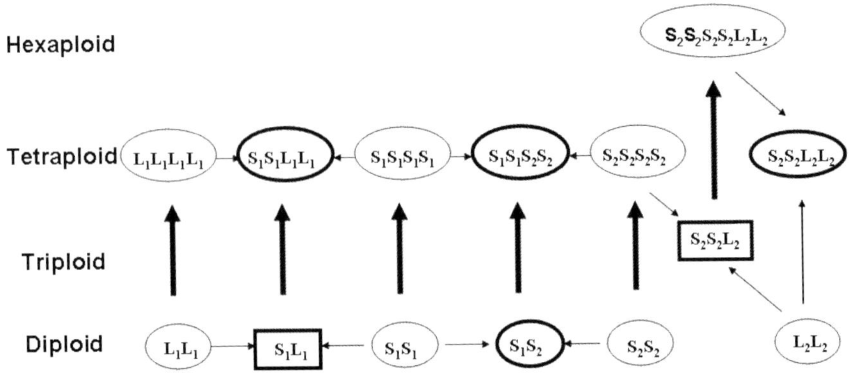

Fig. 2.10. Pathways of polyploidy breeding of *Phalaenopsis* orchids. S = Small genome; L = Large genome; ⟶ Hybridization; ➡ Chromosome doubling; ▭ Sterile generation; ⬭ Fertile generation.

different chromosome sizes, and they can be improved through the restoration of fertility at the tetraploid level. Even the triploid hybrids can be further improved after chromosome doubling to hexaploid and backcrossing to diploid species. The probability of selecting superior genotypes will be greatly increased because of the restoration of fertility in the sterile hybrids and the production of a large number of progeny for selection. Figure 2.10 shows different pathways in a comprehensive breeding program of *Phalaenopsis* at the polyploid level.

2.5. Conclusion and Prospects

Polyploidy is important in the improvement of *Phalaenopsis* varieties, especially in the "Standard Big Flower" group. Due to the increase in the number of sets of genes in the cell, the accumulation of the additive genetic and heterotic effects of these genes makes the continuous improvement of the varieties possible. In the past 10 years, one of the "Standard Big White Flower" hybrids, *P.* Sogo Yukidian "V3", has been well accepted by the market. At present, it is still the most popular hybrid in this group for export from Taiwan. There is currently no sign of its being replaced. This is another indication of the narrow genetic base of this group of hybrids, which limits their further improvement. The aims of this research project are to provide a database of the nuclear DNA content of *Phalaenopsis* species, to generate efficient technologies to produce polyploid plants, and to set up a germplasm bank of polyploid *Phalaenopsis* species. With the above technologies, it is possible to effect a full-scale upgrade of the breeding program of *Phalaenopsis* to the polyploid level. Consequently, the present tetraploid hybrids will be further improved due to the introgression of a wide range of tetraploid germplasm. Novel varieties will also be created because polyploidy may minimize the problem of sterility encountered by certain combinations of hybridization. It is expected that the production of new and unique hybrids will greatly impact the global market when these technologies and polyploid germplasm are fully utilized.

Acknowledgments

This work was supported by the National Science Council, Taiwan (NSC-98-2324-B-390-002-CC1). The authors would like to thank the Taiwan Sugar Cooperative for providing diploid *Phalaenopsis* species for this research and for facilitating the setting up of the polyploid *Phalaenopsis* germplasm bank.

References

Alvarez MR. (1968) Quantitative changes in nuclear DNA accompanying postgermination embryonic development in *Vanda* (Orchidaceae). *Amer J Bot* **55**: 1036–1041.

Arends JC. (1970) Cytological observations on genome homology in eight interspecific hybrids of *Phalaenopsis*. *Genetica* **41**: 88–100.

Barow M. (2006) Endopolyploidy in seed plants. *BioEssays* **28**: 271–281.

Barow M, Meister A. (2003) Endopolyploidy in seed plants is differently correlated to systematics, organ, life strategy and genome size. *Plant Cell Environ* **26**: 571–584.

Bolanos-Villegas B, Chen FC. (2007) Cytological identification of chromosomal rearrangements in *Doritaenopsis* and *Phalaenopsis*. *J Int Cooperation* **2**(1): 1–11.

Cardi T, Carputo D, Frusciante L. (1992) *In vitro* shoot regeneration and chromosome doubling in 2x and 3x potato clones. *Am J Potato Res* **69**: 1–12.

Chauvin JE, Souchet C, Dantec JP, Ellissèche D. (2003) Chromosome doubling of 2x Solanum species by oryzalin: Method development and comparison with spontaneous chromosome doubling *in vitro*. *Plant Cell Tiss Org Cult* **73**: 65–73.

Chen JT, Chang WC. (2004) Induction of repetitive embryogenesis from seed-derived protocorms of *Phalaenopsis amabilis* var. *formosa* Shimadzu. *In Vitro Cell Dev Biol-Plant* **40**: 290–293.

Chen JT, Chang WC. (2006) Direct somatic embryogenesis and plant regeneration from leaf explants of *Phalaenopsis amabilis*. *Biologia Plantarum* **50**: 169–173.

Chen WH, Chen YH, Chyou MS, Fu YM, Lin YS, Hsiau PTW. (1999) Development of white Taisuco *Phalaenopsis*. In: Clark J, Elliott WM, Tingley G, Biro J (eds.), *Proceedings of the 16th World Orchid Conference*, Vancouver, Canada, pp. 272–278.

Chen WH, Tang CY, Kao YL. (2009) Ploidy doubling by *in vitro* culture of excised protocorms or protocorm-like bodies in *Phalaenopsis* species. *Plant Cell Tiss Org Cult* **98**: 229–238.

Chen YC, Chang C, Chang WC. (2000) A reliable protocol for plant regeneration from callus culture of *Phalaenopsis*. *In Vitro Cell Dev Biol-Plant* **36**: 420–423.

Chuang HT, Hsu ST, Shen TM. (2008) Breeding barriers in yellow *Phalaenopsis* orchids. *J Taiwan Soc Hort Sci* **54**: 59–66.

Comai L. (2005) The advantages and disadvantages of being polyploid. *Nature Rev Genet* **6**: 836–846.

De Schepper S, Leus L, Eeckhaut T *et al.* (2004) Somatic polyploid petals: Regeneration offers new roads for breeding Belgian pot azaleas. *Plant Cell Tiss Org Cult* **76**: 183–188.

Galbraith DW, Harkins KR, Knapp S. (1991) Systemic endopolyploidy in *Arabidopsis thaliana*. *Plant Physiol* **96**: 985–989.

Griesbach RJ. (1981) Colchicine-induced polyploidy in *Phalaenopsis* orchids. *Plant Cell Tiss Org Cult* **1**: 103–107.

Griesbach RJ. (1985) Polyploidy in *Phalaenopsis* orchid improvement. *J Hered* **76**: 74–75.

Griesbach RJ. (2002) Development of *Phalaenopsis* orchids for the mass-market. In: Janick J, Whipkey A (eds.), *Trends in New Crops and New Uses*, pp. 458–465. ASHS Press, Alexandra, VA.

Hermsen JGT, Ramanna MS, Roest S, Bokelmann GS. (1981) Chromosome doubling through adventitious shoot formation on *in vitro* cultivated leaf explants from diploid interspecific potato hybrids. *Euphytica* **30**: 239–246.

Huang CY. (2001) Studies on somatic embryogenesis and genetic transformation of *Phalaenopsis* orchid. National Chen Kung University, Tainan, Taiwan, Master's Thesis, 93 pp.

Inze D, De Veylder L. (2006) Cell cycle regulation in plant development. *Annu Rev Genet* **40**: 77–105.

Ishii Y, Takamura T, Goi M, Tanaka M. (1998) Callus induction and somatic embryogenesis of *Phalaenopsis*. *Plant Cell Rep* **17**: 446–450.

Jacobsen E. (1981) Polyploidization in leaf callus tissue and in regenerated plants of dihaploid potato. *Plant Cell Tiss Org Cult* **1**: 77–84.

Jean GT. (2008) Morphological observation and nuclear DNA content analysis in ovule and seed development of *Phalaenopsis aphrodite* subsp. *formosana*. National Kaohsiung University, Kaohsiung, Taiwan, Master's Thesis, 98 pp.

Jones WE, Kuehnle AR. (1998) Ploidy identification using flow cytometry in tissues of *Dendrobium* species and cultivars. *Lindleyana* **13**: 11–18.

Kao YY, Chang SB, Lin TS *et al.* (2001) Differential accumulation of heteochromatin as a cause for karyotype variation in *Phalaenopsis* orchids. *Ann Bot* **87**: 387–395.

Karp A, Risiott R, Jones MGK, Bright SWJ. (1984) Chromosome doubling in monohaploid and dihaploid potatoes by regeneration from cultured leaf explants. *Plant Cell Tiss Org Cult* **3**: 363-373.

Kasperbauer MJ, Collins GB. (1972) Reconstitution of diploids from leaf tissue of anther-derived haploids in tobacco. *Crop Sci* **12**: 98–101.

Larkins BA, Dilkes BP, Dante RA *et al.* (2001) Investigating the hows and whys of DNA endoreduplication. *J Exp Bot* **52**: 183–192.

Lee HC, Chen YJ, Markhart AH, Lin TY. (2007) Temperature effects on systemic endoreduplication in orchid during floral development. *Plant Sci* **172**(9): 588–595.

Lee HC, Chiou DW, Chen WH *et al.* (2004) Dynamics of cell growth and endoreduplication during orchid flower development. *Plant Sci* **166**: 659–667.

Lim WL, Loh CS. (2003) Endopolyploidy in *Vanda* Miss Joaquim (Orchidaceae). *New Phytol* **159**: 279–287.

Lin S, Lee HC, Chen WH *et al.* (2001) Nuclear DNA contents of *Phalaenopsis* sp. and *Doritis pulcherrima*. *J Amer Soc Hort Sci* **126**: 195–199.

Murashige T, Nakano R. (1966) Tissue culture as a potential tool in obtaining polyploidy plants. *J Hered* **57**: 115–118.

Murdad R, Hwa KS, Seng CK *et al.* (2006) High frequency multiplication of *Phalaenopsis gigantea* using trimmed bases protocorms technique. *Scientia Horticulturae* **111**: 73–79.

Nakasone HY, Kamemoto H. (1961) Artificial induction of polyploidy in orchids by the use of colchicine. *Hawaii Agri Exp Stn Tech Bull* **42**, 27 pp.

Nontaswatsri C, Fukai S. (2005) Regenerative callus of *Dianthus* "Telstar Scarlet" showing mixoploidy produce diploid plants. *Plant Cell Tiss Org Cult* **83**: 351–355.

Park YS, Kakuta S, Kano A, Okabe M. (1996) Efficient propagation of protocorm-like bodies of *Phalaenopsis* in liquid medium. *Plant Cell Tiss Org Cult* **45**: 79–85.

Tanaka M. (1987) Studies on the clonal propagation of *Phalaenopsis* through *in vitro* culture. *Memo Fac Agri Kagawa Univ* **49**: 1–85.

Tanaka M, Nishimura M, Goi M. (1984) Studies on the clonal propagation of monopodial orchids through tissue culture. (6) Differences in PLB formation capacity of *Phalaenopsis* leaf segments in culture. *Abstr Jpn Soc Hort Sci Autumn Meet* **53**: 366–367.

Tang CY, Chen WH. (2007) Breeding and development of new varieties in *Phalaenopsis*. In: Chen WH, Chen HH (eds.), *Orchid Biotechnology*, pp. 1–22. World Scientific, New Jersey.

Tokuhara K, Mii M. (1993) Micropropagation of *Phalaenopsis* and *Doritaenopsis* by culturing shoot tips of flower stalk buds. *Plant Cell Rep* **13**: 7–11.

Vaughn L, Vaughn V. (1973) The ascendancy of white phalaenopsis. *Amer Orchid Soc Bull* **42**: 231–237.

Yang M, Loh CS. (2004) Systemic endopolyploidy in *Spathoglottis plicata* (Orchidaceae) development. *BMC Cell Biol* **5**: 33–39.

CHAPTER 3

Growth Responses and Changes of Active Components as Influenced by Elevations and Orchid Mycorrhizae on *Anoectochilus formosanus* Hayata

Shu-Fen Cheng† and Doris Chi-Ning Chang†,‡

Two micropropagated lines, B and R, of *Anoectochilus formosanus* Hayata, were separately inoculated with orchid mycorrhizal fungi (OMF), *Rhizoctonia* sp. R02[a bi-nucleated isolate of *Rhizoctonia* sp. (*Ceratobasidium* sp. AG-A)] and R04[a multi-nuclei isolate of *Rhizoctonia solani*, AG-6], and grown by plastic bag cultivation method (PBCM) at three elevations above sea level, including National Taiwan University (NTU, 10 m elevation), Xindian (500m elevation) and Wufong (1000 m elevation) for seven months. Results showed that the survival rates for *ex vitro* growth were more than 80%, and plant production was significantly increased and there was no need to apply any agrichemical. If this orchid was not grown by PBCM then after 3–4 months of cultivation, all plants would die if pesticide or fungicide were not sprayed for every one or two weeks in greenhouse, and the cultivation period was shortened to 1–2 months compared with traditional cultured method. Plants grew in Wufong achieved the best growth performance among three elevations, the fresh weight of mycorrhizal *A. formosanus* was significantly higher than the non-mycorrhizal (NM) control. In Wufong, R04 showed better growth-enhanced effect on line B, while R02 stimulated growth enhancement for line R. For both lines of *A formosanus* cultivated at National Taiwan University (NTU), R02 inoculated plants contained higher level of phenolic compounds and hepatoprotective agent AFEE (*A. formosanus* extraction with ethyl acetate) compared with the non-mycorrhizal (NM) control. Analyses and measurements of antioxidant

*This paper was orginally published in Botanical Studies (50:459-466, 2009).
†Department of Horticulture, National Taiwan University, Taipei 10617, Taiwan.
‡Corresponding author. E-mail: hyha@ntu.edu.tw

capacity by Trolox Equivalent Antioxidant Capacity (TEAC) showed that most of the antioxidant index of mycorrhizal plants were significantly higher than the non-mycorrhizal control. PBCM was proven to be a very labour saving, i.e. this cultural method can save all the human cost of irrigation and fertilization during the rest of cultivation period, and this is an effective method for mass production of agrichemical free *A formosanus* plants. Inoculation of orchid mycorrhizal fungi such as R02 or R04 can significantly increase the production of this orchid with higher antioxidant capacity and hepatoprotective agent content for medicinal use.

3.1. Introduction

3.1.1. *The medicinal effect of A. formosanus*

Anoectochilus formosanus Hayata is a terrestrial orchid species from wild forests in Taiwan. It is an ornamental plant with medicinal value and is widely used for treating high blood pressure and hyperliposis, lowering blood sugar level, and improving control of diabetes, liver diseases, heart diseases and lung diseases (Yen *et al.* 1996). It also displays anticancer and antivirus effects (Lin, 2007; Wu, 2007). Due to its multitudinous medicinal effects, the number of *A. formosanus* plants in the wild has been greatly reduced by intentional picking and uprooting.

3.1.2. *The propagation of A. formosanus*

In recent years, with the advance of tissue culture techniques, large-scale propagation can be achieved by using seedlings and micropropagated plantlets. Future prospects for the development and application of these techniques to *A. formosanus* are therefore very bright. However, traditional intensive culture has several drawbacks. Survival rates of transplanted tissue-cultured plantlets are low, plants tend to grow slowly, and the cultivation period is long. Additionally, the occurrence of diseases such as root and stem rot caused by *Fusarium oxysporum* and *Pythium ultimum*, bacterial soft rot disease, and pest infections by red spider mites and scale insects often result in the continuous usage of pesticides and fungicides during cultivation

(Chang, 1999; Yen *et al.*, 1996). How to raise the survival rate of the seedlings or plantlets for *ex vitro* growth and control the cultivating environment to shorten the growth period as well as reducing the chances of harmful infection are the major concerns for the cultivation of *A. formosanus*.

3.1.3. *PBCM combined with the inoculation of OMF*

Studies have applied the plastic bag cultivation method (PBCM) (Chang, *et al.* 2007b), and combined with the inoculation of *Rhizoctonia* spp. of orchid mycorrhizal fungi (OMF) (R02 and R04) for the cultivation of *A. formosanus*. The PBCM offers several advantages such as lowering cultivation times, reducing labor, and eliminating the need for pesticides or fungicides. Appropriate ecological conditions are important determinants for the successful artificial culture of *A. formosanus*. Lowering temperature has been known to favor the growth of *A. formosanus* and to reduce pathogen occurrence (Chang *et al.*, 2007a; Lee, 2001). However, many people have for practical reasons tried to grow this orchid at low elevations. Therefore, we decided to conduct our experiments at three locations of different elevations and growth temperatures — NTU campus (10m above sea level), Shihzihtou Mountain in Xindian (500m elevation) and Wufong (1,000m elevation) — to determine the most suitable site for the cultivation of *A. formosanus*.

The tissue-cultured *A. formosanus* plants are considered to be less medicinally effective than the wild plants, which generally have three- to five-fold higher prices on the market. Previous studies have demonstrated that *A. formosanus* inoculated with OMF exhibits higher levels of superoxide dismutase (SOD), acid phosphatase (ACP) and alkaline phosphatase (AKP) activities, and higher components of ascorbic acid, phenolic compounds, flavonoids, polysaccharides and phosphoric acid, and as a result they are more effective as a medicinal source (Chang *et al.*, 2007a; Chou, 2004).

In this study, we aim to achieve three goals: (i) to evaluate the possibility of using the PBCM for the mass production of *A. formosanus*,

(ii) to understand the interactions between the cultivating lines and environmental factors such as elevation and temperature, and (iii) to analyze and determine the antioxidant capacity and hepato-protective activity among the mycorrhizal and non-mycorrhizal *A. formosanus*.

3.2. Results

3.2.1. *Plastic bag cultivation method (PBCM)*

PBCM has been applied for the cultivation of *A. formosanus* (Fig. 3.1). This method offers several advantages. No watering or fertilization is needed during the whole cultivation period, and the potential threat of pest attack is eliminated, and labor requirements for cultivation management are greatly reduced (Chang *et al.*, 2007b). In this study, PBCM

Fig. 3.1. Plastic bag cultivation method (PBCM) for *A. formosanus*. The metal wire (MW) prevents the collapse of the bag.

was applied to large-scale cultivation of *A. formosanus* at three sites of different elevations, and a survival rate of more than 80 percent was attained. For greenhouse cultivation of *A. formosanus*, an intensive chemical spray program is needed and must be carried out regularly. Without the regular use of pesticides, fungicides and chemicals to kill mites, most plants will be attacked. Using the PBCM, however, the survival rate of the *ex vitro* grown plantlets and thus plant production are greatly increased without applying pesticides, fungicides or any other agrichemicals. The growth period is also shortened. As a result of our studies, the PBCM has now been promoted in the Wufong area and applied to large-scale, year-round commercial cultivation for more than 100,000 plants, and the survival rates have been more than 90 percent.

3.2.2. *Inoculation of OMF for the cultivation of A. formosanus*

The PBCM was adopted for the vessel cultivation of line B (wild type) and line R (hybrid of wild type and red stem type) of *A. formosanus* for both mycorrhizal and non-mycorrhizal control plants. After seven months of growth at three locations, the best plant growth was achieved in Wufong (1,000 m elevation). The number of leaves, fresh weight and chlorophyll meter reading of the plants at this location were all higher than those at other sites (Figs. 3.2–3.4). In addition, among the plants in Wufong, those inoculated with OMF [either R02, a bi-nucleated isolate of *Rhizoctonia* sp. (*Ceratobasidium* sp. AG-A), or R04, a multi-nuclei isolate of *Rhizoctonia solani* (AG-6)] attained better growth than the non-mycorrhizal control. Based on the fresh weight, R04 showed better growth-promoting effects for line B, while R02 stimulated better growth for line R (Figs. 3.2 and 3.3). Thus, different lines of *A. formosanus* plants appears to react differently to different OMF, and OMF inoculation does influence plant growth. Compared to the non-mycorrhizal control, OMF inoculation clearly promoted plant growth and development. The fresh weight of mycorrhizal plants was consistently higher at all three elevations (Figs. 3.2 and 3.3). Comparing the three locations, we also found that the fresh

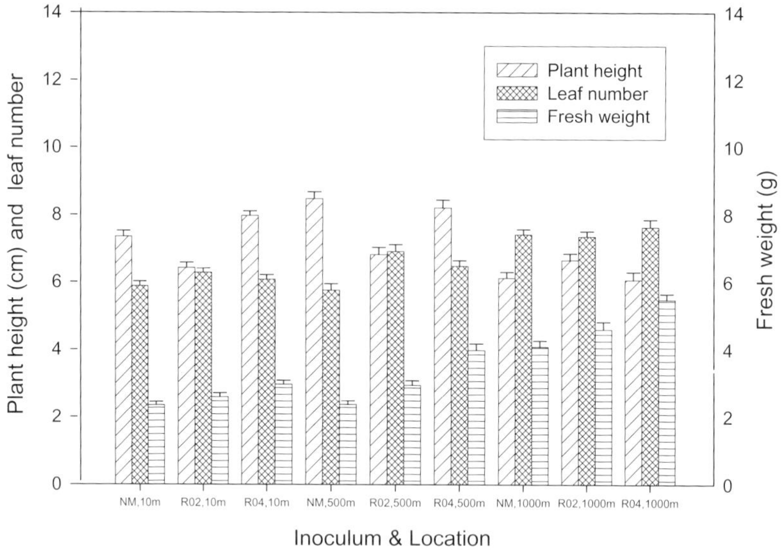

Fig. 3.2. Growth of orchid mycorrhizal fungi (*Rhizoctonia* sp. R02 and R04) inoculated and non-mycorrhizal (NM) plants of *A. formosanus* (line B) using the plastic bag cultivation method (PBCM) at National Taiwan University (NTU, 10m elevation), Xindian (500 m) and Wufong (1,000m) for seven months.

weight of plants in Wufong was the heaviest even though taller plants were found at the two lower elevations, NTU and Xindian (Figs. 3.2 and 3.3).

Further examination revealed that the plants in Wufong displayed shorter internodes but thicker stems, while the plants at NTU generally showed the opposite features, with longer internodes and thinner stems. We speculate that the lower temperature was the likely cause of the heavier fresh weight of the plants at Wufong; the average temperatures of the NTU and Xindian were about 23–30°C and 18–27°C respectively, while the Wufong site was around 15–25°C. When comparing *A. formosanus* plants cultivated in a phytotron, those plants cultivated under controlled temperatures of 20/15°C (day/night) had shorter internodes and thicker stems; samples kept at 30/25°C (day/night) in a phytotron grew taller and had thinner stems (unpublished data).

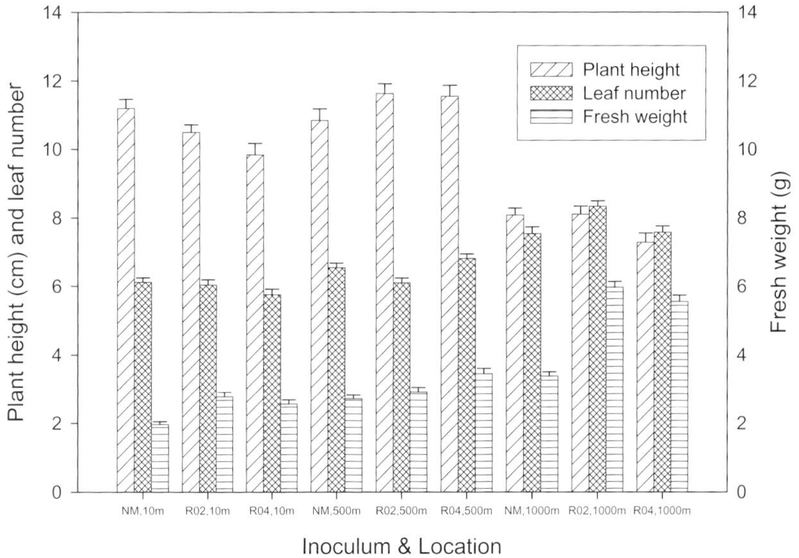

Fig. 3.3. Growth of orchid mycorrhizal fungi (*Rhizoctonia* sp. R02 and R04) inoculated and non-mycorrhizal (NM) plants of *A. formosanus* (line R) using the plastic bag cultivation method (PBCM) at National Taiwan University (NTU, 10m elevation), Xindian (500m) and Wufong (1,000m) for seven months.

From our studies, we conclude that temperature plays a critical role in the growth of *A. formosanus* and is the main reason for the higher fresh weight of the plants grown in Wufong. Therefore, the ideal growing location for *A. formosanus* is at higher elevations where temperatures are lower and there is a greater day/night temperature difference. Previous studies have reported that when line C and line T of *A. formosanus* was inoculated either with *Rhizoctonia* sp. R02 or R04, growth-promoting effects were clearly observed (Chang *et al.*, 2007b). In our study, two different lines, line B and line R, were chosen and also inoculated with R02 or R04, and similar growth-enhancing effects were observed. The results indicate that OMF inoculation can promote the growth of *A. formosanus* both in the cultivation vessel and after transplantation. The absorbing areas of roots are increased as a result of OMF inoculation, and therefore the plantlets can efficiently

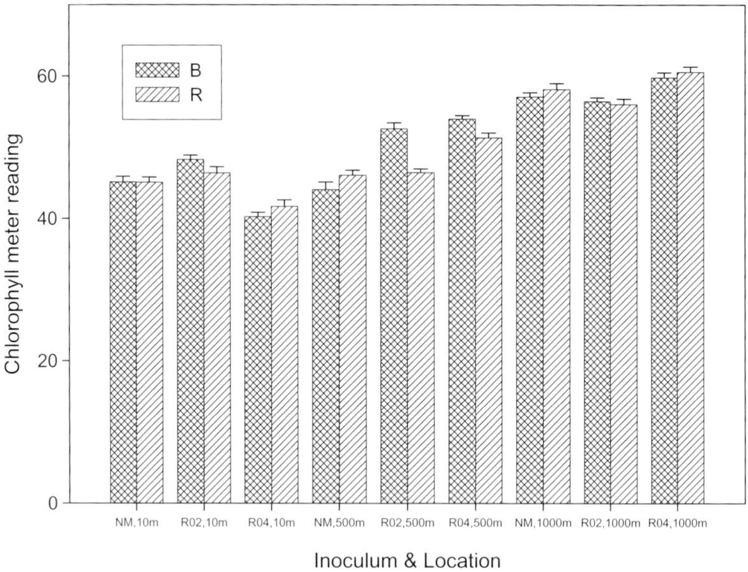

Fig. 3.4. Chlorophyll meter reading value of orchid mycorrhizal fungi (*Rhizoctonia* sp. R02 and R04) inoculated and non-mycorrhizal (NM) plants of *A. formosanus* (line B and line R) using the plastic bag cultivation method (PBCM) at National Taiwan University (10m elevation), Xindian (500m) and Wufong (1,000m) for seven months.

absorb nutrients and water. It has also been reported that the quantity of chlorophyll as well as the efficiency of photosynthesis are increased due to OMF inoculation (Chu, 2000). In our experiments, chlorophyll meter readings among mycorrhizal plants were generally higher than the controls (Fig. 3.4). However, when fresh weight and chlorophyll meter readings were cross-referenced, a positive correlation was not found. This may be due to the fact that plant growth is indeed influenced by multiple factors.

3.2.3. *Analysis of the antioxidant capacity*

3.2.3.1. *Comparison and analysis of phenolic compounds*

Phenolic compounds are widely distributed in the plant kingdom. They are plant secondary metabolites that are often used as an index for antioxidant capacity (Castellucio *et al.*, 1995). Our analyses revealed that *A. formosanus* (line B) grown in Wufong contained a higher level of phenolic compounds; however, the levels of phenolic compounds in the NM control and mycorrhizal plants at this location were not significantly different (Fig. 3.5a). The levels of phenolic compounds in mycorrhizal plants at NTU and Xindian were higher than the non-mycorrhizal controls (Fig. 3.5a). The levels of phenolic compounds among line R plants inoculated with R04 at the three locations were similar (Fig. 3.5a). For the NM and R02-inoculated plants, the levels of phenolic compounds were higher at NTU and Wufong; the R02-inoculated group had higher levels of phenolic compounds, with 5.06 mg/g in the NTU plants and 4.96 mg/g in those at Wufong. Considering the dilution effect of fast-growing plants, we converted the levels of phenolic compounds to a whole-plant basis as shown in Fig. 3.5b. The results showed that the levels of phenolic compounds were highest among the plants grown in Wufong both for line B and line R, and the phenolic compound levels found in the OMF inoculated plants were higher than the NM plants (Fig. 3.5b). Previous studies have indicated that the cultivation environment affects the types and quantity of phenolic compounds in a plant (Dumas *et al.*, 2003; Roussous *et al.*, 2007). Our results demonstrate that a cooler area such as Wufong was more suitable for the growth of *A. formosanus* and that the levels of phenolic compounds found in these plants were higher.

3.2.3.2. *Test of Trolox equivalent antioxidant capacity*

In previous experiments, the antioxidant capacity of *A. formosanus* was tested using the TEAC (Trolox equivalent antioxident capacity), FRAP (Ferric reducing antioxidant power) and DPPH (1,1-diphenyl-2-picrylhydrazyl) methods. In this study, TEAC was chosen to determine the antioxidant capacity due to the method's comparative stability and the significant differences in readings between mycorrhizal and non-mycorrhizal plants.

Our results showed that the TEAC readings of the mycorrhizal line B plants at NTU and at Xindian were higher than the NM plants. In line

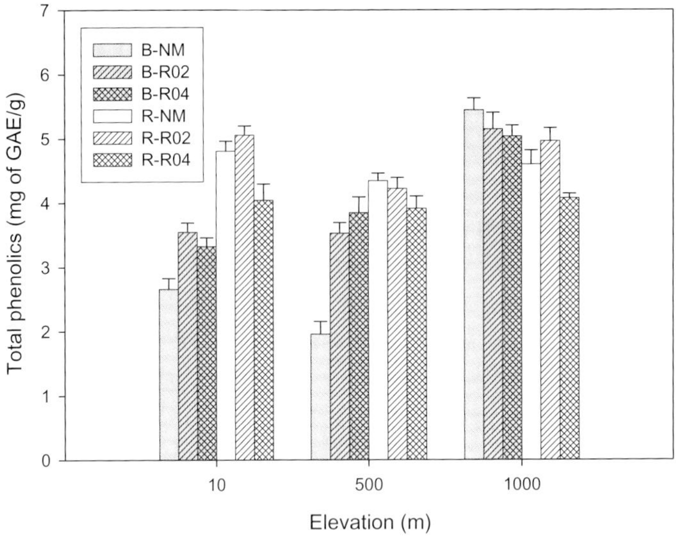

(a)

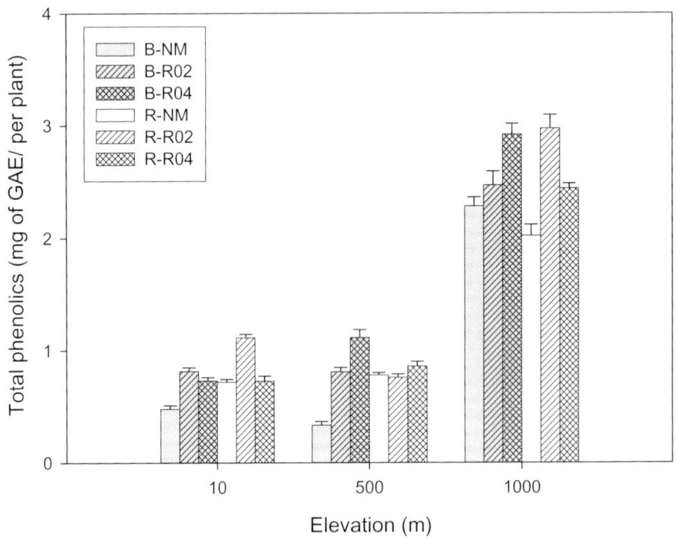

(b)

Fig. 3.5. Levels of phenolic compounds for two lines of *A. formosanus* inoculated with *Rhizoctonia* sp. R02 and R04 using the PBCM at National Taiwan University (10m elevation), Xindian (500m) and Wufong (1,000m) for seven months: (a) level of phenolic compounds per gram, and (b) level of phenolic compounds per plant.

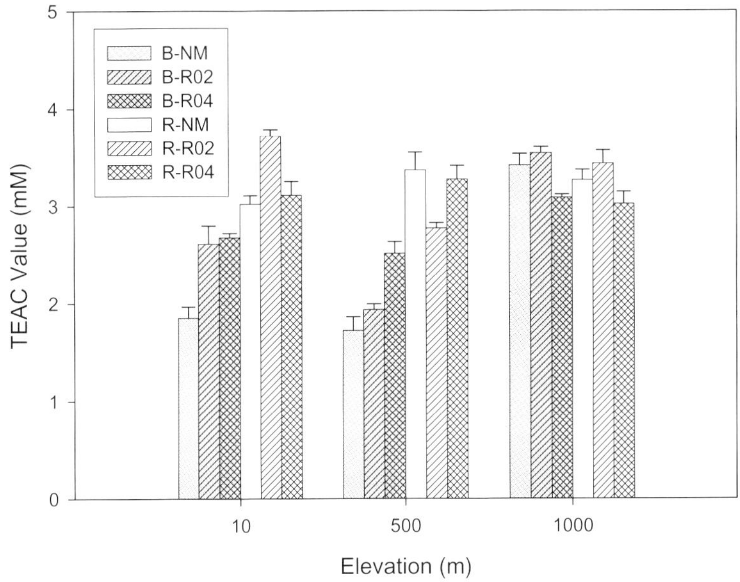

Fig. 3.6. Analyses of Trolox equivalent antioxidant capacity (TEAC) values for two lines of *A. formosanus* inoculated with *Rhizoctonia* sp. R02 and R04 and cultivated at NTU (10m elevation), Xindian (500m) and Wufong (1,000m) for seven months.

R plants, the R02-inoculated group at NTU and at Wufong had higher TEAC readings (Fig. 3.6).

Previous studies have shown that *A. formosanus* extract displays free radical scavenging activity (Wang *et al.*, 2002). Under the treatment of *A. formosanus*, HepG22 cells demonstrated stronger solvency over free radicals and hydrogen peroxide and maintained an effective antioxidant capacity (Hsieh, 2001). *A. formosanus* also contains various enzymes including superoxide dismutase (SOD), catalase, peroxidase and ascorbate peroxidase (Wang, 1999). In 2004, analyses of the enzyme activities and the plant components demonstrated that a stronger antioxidant capacity was found in the mycorrhizal *A. formosanus* plants. Our results showed that mycorrhizal line B plants at NTU and at Xindian exhibited higher TEAC readings than the NM control plants. R02-inoculated line R plants cultivated at NTU and Wufong had higher TEAC readings as well. The results demonstrated that the

4.1. Introduction

Of all orchids, potted *Phalaenopsis* orchids are currently the most commercialized and command the highest overall production value. *Phalaenopsis* orchids are commonly grown in sphagnum moss or a bark-based medium. Taiwan is a major exporter of *Phalaenopsis* plants, and the medium of choice among Taiwanese growers and other exporting countries in the region is sphagnum moss. The use of sphagnum moss to cultivate *Phalaenopsis* is also gaining worldwide importance and *Phalaenopsis* orchids potted in sphagnum are exported in large numbers to many countries.

Sphagnum moss is a group of mosses belonging to the genus *Sphagnum*. Dead cells of sphagnum moss have a large volume with thin but firm cell walls, which are excellent for transmitting water and maintaining its shape (Puustjarvi, 1977). Compared with peat-based media, which are widely used for potted plants, sphagnum moss has very high air-filled porosity (v/v), ranging from 15% to 26% under general conditions (Hwang and Jeong, 2007; Yao and Chang, unpublished data). Sphagnum moss also has a high cation exchange capacity (CEC), ranging from 26–120 meq/100 g, as the pH increases from 3 to 7 (Kubota *et al.*, 1993), whereas the CEC ranges of bark and peatmoss are lower at 8–60 and 7–13 meq/100 g, respectively (Nelson, 2003). These properties make sphagnum moss an excellent medium for epiphytic orchids. For potting *Phalaenopsis*, the sphagnum is commonly packed at a bulk density of approximately $50\ mg \cdot cm^{-3}$.

For proper fertilization management of crops, the nutritional status of the root zone needs to be accurately monitored via testing of the medium solution. The term "medium solution" is equivalent to the term "soil solution", and is applicable to plants growing in soilless media. Chemical analyses of soilless media can be classified into saturated media extract (SME), suspensions, and displaced soil solution methods (Bunt, 1986). Saturated media extract (Warncke, 1986), 1 substrate: 2 water (v/v) suspension (Sonneveld, 1990), and pour-through (PT; Wright, 1986) are the most commonly used procedures, respectively, for the three categories. The SME method is not applicable to sphagnum moss, because the saturation endpoint cannot be determined due to the structure of sphagnum moss. The 1:2 suspension can be adapted to sphagnum moss but it is a destructive method. Pour-through is a non-destructive, inexpensive, and quick medium extraction method and has been used for peat-based medium by growers producing floral crops. The application of PT to *Phalaenopsis* has been suggested (Wang *et al.*, 2005), but no scientific data were provided in that technical report. Due to some special characteristics of sphagnum moss, such as high air-filled porosity and CEC, it is uncertain whether it is feasible to apply PT to sphagnum moss.

4.2. Validation of the Pour-through Method for *Phalaenopsis* Cultivation with Sphagnum Moss

The principle of PT is solution displacement. The high air-filled porosity of sphagnum moss may interfere with the displacement process, hence the PT method needs to be validated for use on sphagnum moss substrate before it is recommended for wider application. In highly porous substrates, some of the displacing solution may bypass finer pores and instead, flow through channels formed by larger pores, such that the displacing solution becomes part of the leachate. To investigate if this channeling effect is significant at the usual packing density of sphagnum moss in potted *Phalaenopsis*, the authors conducted an experiment to study the effects of various displacing solutions on the physicochemical properties of the leachate.

Vegetatively propagated *Phalaenopsis* Sogo Yukidian "V3" plants in sphagnum substrate in 8.5-cm pots were purchased and transplanted to 10.5-cm pots. Sphagnum moss from Chile was used as the potting medium. The moss was soaked in water overnight, and then wrung with a spin-drying machine until just moist. The moss was carefully wrapped around the roots and packed into each pot at a bulk density of around 50 $mg \cdot cm^{-3}$. During the experimental period, the plants were grown in a greenhouse with a pad and fan cooling, where the photosynthetically active radiation at noon was approximately 400 $\mu mol \cdot m^{-2} \cdot s^{-1}$.

Seven displacing solutions with different electrical conductivities (ECs) were prepared, as follows: double-distilled water, Taisun Pure Water (Taisun Enterprise Co., Taipei, Taiwan), tap water, Y.E.S. Mineral Water (Yes! Co., Ilan, Taiwan), and 20N-8.7P–16.6K fertilizer solutions

Table 4.1 Leachate Electrical Conductivity (EC) and pH Values of Sphagnum Moss as Affected by EC Values of the Displacing Solution*

Solution Applied	EC of the Applied Solution ($dS{\cdot}m^{-1}$)	pH of the Applied Solution	Leachate EC ($dS{\cdot}m^{-1}$)	Leachate pH
Double-distilled water	0.00	6.50	0.73 b^{Y}	3.51 a
Taisun Pure Water	0.00	7.33	0.70 b	3.71 a
Tap water	0.09	7.48	0.70 b	3.63 a
Y.E.S Mineral Water	0.20	7.40	0.71 b	3.46 a
100 $mg{\cdot}L^{-1}$ N^{Z}	0.51	6.60	0.64 b	3.61 a
200 $mg{\cdot}L^{-1}$ N^{Z}	0.93	6.48	0.75 b	3.52 a
5000 $mg{\cdot}L^{-1}$ N^{Z}	16.74	4.98	1.00 a	3.53 a

*From Yao *et al., HortScience* (2008) **43**: 2167–2170. The authors wish to thank *HortScience* for providing the figures and tables cited in the text.
[Z]Different concentrations of 20N-8.7P-16.6K fertilizer (Peters Professional 20-20-20).
[Y]Means in the same column followed by a different letter are significantly different at $P \leq 0.05$ by LSD test ($n = 15$).

(Peters Professional 20-20-20; The Scotts Co., Marysville, OH) at 100, 200, and 5,000 $mg{\cdot}L^{-1}$ N (as shown in Table 4.1). Plants were first fertigated with a 20N-8.7P–16.6K fertilizer solution at 200 $mg{\cdot}L^{-1}$ N. After 1 h, 60 mL of a displacing solution was applied to each pot. The leachate was collected and tested for pH and EC with a pH and EC meter (IQ170; IQ Scientific Instruments, Carlsbad, CA). There were 15 single-plant replications for every treatment in a completely randomized design.

The authors found that the EC values of 0.001–0.93 $dS{\cdot}m^{-1}$ of a displacing solution did not affect either the EC or pH of the leachate (Table 4.1). One displacing solution which had an unusually high EC value of 16.7 $dS{\cdot}m^{-1}$ was found to have raised the leachate EC. It is reasonable for the leachate EC value to increase when a displacing solution with a very high EC is used, because a little contamination from the displacing solution (e.g. the solution dripping along the pot edge) can have a significant influence on the leachate properties. The addition of as little as 0.8 mL of a 16.7 $dS{\cdot}m^{-1}$ solution to the leachate was sufficient to cause the observed rise in the leachate EC value (0.3 $dS{\cdot}m^{-1}$; calculated from Table 4.1). However, a displacing solution with such a high EC value would never be used in practice. The results shown in Table 4.1 indicate that the displacing solution could successfully press out the medium solution through the sphagnum moss medium without significantly affecting the leachate. In addition, the authors' experiment provides evidence for the first time in the literature on the PT method, that various solutions such as tap water or fertilizer solutions can be used as displacing solutions with greenhouse practice, even though distilled water is suggested for research.

PT was shown to be a reliable medium test method with peat-based and bark-based media, and high correlation coefficients were found between EC levels obtained by PT and SME (Cavins *et al.*, 2004; Yeager *et al.*, 1983). But as has been mentioned earlier, it is difficult to adapt the SME method to sphagnum moss; hence, the SME method is not suitable for direct comparison with the PT method in this case. There is, however, another medium extraction method that can be adapted for testing sphagnum moss medium — the press extraction (PE) method. The PE method was invented for plugs, whereby the medium solution is obtained by application of squeezing pressure (Scoggins *et al.*, 2001). With PE, we can squeeze out the medium solution that authentically represents the nutrient status of the rhizosphere. It was shown that extraction results obtained by PE were highly correlated with those by SME on peat-based medium (Scoggins *et al.*, 2002). Therefore, to further test the feasibility of PT for sphagnum substrates, the authors conducted an experiment to investigate the correlation between the PT and PE methods. Since the plants were grown in soft plastic pots, the PE method was modified by application of pressure to the pot edges by hands until $\approx$30 mL of the medium solution was collected. Newly transplanted *Phalaenopsis* in 10.5-cm pots were placed in a greenhouse for about two weeks and then randomly divided into two groups of 82 plants each. Fertilizer solutions (20N-8.7P–16.6K) with concentrations of 0–400 $mg{\cdot}L^{-1}$ N and EC values of 0–1.6 $dS{\cdot}m^{-1}$, were applied to the plants in both groups. The substrate EC was then tested with PT in one group, and with PE in the other. The data obtained were subjected to a correlation analysis.

The experiment showed that the PT and PE methods produced highly correlated substrate EC levels, with a high correlation coefficient

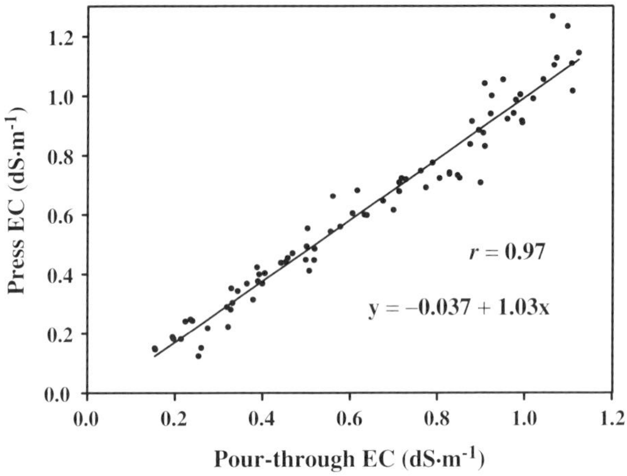

Fig. 4.1. Relationship between substrate electrical conductivity (EC) tested by the pour-through method and by the press method. The EC of the fertilizer solutions applied ranged 0–1.6 dS·m^{-1}. (From Yao *et al.*, *HortScience* (2008) **43**: 2167–2170. The authors wish to thank *HortScience* for providing the figures and tables cited in the text.)

between the two methods (r = 0.97; Fig. 4.1). Another notable observation was that the application of fertilizer solutions with EC range 0–1.6 dS·m^{-1} corresponded to a substrate EC range of 0.2–1.1 dS·m^{-1}. This demonstrates that when a low fertilizer concentration was applied, some nutrients were released from the sphagnum moss into the medium solution, while with higher fertilizer concentrations, some nutrients were adsorbed. This observation indicates that sphagnum moss has a strong buffering capacity for nutrients. The substrate can hold some of the nutrients, such that they cannot be extracted either by PT or PE. The adsorbed nutrients could be labile, i.e. adsorbed from the medium solution when the fertilizer concentration is high and released back into the medium solution when the concentration is low. This also demonstrates that the EC of the fertilizer solution applied is not necessarily equal to that of the medium solution. After the fertilizer solution has been applied to the medium, it undergoes equilibration before becoming a part of the medium solution. Hence, simply measuring the EC level of the fertilizer solution before application as practiced by some growers is insufficient, and monitoring the EC level of the medium solution is more appropriate. Despite the concern that the high porosity of sphagnum moss may affect the feasibility of PT, the observation that substrate EC values tested by PT resembled the results from PE is further proof of the effectiveness of PT for sphagnum moss medium (Fig. 4.1).

4.3. Establishment of a Standard Pour-through Procedure for Fertility Monitoring of Sphagnum Moss

The PT procedure described by Wright (1986) recommended adding 100 mL of displacing water and collecting 50 mL of leachate, and was based on 3-L growing containers. The volume of the displacing solution used varied with different workers when using containers of different sizes. For example, 45 mL of displacing water was applied to marigolds grown with 1 part peat: 1 part perlite medium in 0.5-L pots (Tolman *et al.*, 1990). Wang (1998) collected ≈30 mL of leachate when applying PT to *Phalaenopsis* grown in bark or 4 part bark: 1 part peat medium. On the other hand, Wright and Hinesley (1991) poured 350 mL of displacing water and collected 100–150 mL of leachate when working with red cedar grown in 5 part bark: 1 part sand medium in 20-L containers. Hence the authors investigated suitable volumes of the displacing solution for *Phalaenopsis* grown in 8.5- and 10.5-cm pots using sphagnum moss medium.

In the first experiment, plants in 10.5-cm pots were fertigated with 20N-8.7P-16.6K fertilizer at 200 $mg \cdot L^{-1}$ N. After 1 h, different volumes of reverse osmosis (RO) water were applied to the surface of the medium, ranging from 40 to 120 mL/pot (Table 4.2). The experiment was conducted with a completely randomized design, with 20 single-plant replications for every treatment. The leachate was collected, measured for volume, and then tested for pH, EC and concentrations of some ions within. The concentrations of ions were measured as follows: NO_3^--N and NH_4^+-N using the magnesium oxide-Devarda alloy method (Page *et al.* 1986); phosphorus using molybdenum-blue spectrophotometry (Murphy and Riley, 1962); and potassium, calcium, and magnesium using inductively-coupled plasma spectrometry (ICP-AES; Optima 2000DV, Perkin Elmer, Wellesley, MA).

Increasing the volume of the displacing water from 40 to 70 mL in the experiment did not affect the extract EC or pH values (Table 4.2). The authors obtained consistent concentrations of leachate NO_3^--N, phosphorus, potassium, calcium, and magnesium when 40–60 mL of displacing water was used (Table 4.2). The leachate EC and concentrations of some mineral ions decreased slightly when >70 mL of water was used (Table 4.2), indicating that over-application of displacing water diluted the leachate. Application of 60 mL of displacing water resulted in ≈50 mL of leachate, which is the volume of leachate recommended by Wright (1986) to be collected for the PT procedure. Therefore, the authors suggest using 60 mL of the displacing solution for PT on sphagnum moss medium for *Phalaenopsis* grown in 10.5-cm pots. This volume is within the range that does not produce a diluting effect on the leachate. Ammonium appears to be a more difficult ion to extract compared with the others, and its concentration in the leachate was lower when 60 mL of displacing water was applied, as compared with using 40 mL; the concentration of all other ions did not change when 60 mL of displacing solution was applied (Table 4.2). When the PT method was compared with the SME technique using peat medium, much-lower levels of NH_4^+-N were observed with SME than with PT, and fixation of NH_4^+ by medium components was suggested as a possible cause (Wright *et al.*, 1990). The causes for this phenomenon are not known; perhaps NH_4^+-N binds more strongly to the substrate. Even though the concentration of NH_4^+-N in the leachate was lower when 60 mL of water was applied, compared with that when 40 mL was used, the EC value was consistent, indicating that the leachate still well represented the status of other nutrients at that volume of displacing water.

In the second experiment, *Phalaenopsis* plants in 8.5-cm pots were grown in a greenhouse. On the day of the experiment, the authors fertigated the plants, and 1 h later applied 30–70 mL of RO water. The leachate was collected, measured for volume, and then tested for pH and EC. The experiment was conducted with a completely randomized design with 18 single-plant replications in each treatment. The results showed a similar trend to that of the 10.5-cm-pot experiment (Tables 4.2 and 4.3). Although the leachate EC values did

Table 4.2 Leachate Status of Sphagnum Moss as Affected by the Displacing Water Volumes (in 10.5-cm Pots)*

Volume Applied (mL)	Volume Collected[z] (mL)	Leachate EC ($dS \cdot m^{-1}$)	Leachate pH	NH_4^+-N ($mg \cdot L^{-1}$)	NO_3^--N ($mg \cdot L^{-1}$)	P ($mg \cdot L^{-1}$)	K ($mg \cdot L^{-1}$)	Ca ($mg \cdot L^{-1}$)	Mg ($mg \cdot L^{-1}$)
40	34	0.64 a[Y]	4.01 bc	19.7 a	19.8 a	31.9 a	62.8 ab	5.0 ab	4.9 abc
50	43	0.63 ab	3.99 c	18.0 ab	19.7 a	31.1 ab	65.4 a	5.6 a	5.3 a
60	51	0.62 b	3.99 bc	16.5 bc	19.8 a	31.3 ab	65.3 a	5.4 a	5.2 ab
70	56	0.62 ab	3.99 c	16.7 bc	18.2 ab	30.0 bc	62.0 abc	5.4 a	5.1 ab
80	64	0.61 c	3.98 c	16.9 bc	18.0 ab	28.8 c	59.6 bc	4.2 bc	4.6 bc
90	74	0.59 cd	4.08 ab						
100	77	0.58 de	4.12 a	14.8 c	17.1 ab	28.7 c	57.0 c	4.4 bc	4.7 bc
110	91	0.58 de	4.11 a						
120	106	0.58 e	4.12 a	16.5 bc	16.2 b	26.2 d	58.8 bc	3.9 c	4.4 c

*From Yao *et al.*, *HortScience* (2008) **43**: 2167–2170. The authors wish to thank *HortScience* for providing the figures and tables cited in the text.

[z] Linear regression of the displacing water volumes (x) vs. leachate volumes (y): $y = -1.136 + 0.841x$; $r^2 = 0.98$, $P < 0.0001$.

[Y] Means in the same column followed by a different letter are significantly different at $P \leq 0.05$ by LSD test ($n = 20$).

Table 4.3 Leachate Electrical Conductivity (EC) and pH of Sphagnum Moss as Affected by the Displacing Water Volumes (in 8.5-cm Pots)*

Volume Applied (mL)	Volume Collected[z] (mL)	Leachate EC ($dS{\cdot}m^{-1}$)	Leachate pH
30	28	1.18 a[y]	2.73 b
40	38	1.16 a	2.91 a
50	49	1.18 a	2.76 b
60	56	1.14 a	2.70 b
70	64	1.08 b	2.78 b

*From Yao *et al.*, *HortScience* (2008) **43**: 2167–2170. The authors wish to thank *HortScience* for providing the figures and tables cited in the text.
[z]Linear regression of the displacing water volumes (x) vs. leachate volumes (y): $y = 2.007 + 0.900x$; $r^2 = 0.996$, $P < 0.0001$.
[y]Means in the same column followed by a different letter are significantly different at $P \leq 0.05$ by LSD test ($n = 18$).

not change when 30–60 mL of displacing water was used, it was noted that leachate dilution occurred when >60 mL of water was applied (Table 4.3). Because the volume of the displacing water that was proposed for 10.5-cm pots (60 mL) was close to the volume at which the leachate might be diluted, the authors propose using a volume of 40 mL for 8.5-cm pots.

Wright (1986) suggested that PT be carried out 1 h after fertigation. Cavins *et al.* (2004) demonstrated that timing the sampling at 13–240 min after fertigation did not result in different leachate EC values. The authors conducted an experiment to determine how soon sampling can be done after fertigation for *Phalaenopsis* grown in sphagnum moss. Fertigation of the plants was carried out in 10.5-cm pots with 15N-2.2P-12.5K fertilizer (Peters Excel 15-5-15 Cal-Mag; The Scotts Co.) at 200 $mg{\cdot}L^{-1}$ N and 60 mL of RO water was applied to displace the medium solution at 0, 20, 40, 60, 80, 100, 120, 140, and 160 min after fertigation. Sampling at 0 min was actually done 0 to 1 min after fertigation because of technical difficulties. The experiment was arranged in a completely randomized design with 20 single-plant replications per treatment.

The authors found that timing of sampling at 20–160 min after fertigation did not result in different EC values (Table 4.4). However,

Table 4.4 Leachate Electrical Conductivity (EC) and pH of Sphagnum Moss as Affected by Leachate Collection Time after Fertigation*

Collection Time (min)	Leachate EC ($dS \cdot m^{-1}$)	Leachate pH
0	1.22 a[z]	3.46 bcd
20	1.05 b	3.68 ab
40	1.00 b	3.64 abc
60	1.06 b	3.56 abcd
80	0.98 b	3.65 abc
100	1.03 b	3.74 a
120	0.99 b	3.62 abcd
140	1.05 b	3.43 cd
160	1.00 b	3.40 d

*From Yao *et al.*, *HortScience* (2008) **43**: 2167–2170. The authors wish to thank *HortScience* for providing the figures and tables cited in the text.
[z]Means in the same column followed by a different letter are significantly different at $P \leq 0.05$ by LSD test ($n = 20$).

leachate sampled at 0 min after fertigation had a higher EC value than other treatments (Table 4.4), and this level was close to that of the fertilizer solution (1.32 $dS \cdot m^{-1}$). The results show that equilibration of nutrients between the added fertilizer and the substrate takes approximately 20 min. In another experiment, the authors tested how long the sampling time could be delayed after fertigation, and found that as long as the medium was maintained at a saturated condition after fertigation, the sampling time could be delayed for up to 10 days with no significant change in leachate EC (Yao *et al.*, 2008). For *Phalaenopsis* grown in sphagnum moss, a standard PT sampling time of 1 h after fertigation, the same as timing suggested by Wright (1986), is recommended, but collecting the leachate anytime between 20 min and 24 h after fertigation is acceptable for practical application.

Based on the results of the experiments described above, the authors suggest a standard procedure for PT for *Phalaenopsis* grown in sphagnum moss as follows (illustrated in Fig. 4.2):

1. Fertigate the medium to its maximum water-holding capacity.
2. Wait about 1 h for the nutrients to equilibrate between the medium solution and the substrate.

Fig. 4.2. Recommended pour-through procedure for *Phalaenopsis* cultivation with sphagnum moss.

3. Pour an appropriate amount of displacing solution (60 mL for 10.5-cm pots or 40 mL for 8.5-cm pots).
4. Collect the leachate for testing.

4.4. Practical Application and Some Considerations

The EC of the medium can be used as an estimate of the nutrient content of the growing medium. Maintaining the medium EC within an optimal range is important for vigorous plant growth. The optimal medium EC varies according to the crop, and so far, no EC range has been recommended for *Phalaenopsis* cultivation using sphagnum moss. The author sampled *Phalaenopsis* plants from 10 nurseries in Taiwan to study the relationship between leachate EC levels and quality of roots (Yao, 2007). In addition, experiments were conducted to determine the optimal EC range for *Phalaenopsis* cultivated using sphagnum moss in which different concentrations of 15N-2.2P-12.5K fertilizer were used. The leachate EC that corresponded to good growth and flowering was then determined. It was found that a fertilizer concentration of 100 $mg{\cdot}L^{-1}$ N produced balanced root and shoot growth and was adequate for healthy growth and flowering (Yao, 2007). These investigations showed that the optimal leachate EC range was 0.6–1.0 $dS{\cdot}m^{-1}$, while EC levels of >1.5 $dS{\cdot}m^{-1}$ were detrimental to root growth. The recommended EC range was tested on thousands of plants in nurseries (data not presented). The decision-making procedure for fertilization management of *Phalaenopsis* potted using sphagnum moss by the PT method is shown in Table 4.5. The authors recommend that growers aim for an EC value of 1.0 $dS{\cdot}m^{-1}$. When leachate EC values of >1.5 $dS{\cdot}m^{-1}$ are found, immediate leaching is suggested, while a higher fertilizer concentration at the next application is necessary when the tested EC is <0.6 $dS{\cdot}m^{-1}$.

As *Phalaenopsis* roots are sensitive to salinity and are easily damaged by elevated salt concentrations, the authors used distilled or RO water for the experiments. Therefore, the optimum range of leachate EC as recommended by the authors represents EC levels that are mainly due to nutrients from the fertilizer and not to other salts from the water source. Use of RO water for irrigation and fertigation in practice is

Table 4.5 Fertilization Management in *Phalaenopsis* Cultivation by the Pour-through Method

Leachate EC ($dS{\cdot}m^{-1}$)	Management
>1.5	Immediately leach with water
1.0–1.5	Reduce fertilizer concentration at the next fertigation
0.6–1.0	EC in an optimal range; maintain regular fertilization program
<0.6	Increase fertilizer concentration at the next fertigation

recommended, because if the source water has a high EC, it will be more difficult to provide an adequate level of nutrients without exceeding the optimum EC range.

An incorrect sampling method can introduce errors in the analysis of medium solution with the PT technique. For example, sampling pots from sides of benches can be used for convenience, but pots on the sides tend to dry out more quickly than pots located in bench centers. Another example is the tendency of some growers to choose inferior plants for testing. Such practices can introduce significant error into the analysis, and the importance of representative samplings of materials cannot be overemphasized.

The medium solution can also be directly measured with commercially available EC meters, which contain probes to be inserted into the medium. The authors tested the utility of two such devices for measuring EC levels of medium solutions. In total, 82 *Phalaenopsis* plants in 10.5-cm pots were used in that experiment. Fertilizer solutions (20N-8.7P-16.6K) with concentrations of 0–400 $mg{\cdot}L^{-1}$ N were applied to the plants. The medium solution of each pot was tested with a Delta T WET sensor (Delta-T Corporation, Williamsburg, VA), with a Decagon EC sensor (Decagon Devices, Inc., Pullman, WA), and with the PT method. Data obtained from each device were subjected to a correlation analysis with the PT method.

EC values of medium solutions obtained with both the Delta T WET sensor and Decagon EC sensor were highly correlated with those obtained with the PT method, with respective correlation coefficients

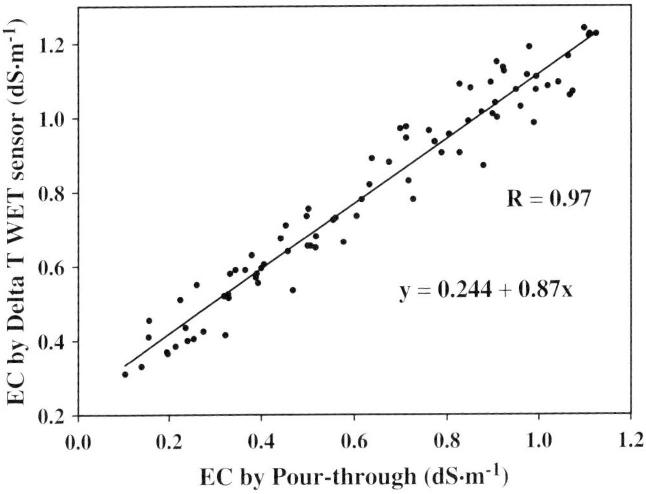

Fig. 4.3. Relationship between the medium electrical conductivity (EC) tested by the pour-through method and by a Delta T WET sensor.

of 0.97 and 0.92 (Figs. 4.3 and 4.4). Data obtained with the Delta T WET sensor were uniformly correlated with those obtained with the PT method over the tested EC range (Fig. 4.3), while values obtained with the Decagon EC sensor had a better correlation with PT data at lower EC levels, but exhibited greater scattering at a higher EC range (Fig. 4.4). The results indicate that the two devices can reliably be used and are comparable to the PT method for monitoring the fertilizer content of sphagnum moss in *Phalaenopsis* cultivation. Using the graphical equations in Figs. 4.3 and 4.4, the optimal range of EC readings of each device can be calculated from that obtained with the PT method (i.e. 0.6–1.0 dS·m^{-1}). These devices are convenient to use, but have a few disadvantages. They are expensive, and insertion of the probes into the medium may damage the roots. In addition, the same pot that is being sampled may give different EC values at different insertion points in the substrate. This is because the probe may come into contact with air pockets or sap from injured roots in addition to the medium solution being measured.

occasional under- or over-fertilization. This result is not surprising considering the high CEC of sphagnum moss.

Bark is another commonly used medium for growing orchids, either used alone or as a component in potting mixtures. Cultivators of epiphytic orchids typically use coarser grades of bark. The PT procedure was originally developed for bark-based media using very fine grades of bark of smaller than 6.4 mm (Yeager *et al.*, 1983), but the application of PT for coarse bark has not been studied in detail. Since coarse bark has very large pores between the particles, the high porosity may interfere with the displacement of the medium solution by the PT method. The moisture distribution in coarse bark at the maximum container capacity is likely to be heterogeneous due to the effect of gravity; hence, over-application of the displacing solution can easily result in dilution of the leachate. Channeling effects may be significant because the pore spaces between bark particles are even larger than those of sphagnum moss at a bulk density of $\approx 50\ \mathrm{mg \cdot cm^{-3}}$. Because of these characteristics, the feasibility of using PT for coarse bark medium needs to be investigated.

4.6. Conclusion and Perspectives

The authors' study showed that the PT method can be successfully used for sphagnum moss at a bulk density of around $50\ \mathrm{mg \cdot cm^{-3}}$; therefore, PT can be safely recommended for wide application with *Phalaenopsis* and other orchids which are cultivated in sphagnum moss medium at similar bulk densities. However, the feasibility of applying PT to loosely packed sphagnum moss medium is still unknown. Further studies should be done to determine the minimum bulk density of sphagnum moss required for successful application of PT.

An optimal leachate EC range based on the PT method for *Phalaenopsis* grown in sphagnum moss was determined in this study. Investigations of the optimum EC ranges for other major orchid crops such as *Oncidium* or *Dendrobium* in sphagnum moss or other commonly used growing media are logical topics for further research.

Compared with other extraction procedures, the PT method can be performed relatively quickly. However, *Phalaenopsis* is typically mass-produced in large nurseries; therefore, automation will greatly increase the efficiency as compared with manual sampling. For example, the displacing solution can be dispensed mechanically. Other than solution displacement, the authors propose other methods for obtaining medium solution samples, such as by gentle suction with a vacuum or by gently shaking a pot along its axis until the medium solution exits through the drainage holes due to inertia. These methods should be highly amenable to mechanization. Leachate EC measurements and subsequent fertilizer concentration adjustments can automatically be controlled by a computerized system.

Early on, orchid growing was shrouded in mystery, and success among hobbyists and commercial growers alike was mostly based on trial and error. Orchid cultivation has come a long way since that time, and workers in orchid-related sciences have endeavored to achieve better knowledge about this fascinating group of plants. It is hoped that this study has contributed in one way or another towards this pursuit, and that it may benefit hobbyists, growers, and the scholarly community alike.

Acknowledgments

The authors are grateful for the financial support provided by the Council of Agriculture, Executive Yuan, Taiwan (97AS-1.2.1-ST-a1; 98AS-1.2.1-ST-a1) to the first author. They would also like to thank Hadi Şusilo for his critical reading on the manuscript and Hsuan Pai for her illustration of Fig. 4.2.

References

Bunt AC. (1986) Problems in the analysis of organic and lightweight potting substrates. *HortScience* **21**: 229–231.

Cavins TJ, Whipker BE, Fonteno WC. (2004) Establishment of calibration curves for comparing pour-through and saturated media extract nutrient values. *HortScience* **39**: 1635–1639.

Hwang SJ, Jeong BR. (2007) Growth of *Phalaenopsis* plants in five different potting media. *J Jpn Soc Hort Sci* **76**: 319–326.

Kubota S, Kato T, Yoneda K. (1993) The effects of the concentration of fertilizer application and the physico-chemical properties of sphagnum moss and clay pots on the growth of *Phalaenopsis*. *J Jpn Soc Hort Sci* **63**: 601–609.

Murphy J, Riley JP. (1962) A modified single solution method for the determination of phosphorus in natural water. *Anal Chim Acta* **27**: 31–36.

Nelson PV. (2003) *Greenhouse Operations and Management.* Prentice Hall, Upper Saddle River, NJ.

Page AL, Miller RH, Keeney DR. (1986) *Methods of Soil Analysis.* American Society of Agronomy, Madison, WI.

Puustjarvi V. (1977) *Peat and Its Use in Horticulture.* Turveteollisuusliitto, Helsinki, Finland.

Scoggins HL, Bailey DA, Nelson PV. (2001) Development of the press extraction method for plug substrate analysis: Quantitative relationships between solution extraction techniques. *HortScience* **36**: 918–921.

Scoggins HL, Bailey DA, Nelson PV. (2002) Efficacy of the press extraction method for bedding plant plug nutrient monitoring. *HortScience* **37:** 108–112.

Sonneveld C. (1990) Estimating quantities of water-soluble nutrients in soils using a specific 1:2 by volume extract. *Commun Soil Sci Plant Anal* **21**: 1257–1265.

Tolman DA, Niemiera AX, Wright RD. (1990) Influence of plant age on nutrient absorption for marigold seedlings. *HortScience* **25**: 1612–1613.

Wang YT. (1998) Impact of salinity and media on growth and flowering of a hybrid *Phalaenopsis* orchid. *HortScience* **33**: 247–250.

Wang YT, Blanchard M, Lopez R, *et al.* (2005) The orchid grower, Part II. *Greenhouse Grower* **23**(8): 70–72.

Warncke DD. (1986) Analyzing greenhouse growth media by the saturation extraction method. *HortScience* **21**: 223–225.

Wright RD.(1986) The pour-through nutrient extraction method procedure. *HortScience* **21**: 227–229.

Wright RD, Grueber KL, Leda C. (1990) Medium nutrient extraction with the pour-through and saturated medium extract procedures for poinsettia. *HortScience* **25**: 658–660.

Wright RD, Hinesley LE. (1991) Growth of containerized eastern redcedar amended with dolomitic limestone and micronutrients. *HortScience* **26**: 143–145.

Yao HY. (2007) Applying pour-through medium solution testing method on *Phalaenopsis* grown with sphagnum moss. MS Thesis, National Taiwan University, Taipei (in Chinese with English abstract).

Yao HY, Chung RS, Ho SB *et al.* (2008) Adapting the pour-through medium extraction method to *Phalaenopsis* grown in sphagnum moss. *HortScience* **43**: 2167–2170.

Yeager TH, Wright RD, Donohue SJ. (1983) Comparison of pour-through and saturated pine bark extract N, P, K, and pH levels. *J Amer Soc Hort Sci* **108**: 112–114.

CHAPTER 5

Compositional Variation in the Polysaccharides of the Pseudobulb of *Oncidium* Gower Ramsey During Different Stages of Inflorescence Development*

Kai-Wun Yeh[†,§], Heng-Long Wang[‡] and Chin-Hui Shen[†]

Oncidium Gower Ramsey is a popular and profitable cut flower in the Asian flower market. Its flowering time is usually irregular. The pseudobulb is a storage organ for nutrition, minerals and water during vegetative development, and provides flowers nutrients during the reproductive stage. The carbohydrate mobilization in the pseudobulb during the vegetative and reproductive stages is investigated in this chapter. The results reveal that polysaccharides such as mannan and pectin are degraded and interconverted. The components of polysaccharides are also analyzed.

5.1. Introduction

Oncidium Gower Ramsey is a thin-leaved epiphytic C_3 orchid extensively cultivated for its exuberant yellow flowers. One of the characteristic features of this orchid is the presence of a pseudobulb, which is an enlarged portion of the internode below leaf-2 (Fig. 5.1).

*This chapter is dedicated to Dr. Shripathi Venkatagiri, a visiting professor from Karnatak University, Dharwad, India (2007–2008).

[†]Institute of Plant Biology, National Taiwan University, Taipei 106, Taiwan.
[‡]Department of Life Science, National University of Kaohsiung, Kaohsiung, Taiwan.
[§]Corresponding author. E-mail: ykwbppp@ntu.edu.tw

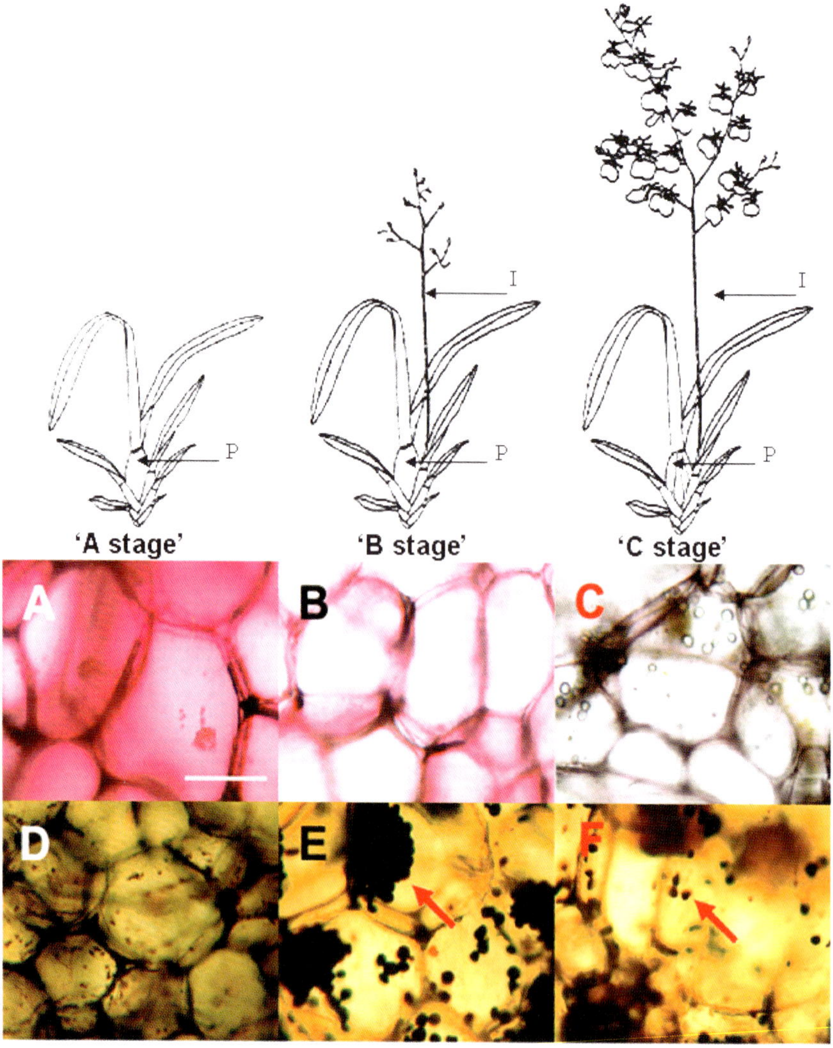

Fig. 5.1. Transverse sections of the pseudobulb stained with ruthenium red (A, B and C) for pectin, and iodine (D, E and F) for starch at different stages of inflorescence development. A and D, before inflorescence initiation stage (**A stage**); B and E, middle inflorescence development stage (**B stage**); and C and F, late inflorescence stage (**C stage**). I, inflorescence; P, pseudobulb. Embedded arrows indicate starch grains. [From Wang et al., *Planta* (2008) **227**: 1063–1077. The authors wish to thank *Planta* for providing the figures cited in the text.]

The function of the pseudobulb is manifold: from acting as a storage organ for water and nutrients to being responsible for supplying energy during initiation and development of inflorescence (Hew and Ng, 1996; Wang *et al.*, 2003). Nutrient storage in the orchid pseudobulb may be linked to vegetative growth as well as flowering (Zimmerman, 1990). Mobilization of water soluble polysaccharides has been observed in the pseudobulbs of epiphytic orchids during water deficit stress (Stancato *et al.*, 2001). The function of the orchid pseudobulb in the context of the relation between water, nutrients and carbohydrates has received considerable attention (Hew and Ng, 1996; Yong and Hew, 1995). It appears that there is a direct link between the flowering process and biochemical changes related to carbohydrate metabolism in the pseudobulb (Hew and Ng, 1996; Wang *et al.*, 2003).

In addition to their role as reserve food, a variety of other functions has also been described for carbohydrates. Sugars and oligosaccharides function as signaling molecules in plant growth and development (Baier *et al.*, 2004; Dumville and Fry, 2000; Gibson, 2005; Koch, 1996; Koch, 2004; Smeekens, 2000). Compositional variation in polysaccharides has been observed in plants during fruit ripening (Koh and Melton, 2002), seed development (Chanda, 2002), pod development (Stolle-Smiths *et al.*, 1999) and phloem fiber development (Gorshkova *et al.*, 1996), indicating that polysaccharides play a role in all these processes. The present study has been undertaken to examine compositional variation in polysaccharides in the pseudobulb of *Oncidium* during the flowering process.

Apart from their physiological role in plants *in vivo*, medicinal applications of plant polysaccharides in the area of dermatology — especially as antimicrobial, antioxidant and wound healing agents — have also been extensively described in recent years (Paulsen, 2001). The use of polysaccharides for improved delivery of therapeutic agents (Ghelardi *et al.*, 2000; Mehvar, 2003) is another area of biomedical application of plant-derived polysaccharides. Our study also deals with the isolation and characterization of polysaccharides from the pseudobulb of *Oncidium* "Gower Ramsey" that can be exploited for biomedical applications.

5.2. Biochemical Characterization of Polysaccharides in the Pseudobulb

The pseudobulbs selected for the extraction and analysis of polysaccharides are classified into four stages based on the growth and developmental conditions. They are: the A′ stage, the young pseudobulb about one month before the initiation of inflorescence; the A stage, just before the initiation of inflorescence; the B stage, when the inflorescence is about 30 cm long; and the C stage, when the pseudobulb is sustaining the inflorescence in full bloom.

5.2.1. *Extraction of polysaccharide components at various growth stages of the pseudobulb*

5.2.1.1. *Extraction of mannans*

Mannans were isolated from the pseudobulb following the method described by Mulimani and Prashanth (2002) with a little modification. The pseudobulb was cut into small pieces and ground into fine powder in liquid nitrogen using mortar and pestle. The powder was dispensed in hot water (10 ml/g tissue), stirring at 60°C for 5 h, filtered through two layers of Miracloth, and extracted repeatedly. The filtrates were combined and trichloro acetic acid (TCA) was added to a final concentration of 5 percent (w/v) while slowly stirring the solution. The contents were kept at 4°C for 1 h for the precipitation of proteins and subsequently centrifuged at 20,000 × g for 30 min. The supernatant was collected and filtered through two layers of Miracloth to avoid solid particles getting into the clear solution. The mannans were precipitated from the clear solution by adding chilled ethanol to a final concentration of 80 percent (v/v) and kept at 4°C overnight. The precipitate was collected by centrifuging at 20,000 × g for 30 min at 4°C. The pellet was washed with a mixture of ethanol and ether (1:1, v/v) and dried *in vacuo*. The amount was quantified gravimetrically.

5.2.1.2. *Extraction of pectins*

The pseudobulb was first chopped and ground into fine powder in liquid nitrogen using mortar and pestle. The fine powder was dispensed in 80 percent ethanol (5 ml/g tissue), boiled for 40 min, and filtered. The residue was washed twice with 80 percent ethanol followed by acetone and air-dried to get alcohol-insoluble residue (AIR). Starch was removed from the AIR as per the method described by Stolle-Smiths *et al.* (1999) with a slight modification. The AIR was suspended in 90 percent (v/v) dimethylsulfoxide (DMSO, 10 ml/g tissue) for 16 h at 20°C and centrifuged at 20,000 × g for 20 min. The pectic polysaccharide was extracted from the starch-free AIR essentially following the method described by Western *et al.* (2001). The starch-free AIR was stirred in 0.5 percent ammonium oxalate (25 ml/g AIR) at 80°C for 1 h and then centrifuged at 20,000 × g for 20 min. The supernatant was collected and ethanol was added to a final concentration of 80 percent (v/v) to precipitate pectic polysaccharides. The precipitate was collected by filtration through four layers of Miracloth and vacuum dried. The amount was quantified gravimetrically. The ammonium oxalate fraction was tested for mannan contamination by adding Fehling's solution, which precipitates any mannans present.

5.2.1.3. *Polysaccharide hydrolysis procedures and high performance anion exchange chromatography with pulsed amperometric detection (HPAEC-PAD) analysis*

Acid hydrolysis of pectins and mannans was done as per the method described by Suda *et al.* (2003) with a little modification. Ten milligrams of pectin/mannan was pre-incubated in 0.1 ml of 72 percent H_2SO_4 at 30°C for 30 min and then the mixture was diluted to 2N H_2SO_4 by the addition of 1.1 ml of Milli-Q water. The hydrolysis was carried out at 120°C in an autoclave for 1 h. After acid hydrolysis, the mixture was neutralized with 2N NaOH (prepared from 50 percent NaOH, Fisher Scientific) before HPAEC analysis.

The resulting monosaccharides were analyzed by HPAEC-PAD (Dionex, Sunnyvale, USA) with an ED 50 (Dionex) detector and CarboPak PA-10 analytical column (2 × 250 mm) with a PA-10 guard column (Dionex). The acid hydrolysates of mannans were analyzed by eluting the sugars isocratically with 20 mM NaOH for 20 min followed by 200 mM NaOH for 10 min with a flow rate of 0.25 ml/min. The column was equilibrated with 20 mM NaOH for 10 min between each run of samples in a sequence.

The acid hydrolysates of pectins were analyzed mainly following the method described by Talaga *et al.* (2002) with modifications for convenience. Solvent A was Milli-Q water, solvent B was 0.2 M NaOH, and solvent C was 0.5 M sodium acetate (NaOAc) in 100 mM NaOH. The gradients of NaOH and NaOAc were used simultaneously to elute the sugars by mixing the three solvents to generate the following gradient of NaOH: 0–15 min, 18 mM; 15–18 min, 18–100 mM; 18–45 min, 100 mM; 45.1–50 min, 18 mM. The simultaneous gradient of NaOAc was 18–35 min, 0–250 mM; 35.1–45 min, 500 mM. The column was equilibrated for 10 min with 18 mM NaOH between each run of samples in a sequence. The PAD was set to the following waveform: 0.00 min, 0.10 V; 0.20 min, 0.10 V (integration = begin); 0.40 min, 0.10 V (integration = end); 0.41 min, −2.00 V; 0.42 min, −2.00 V; 0.43 min, 0.60 V; 0.44 min, −0.10 V; and 0.50 min, −0.10 V. Peaks were identified based on the relative retention time of peaks of authentic monosaccharide standards (Sigma). Sugars were quantified after calibration of PAD response to concentration of different sugars.

5.2.2. *Mannan and pectin profiles in different developmental stages of the pseudobulb*

The total polysaccharide content in the pseudobulb was microscopically observed by tissue section and staining before HPLC analysis. Ruthenium red staining clearly showed that the pectin content decreased from the A′ stage to the C stage, and starch accumulated transiently in the A–B stages (Fig. 5.1). By using HPLC method, the total mannan and pectin content of the pseudobulb was also found to decrease from the A′ stage to the C stage (Figs. 5.2 and 5.3). It appears that the mobilization of mannans and pectins in the pseudobulb could be an essential step in the induction of flowering in *Oncidium*. Expression sequence tag (EST) analysis of genes involved in carbohydrate metabolism in the pseudobulb at different stages of inflorescence development indicated a metabolic network

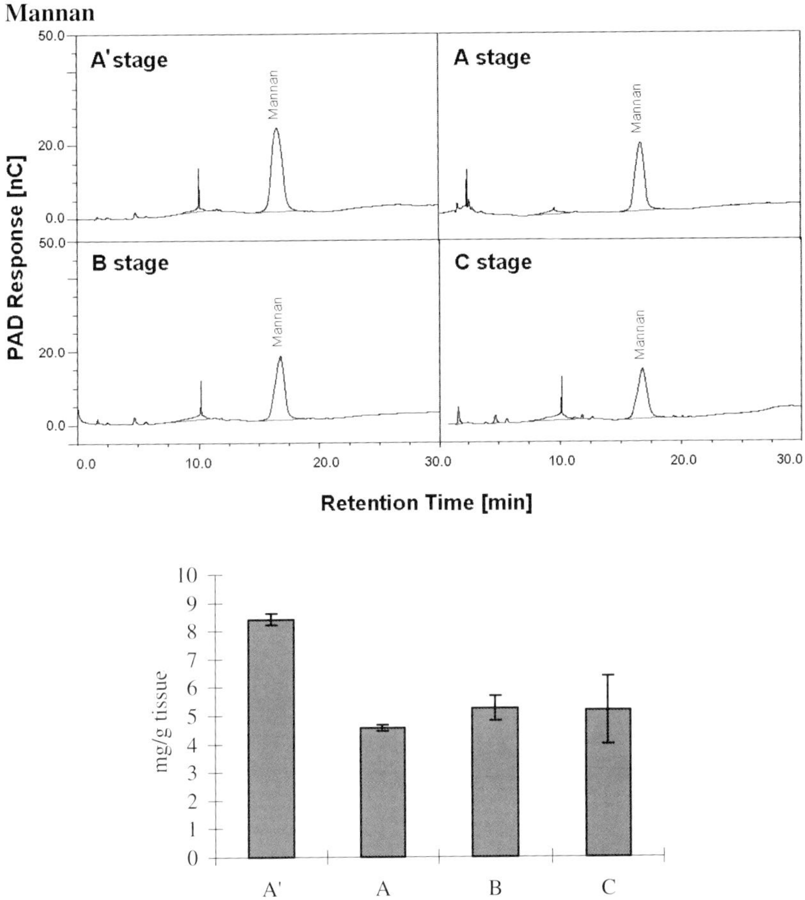

Fig. 5.2. Mannan content of the pseudobulb of *Oncidium* at different stages of inflorescence development. Vertical bars represent standard error of the mean of two independent measurements. [From Wang et al., *Planta* (2008) **227**: 1063–1077. The authors wish to thank *Planta* for providing the figures cited in the text.]

of sucrose-mannan-starch-pectin interconversion in *Oncidium* (Wang *et al.*, 2008).

Mannans are the designated cell wall storage polysaccharides along with xyloglucans and galactans, which have the function of both storage and mechanical support (Buckeridge, 2000). Three different

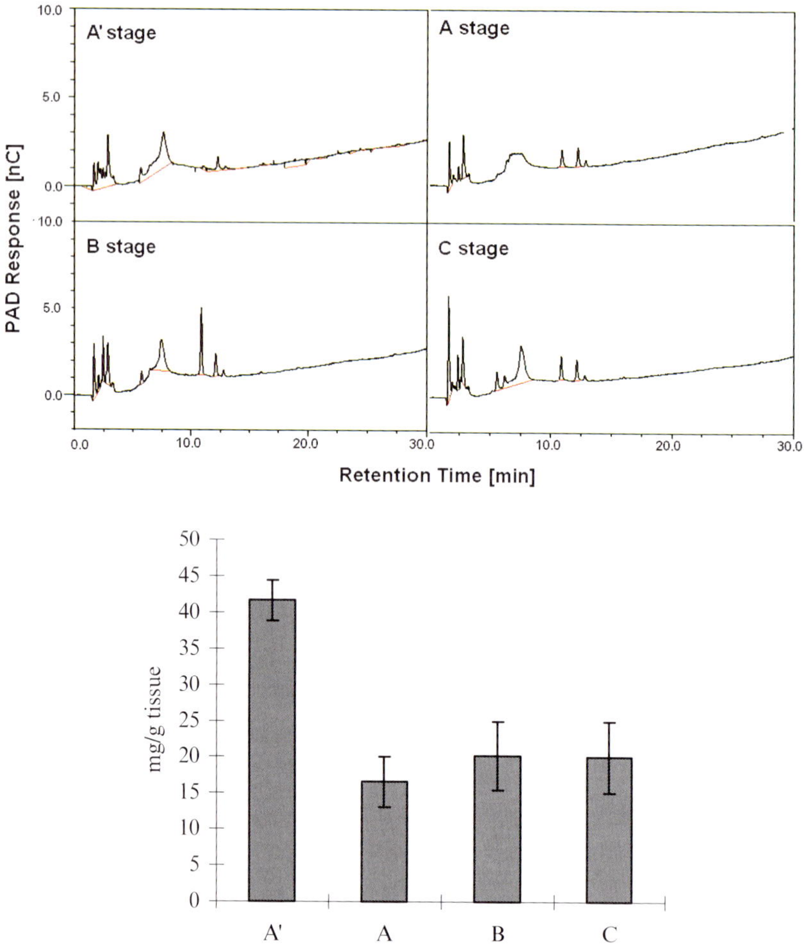

Fig. 5.3. Pectin content of the pseudobulb of *Oncidium* at different stages of inflorescence development. Vertical bars represent standard error of the mean of two independent measurements. [From Wang et al., *Planta* (2008) **227**: 1063–1077. The authors wish to thank *Planta* for providing the figures cited in the text.]

types of mannans have been found to occur in plants: pure mannans, galactomannans and glucomannans (Buckeridge, 2000). Unlike galactomannans and glucomannans, pure mannans are insoluble in water and are considered to contain 90 percent or more mannopyranose with a linear chain of β-(1-4)-linked mannopyranosyl residues and up to 10 percent α-(1-6)-linked galactopyranosyl residues. Galactomannans are structurally similar to mannans and are composed of a linear backbone of β-(1-4)-linked mannopyranosyl residues containing α-(1-6)-linked galactopyranosyl residues in larger proportions than those of pure mannans. Glucomannans have a linear β-(1-4)-linked chain of almost equal numbers of β-glucopyranosyl and β-mannopyranosyl residues with an occasional branch containing α-(1-6)-linked galactopyranosyl residues (Buckeridge, 2000). The water extractability of *Oncidium* mannans indicates that the pseudobulb contains water soluble glucomannans and galactomannans. However, the sugar composition of mannans in the *Oncidium* pseudobulb indicates that mannose accounts for about 90 percent of the total sugars, which is a characteristic feature of pure mannans (Table 5.1). Based on gas chromatography mass spectrometry (GC-MS) analysis and the specific enzyme digestion of mannans isolated from the pseudobulb, it was reported that the pseudobulb predominantly contained (1,4)-β-mannans (Wang *et al.*, 2006). Analyzing the monosaccharide composition of mannans revealed that there is a significant decrease in galactose with an apparent increase in glucose content from the A′ stage to the C stage, indicating both qualitative and quantitative change in mannans in the pseudobulb during the

Table 5.1 Monosaccharide Composition of Mannans of the *Oncidium* Pseudobulb at Different Stages of Inflorescence Development. Values are the Mean of Three Independent Experiments Plus/Minus Standard Error

	Mol percent			
	A′ Stage	**A Stage**	**B Stage**	**C Stage**
Arabinose	3.00 ± 1.20	1.04 ± 0.53	0.54 ± 0.27	0.57 ± 0.28
Galactose	[a]4.59 ± 0.45	[b]2.56 ± 0.30	[c]1.26 ± 0.13	[c]1.27 ± 0.04
Glucose	5.45 ± 0.24	4.24 ± 0.71	6.10 ± 0.69	14.61 ± 6.41
Mannose	86.96 ± 1.73	92.16 ± 0.13	92.10 ± 0.29	83.55 ± 6.15

Numbers which are not indicated by any prefix or indicated by the same letter are not statistically significant; those indicated by different letters are statistically significant.

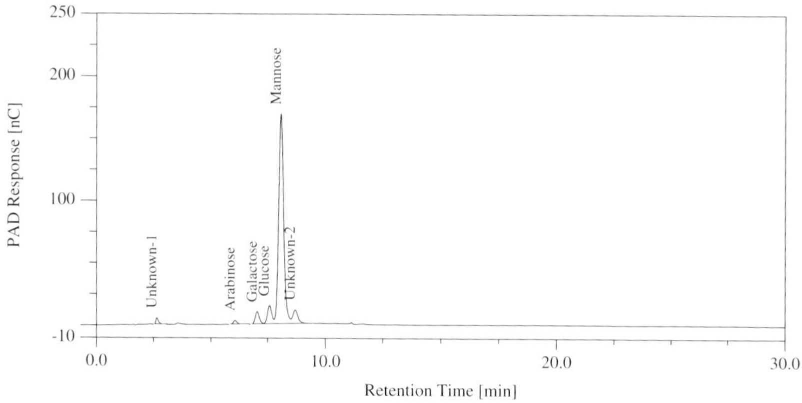

Fig. 5.4. HPAEC chromatogram of the 2N H_2SO_4 hydrolysate of mannans of the *Oncidium* pseudobulb.

flowering process. The presence of arabinose in the mannan extract indicates that *Oncidium* mannans also contain arabinose (Fig. 5.4 and Table 5.1). Since arabinose is not a component of mannans in higher plants, it is quite possible that the current extraction procedure (Mulimani and Prashanth, 2002) might also extract a small fraction of arabinogalactans which occur widely in higher plants (Immerzeel, 2004). Although no direct role for mannans in flowering has been described so far, it appears that mannan metabolism is linked to the generation of antioxidants (Smirnoff, 2001), which are much needed by the plant during the metabolic outburst at the time of flowering. The degree of galactosylation of mannans shifts their function from mechanical support to water related functions (Buckeridge *et al.*, 2000; Mulimani and Prashanth, 2002). Galactomannans have been found to retain water and act as osmoprotectants during times of water stress (Reid and Bewley, 1979). As *Oncidium* is an epiphytic orchid growing in xeric habitats, it might have evolved to store mannans as multifunctional storage polysaccharides in the pseudobulb.

Pectin is a family of complex cell wall polysaccharides containing α-(1-4)-linked D-galactopyranosyl uronic acid residues present in all plant primary walls. Pectins can be extracted from AIR by employing a variety of aqueous solvents (Femenia, 1999). We used 0.5 percent ammonium oxalate solution for the extraction of pectins (Western *et al.*, 2001). A pectin fraction without mannan contamination was obtained using this method. The purity of pectin was ascertained by testing the ammonium oxalate fraction with Fehling's solution, which precipitates any contaminating mannans present.

Three different classes of pectins have been described based on sugar composition and structure: homogalacturonan (HRG), rhamnogalacturonan I (RG-I) and rhamnogalacturonan II (RG-II). The biological activity of pectin and pectic polysaccharides ranges from mechanical support to signaling in various aspects of plant metabolism. Ridley *et al.* (2001) reviewed the pectin structure, biosynthesis and oligogalacturonide-related signaling in plants, while Willats *et al.* (2001) reviewed the cell biology and functional analysis of pectins. The decrease in the total pectin content in the pseudobulb from the A′ stage to the flower production C stage (Fig. 5.3) indicates that pectin is also mobilized in the pseudobulb of *Oncidium* during the flowering process. The solubilization of pectin might be mediated through both enzymatic and nonenzymatic mechanisms. The enzymes involved in pectin degradation are pectin esterase, pectate lyase and polygalacturonase (Dumville and Fry, 2000) and the nonenzymatic mechanism involves the solubilization of pectin by ascorbate (Dumville and Fry, 2003).

Six peaks were identified in the HPAEC profile of the 2N H_2SO_4 hydrolysate of pectins: rhamnose, arabinose, galactose, glucose, xylose and mannose, and galacturonic acid (Fig. 5.5). It was not possible to separate mannose and xylose under the present conditions of chromatographic column development and elution. In fact, mannose and xylose are co-eluted under most chromatographic conditions (Jahnel *et al.*, 1998). The residual signal appearing at 27.5 min is due to a narrow zone of hydroxide ions being displaced from the column by increased acetate gradient (Talaga, 2002). The low peak intensity of galacturonic acid is due to low PAD response to this sugar, as observed in the calibration run (results not shown). The relative percentages of rhamnose, arabinose and galactose were found to decrease in older pseudobulbs, whereas those of glucose and galacturonic acid increased (Table 5.2). A change in the quality

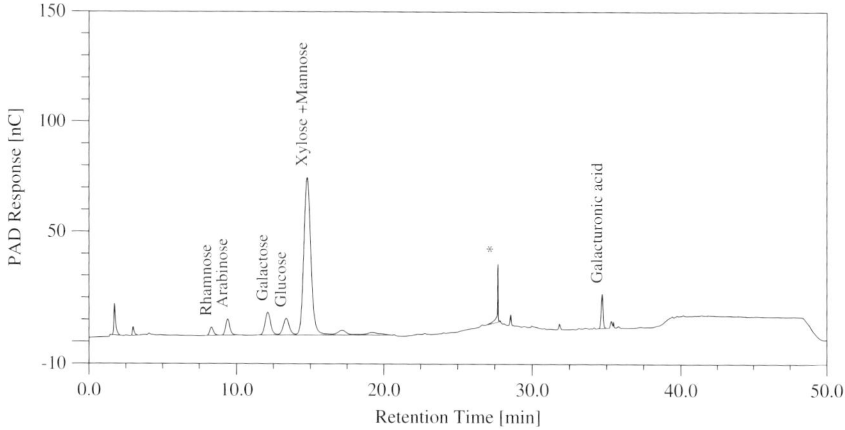

Fig. 5.5. HPAEC chromatogram of the 2N H_2SO_4 hydrolysate of pectins of the *Oncidium* pseudobulb. The peak appearing at around 27.5 min (marked with an asterisk) indicates the hydroxyl ion, which is eluted by the sodium acetate gradient.

Table 5.2 Monosaccharide Composition of Pectins of the *Oncidium* Pseudobulb at Different Stages of Inflorescence Development. Values are the Mean of Four Independent Experiments Plus/Minus Standard Error

	Mol %			
	A′ Stage	A Stage	B Stage	C Stage
Rhamnose	4.23 ± 0.62	[a]5.44 ± 0.37	[b]2.66 ± 0.38	[b]2.71 ± 0.50
Arabinose	[a]9.26 ± 0.66	[a]8.97 ± 0.10	[b]4.28 ± 0.22	[b]4.51 ± 0.08
Galactose	22.61 ± 7.21	[a]24.65 ± 1.35	[b]7.42 ± 0.96	[b]7.91 ± 0.77
Glucose	4.45 ± 0.87	[a]2.91 ± 0.06	[b]12.96 ± 4.07	[b]13.42 ± 0.70
Xylose/Mannose	41.10 ± 9.75	19.12 ± 3.02	32.61 ± 4.90	26.33 ± 4.72
Galacturonic acid	[a]18.34 ± 2.38	[b]38.91 ± 1.20	[b]40.07 ± 2.32	[b]45.12 ± 5.38

Numbers which are not indicated by any prefix or indicated by the same letter are not statistically significant; those indicated by different letters are statistically significant.

of pectins — an increase in the ratio of homogalacturonan to rhamnogalacturonan — was evident in the pseudobulb during the flowering process. An increased galacturonic acid content in the later stage of inflorescence development (Table 5.2) suggests that pectins with fewer side chains of other sugars (homogalacturonans) were generated for a specific reason. In fact, homogalacturonan could be recovered after enzymatic hydrolysis (endo-polygalacturonase plus pectin methyl esterase plus side-chain degrading enzymes) of sugar beet pectin followed by anion-exchange and size exclusion chromatography (Ralet *et al.*, 2005).

Interestingly, we have also observed the increased expression of polygalacturonase, pectinesterase and pectate lyase in the pseudobulb before the initiation of inflorescence (unpublished). This suggests that homogalacturonans are produced *in vivo* in the pseudobulb during the flowering process. Furthermore, homogalacturonan is an important source of oligogalacturonides that appear to have roles as signaling molecules in both plant development and defense (Capodicasa *et al.*, 2004). Oligogalacturonides have been considered as non-traditional plant growth regulators possessing the ability to induce flower formation along with other functions (Creelman and Mullet, 1997). Immunolocalization of homogalacturonan in the apex of the long-day plant *Sinapis alba* at floral transition indicated that pectin content drops dramatically in the first hours of this transition (Sobry, 2005). Therefore, pectin derived signals from the pseudobulb might play a role in inflorescence initiation, and an investigation of the involvement of oligogalacturonides in flower induction in *Oncidium* is underway.

5.3. Conclusions

The pseudobulb of *Oncidium* Gower Ramsey is an organ possessing several functions — from acting as storage organ for water and nutrients to being responsible for initiation and development of inflorescence. The mucilaginous content of the pseudobulb is a rich source of polysaccharides, especially pectins and mannans. The pectin and mannan content of the pseudobulb was found to decrease during the initiation and development of inflorescence. HPAEC-PAD analysis of the pseudobulb mannans revealed that there was a decrease in relative content of galactose with an apparent increase in glucose after the initiation of inflorescence axis. A monosaccharide analysis of pseudobulb pectins revealed that there was a decrease in the relative content of rhamnose, arabinose and galactose, and a corresponding increase in glucose and galacturonic acid, indicating that the ratio of homogalacturonan to rhamnogalacturonan increased during the flowering process. A close relationship between the compositional variation of polysaccharides in the pseudobulb and the flowering process in *Oncidium* was discussed.

5.4. Future Prospects

The pseudobulb of *Oncidium* Gower Ramsey has been found to be a rich source of polysaccharides (5–9 mg of mannans and 15–45 mg of pectins per gram tissue; Figs. 5.2 and 5.3). In addition to this, there is also a qualitative change in these polysaccharides in the pseudobulb at different stages of inflorescence development (Tables 5.1 and 5.2). Plant polysaccharides are gaining importance for their immunostimulatory activity and biomedical applications (Paulsen, 2001). Further studies on the biomedical properties of polysaccharides in the pseudobulb might elucidate *Oncidium* as an important source of novel bioactive polysaccharides in addition to its floricultural importance.

Acknowledgments

The authors thank the National Science Council, Taiwan for the financial support granted to Prof. Kai-Wun Yeh under the grants NSC-93-2317-3-002-012 and NSC-92-2317-B-002-006.

References

Baier M, Hemmann G, Holman R *et al.* (2004) Characterization of mutants in *Arabidopsis* showing increased sugar-specific gene expression, growth and developmental responses. *Plant Physiol* **134**: 81–91.

Buckeridge MS, Santos HP, Tiné MAS. (2000) Mobilization of storage cell wall polysaccharides in seeds. *Plant Physiol Biochem* **38**: 141–156.

Capodicasa C, Vairo D, Zabotina O *et al.* (2004) Targeted modification of homogalacturonan by transgenic expression of a fungal polygalacturonase alters plant growth. *Plant Physiol* **135**: 1294–1304.

Chanda SV. (2002) Changes in cell wall polysaccharides during sunflower seed development. *Bulg J Plant Physiol* **28**: 21–29.

Creelman RA, Mullet JA. (1997) Oligosaccharins, brassinolides, and jasmonates: Nontraditional regulators of plant growth, development, and gene expression. *Plant Cell* **9**: 1211–1223.

Dumville JC, Fry SC. (2000) Uronic acid-containing oligosaccharins: Their biosynthesis, degradation and signaling roles in non-deseased plant tissues. *Plant Physiol Biochem* **38**: 125–140.

Dumville JC, Fry SC. (2003) Solubilization of tomato fruit pectins by ascorbate: A possible non-enzymatic mechanism of fruit softening. *Planta* **217**: 951–961.

Femenia A, Waldron KW, Robertson JA, Selvendran RR. (1999) Compositional and structural modification of the cell wall of cauliflower (*Brassica oleracea* L. Var *botrytis*) during tissue development and plant maturation. *Carbohydr Polym* **39**: 101–108.

Ghelardi E, Tavanti A, Celendroni F *et al.* (2000) Effect of a novel mucoadhesive polysaccharide obtained from tamarind seeds on the intraocular penetration of gentamicin and ofloxacin in rabbits. *J Antimicrob Chemother* **46**: 831–834.

Gibson SI. (2005) Control of plant development and gene expression by sugar signaling. *Curr Opin Plant Biol* **8**: 93–102.

Gorshkova TA, Wyatt SE, Salnikov VV *et al.* (1996) Cell-wall polysaccharides of developing flax plants. *Plant Physiol* **110**: 721–729.

Hew CS, Ng CKY. (1996) Changes in mineral and carbohydrate content in pseudobulbs of C_3 epiphytic orchid hybrid *Oncidium* Goldiana at different growth stages. *Lindleyana* **11**: 125–134.

Immerzeel P, Schols HA, Voragen AGJ, de Vries SC. (2004) Different arabinogalactan proteins are present in carrot (*Daucus carota*) cell culture medium and in seeds. *Physiol Plant* **122**: 181–189.

Jahnel JB, Ilieva P, Frimmel FH. (1998) HPAE-PAD — A sensitive method for the determination of carbohydrates, *Fresenius J Anal Chem* **360**: 827–829.

Koch KE. (1996) Carbohydrate-modulated gene expression in plants. *Ann Rev Plant Physiol Plant Mol Biol* **47**: 509–540.

Koch K. (2004) Sucrose metabolism: Regulatory mechanisms and pivotal roles in sugar sensing and plant development. *Curr Opin Plant Biol* **7**: 235–246.

Koh TH, Melton LD. (2002) Ripening-related changes in cell wall polysaccharides of strawberry cortical and pith tissues. *Postharvest Biol Technol* **26**: 23–33.

Mehvar R. (2003) Recent trends in the use of polysaccharides for improved delivery of therapeutic agents: Pharmacokinetic and pharmacodynamic perspectives. *Curr Pharm Biotechnol* **4**: 283–302.

Mulimani VH, Prashanth SJ. (2002) Investigating plant galactomannans. *Biochem Mol Biol Edu* **30**: 101–103.

Paulsen BS. (2001) Plant polysaccharides with immunomostimulatory activities. *Curr Org Chem* **5**: 939–950.

Ralet M-C, Cabrera JC, Bonnin E *et al.* (2005) Mapping sugar beet pectin acetylation pattern. *Phytochemistry* **66**: 1832–1843.

Reid JSG, Bewley JD. (1979) A dual rôle for the endosperm and its galactomannan reserves in the germinative physiology of fenugreek (*Trigonella foenum-graecum* L.), an endospermic leguminous seed. *Planta* **147**: 145–150.

Ridley BL, O'Neil MA, Mohnen D. (2001) Pectins: Structure, biosynthesis, and oligogalacturonide-related signaling. *Phytochemistry* **57**: 929–967.

Smeekens S. (2000) Sugar-induced signal transduction in plants. *Ann Rev Plant Physiol Plant Mol Biol* **51**: 49–81.

Smirnoff N, Conklin PL, Loewus FA. (2001) Biosynthesis of ascorbic acid in plants: A renaissance. *Ann Rev Plant Physiol Plant Mol Biol* **52**: 437–467.

Sobry S, Havelange A, Liners F, Cutsem PV. (2005) Immunolocalization of homogalacturonans in the apex of the long-day plant *Sinapis alba* at floral transition. The pectin content drops dramatically in the first hours of this transition. *Physiol Plant* **123**: 339–347.

Stancato GC, Mazzafera P, Buckeridge MS. (2001) Effect of a drought period on the mobilization of non-structural carbohydrates, photosynthetic efficiency and water status in an epiphytic orchid. *Plant Physiol Biochem* **39**: 1009–1016.

Stolle-Smiths T, Beekhuizen JG, Kok MTC *et al.* (1999) Changes in cell wall polysaccharides of green bean pods during development. *Plant Physiol* **121**: 363–372.

Suda CNK, Buckedidge MS, Giorgini JF. (2003) Cell wall hydrolases in the seeds of *Euphorbia heterophylla* L. during germination and early seedling development. *Braz J Plant Physiol* **15**: 135–147.

Talaga P, Vialle S, Moreau M. (2002) Development of a high-performance anion-exchange chromatography with pulsed-amperometric detection based quantification assay for pneumococcal polysaccharides and conjugates. *Vaccine* **20**: 2474–2484.

Wang CY, Chiou CY, Wang HL, Krishnamurthy R. (2008) Carbohydrate mobilization and gene regulatory profile in the pseudobulb of *Oncidium* orchid during the flowering process. *Planta* **227**: 1063–1077.

Wang H-L, Chung J-D, Yeh K-W. (2003) Changes in carbohydrate and free amino acid pools in current pseudobulb of *Oncidium* "Gower Ramsey" during inflorescence development. *J Agric Assoc China* **4**: 476–488.

Wang H-L, Yeh K-W, Chen P-R *et al.* (2006) Isolation and characterization of pure mannans from *Oncidium "Gower Ramsey"* current pseudobulb during initial inflorescence development. *Biosci Biotechnol Biochem* **70**: 551–553.

Western TL, Burn J, Tan WL *et al.* (2001) Isolation and characterization of mutants defective in seed coat mucilage secretory cell development in Arabidopsis. *Plant Physiol* **127**: 998–1011.

Willats WGT, McCartney L, Mackie W, Knox JP. (2001) Pectin: Cell biology and prospects for functional analysis. *Plant Mol Biol* **47**: 9–27.

Yong JWH, Hew CS. (1995) The importance of photoassimilate contribution from current shoots and connected back shoots to inflorescence size in the thin-leaved sympodial orchid *Oncidium* Goldiana. *Int J Plant Sci* **156**: 450–459.

Zimmerman JK. (1990) Role of pseudobulbs in growth and flowering in *Catasetum viridiflavum* (Orchidaceae). *Am J Bot* **77**: 533–542.

CHAPTER 6

Coloration and Color Patterning in Floral Tissues of *Oncidium* Gower Ramsey

Chung-Yi Chiou* and Kai-Wun Yeh*,†

This chapter investigates the molecular regulatory mechanism of the anthocyanin-biosynthetic genes involving in floral pigmentation of *Oncidium* Gower Ramsey. The results reveal that the MYB gene is inactive, suppressing the regulation of chalcone isomerase (CHI) and dihydroflavonol reductase (DFR). This results in the elimination of anthocyanin pigments in floral tissues.

6.1. Introduction

In nature, flower pigments such as carotenoids, anthocyanins and betalains are responsible for the natural attractive factors of plants. These three groups of pigments play important ecological functions, such as attracting animal pollinators (Schaefer and Wilkinson, 2004). Among them, anthocyanins have the broadest distribution in flowering plants and their biosynthetic pathway has been well characterized in several plant species (Grotewold, 2006). Anthocyanins are derived from the phenylpropanoid pathway. Chalcone synthase (CHS) is the first key enzyme to produce a tetrahydroxychalcone, which acts as the precursor for all classes of flavonoids. A subsequent enzymatic reaction from chalcone to naringenin is catalyzed by chalcone isomerase (CHI), which is further converted to dihydrokaempferol by flavanone 3-hydroxylase (F3H). Finally, three classes of anthocyanidin end-products are completed by consecutive enzymatic activities, including flavonoid 3′-hydroxylase (F3′H), flavonoid 3′,5′-hydroxylase

*Institute of Plant Biology, National Taiwan University, Taipei 106, Taiwan.
†Corresponding author. E-mail: ykwbppp@ntu.edu.tw

(F3′5′H), and dihydroflavonol 4-reductase (DFR). Anthocyanidin synthase (ANS) catalyzes the reaction from the colorless leucoanthocyanidin to the colored anthocyanidin. Almost all anthocyanidins undergo several modifications, such as glycosylation or methylation, by UDP-glucoside:flavonoid 3-*O*-glucosyltransferase (3GT) and anthocyanin methyltransferase (AMT) (Fig. 6.1a). These water-soluble pigments are eventually accumulated in the vacuoles of the epidermal cell and are

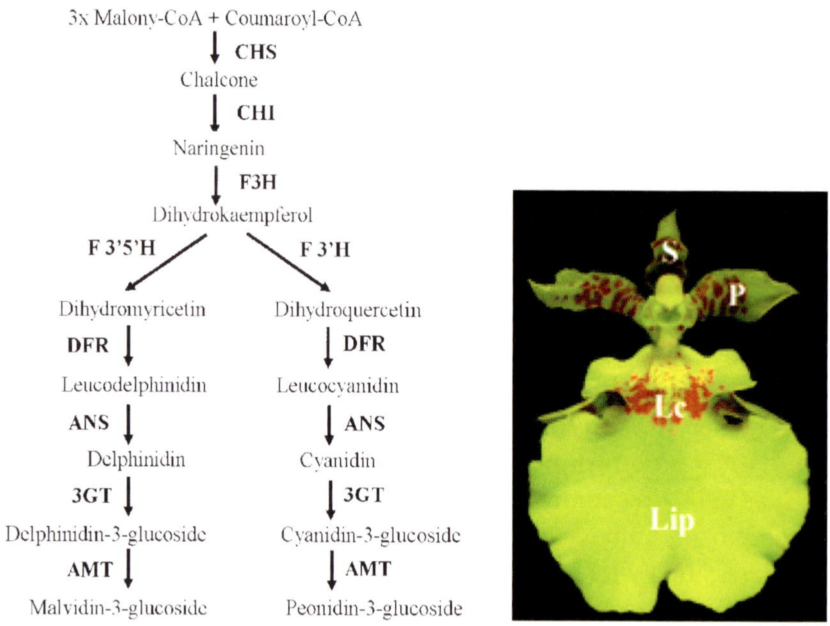

(a) (b)

Fig. 6.1. Flower phenotype of *Oncidium* "Gower Ramsey" and related anthocyanin biosynthetic pathway. (a) Anthocyanin biosynthetic pathway. Names of enzymes are abbreviated in bold capital letters as follows: CHS, chalcone synthase; CHI, chalcone flavanone isomerase; F3H, flavanone 3-hydroxylase; F3′H, flavonoid 3′-hydroxylase; F3′5′H, flavonoid 3′,5′-hydroxylase; DFR, dihydroflavonol 4-redutase; ANS, anthocyanidin synthase; 3GT, UDP-glucoside:flavonoid 3-*O*-glucosyltransferase; AMT, anthocyanin methyltransferase. (b) The different parts of the flower are as follows: S, sepal; P, petal; Lc, Lip crest; Lip, labellum. [From Chiou and Yeh (2008). The authors wish to thank *Plant Molecule Biology* for providing the figures cited in the text.]

responsible for the appearance of color. Delphinidin derivatives usually make colors look purple or dark purple, whereas cyanidin derivatives make flowers appear red in color.

It has been established that the expression of anthocyanin biosynthetic genes is coordinately controlled by two families of transcription regulators, bHLH and MYB proteins (Holton and Cornish, 1995; Koes *et al.*, 2005). bHLH proteins may have overlapping regulatory targets (Zhang *et al.*, 2003; Zimmermann *et al.*, 2004), but MYB proteins are the key components to activate discrete subsets of anthocyanin structural genes (Davies and Schwinn, 2003). However, the regulatory function and mechanism of MYB proteins are not fully understood, and related studies in different plant species have yielded inconsistent findings. Arabidopsis AtPAP1, a MYB protein, up-regulates a number of genes in the anthocyanin biosynthesis pathway from *phenylalanine ammonia-lyase* (*PAL*) to *CHS* and *DFR* (Borevitz *et al.*, 2000; Tohge *et al.*, 2005). By contrast, *MybA* in grapes has a limited ability to activate genes in the very late stages of the anthocyanin biosynthetic pathway, including *3GT* genes (Kobayashi *et al.*, 2002). Interestingly, *FaMYB1* in strawberries activates anthocyanin and flavonol biosynthetic pathways, but it represses these pathways in transgenic tobacco (Aharoni *et al.*, 2001).

Oncidium Gower Ramsey is one of the top-grade cut flowers in the orchid industry. Its unique flower shape and predominant yellow color with mosaic red pigmentation make this hybrid flower highly prized in the market. This orchid is known for its bi-color mixture of carotenoids and anthocyanins, which are localized in sepal, petal and lip crest tissues (Fig. 6.1b). The predominant yellow coloration is characterized by an equal mixture of all-*trans* and 9-*cis* isomers of violaxanthin, with esterification specific to the 9-*cis* isomer (Hieber *et al.*, 2006). Although the red pigments in the plant are ascribed to the coexistence of both carotenoids and anthocyanins (Thammasiri *et al.*, 1986; Matsui and Nakamura, 1988; Matsui *et al.*, 1994), the properties of anthocyanin components in the floral tissues of *Oncidium* are still poorly defined. To identify the anthocyanin compounds, analyzing the expression profiles and coordinate regulation pattern of the relevant biosynthetic genes in the floral tissues of *Oncidium* is the best way to study the molecular mechanism. This chapter aims to unravel the molecular regulation of the anthocyanin-biosynthesis genes that affect the color pattern of *Oncidium* floral tissue.

6.2. Color Pattern in Floral Tissues of *Oncidium* Hybrid

Oncidium Gower Ramsey is a highly profitable cut flower in the Asian marketplace. It is the brilliant yellow-green color in its floral tissues mottled with a brownish-red base in the sepals, petals and lip crest that makes it attractive. Here we aim to unravel the formation of this floral color pattern by a molecular approach.

6.2.1. *HPLC analysis of anthocyanin in Oncidium floral tissues*

Oncidium Gower Ramsey and *Oncidium* Sharry Baby were selected for our analysis of floral pigmentation and the expression pattern of the anthocyanin-biosynthetic gene. The extraction of anthocyanin pigments was carried out following the method of Goodman *et al.* (2004), briefly outlined in the following. *Oncidium* floral tissues were extracted by grinding in an appropriate solvent of 0.1N HCl in methanol. The ground tissues were immediately centrifuged twice to separate the debris. The supernatant was removed and diluted with 5% acetic acid in ratios ranging from 1:1 to 20:1 depending on the pigment concentration. The final solution was then subjected to HPLC analysis, which was performed using a Dionex GP40 gradient pump (Dionex, Sunnyvale, CA) and a Microsorb 100-5 C18 column (Varian, Palo Alto, CA). Pigment separation was carried out by gradient elution with a flow rate of 0.75 ml/min: solvent A, 5% acetic acid; solvent B, acetonitrile, 1 min at 90% A, 10% B; from 90% A, 10% B to 55% A, 45% B in 17.5 min; to 100% B in 2.5 min; at 100% B for 1 min; to 90% A, 10% B in 3 min; at 90% A, 10% B for 3 min. Absorbance was detected at OD_{520} using a model Dionex AD20 detector. Data were collected and analyzed using PEAKNET software (Dionex, Sunnyvale, CA). Anthocyanin compounds, such as malvidin-3-*O*-galactoside, peonidin-3-*O*-glucoside, delphinidin-3-*O*-glucoside and cyanidin-3-*O*-glucoside, were purchased from Fluka and monitored via HPLC as check standards.

The composition of yellow pigment in the *Oncidium* flower has been characterized as the mixture of 9-*cis* violaxanthin, violaxanthin and neoxanthin (Hieber *et al.*, 2006). However, the anthocyanin profile is not clear. To elucidate the anthocyanin components, the extracts from sepal, lip crest, and lip were analyzed separately by HPLC. The results showed that the red portions of the sepal and lip crest contain a mixture of the compounds peonidin-3-*O*-glucoside, delphinidin-3-*O*-glucoside and cyanidin-3-*O*-glucoside (Figs. 6.2a and 6.2b). Interestingly, no anthocyanins were detected in the lip tissue (Fig. 6.2c). The major peak represents cyanidin-3-*O*-glucoside, which accounted for almost 56 percent of the total area in the sepal and lip crest profiles. Other peaks revealed delphinidin-3-*O*-glucoside and peonidin-3-*O*-glucoside, accounting for over 30 percent of the total area in the profiles of the above tissues. The anthocyanin extract of the purple flowers of another *Oncidium* cultivar, called Sharry Baby, was also analyzed by HPLC. Comparing the HPLC profiles of the extracts of cultivar Gower Ramsey and cultivar Sharry Baby, the major peak of the latter changed to delphinidin-3-*O*-glucoside and the content of malvidin-3-*O*-galactoside increased (Fig. 6.2d). The total content of anthocyanin in the flowers of Sharry Baby was 53 times that of Gower Ramsey. These results explain why the flowers of Sharry Baby have a stronger purple color than Gower Ramsey (Fig. 6.2d).

6.2.2. Gene expression pattern of anthocyanin biosynthesis

To investigate the molecular mechanism responsible for the pigmentation pattern and the expression profiles of related anthocyanin biosynthetic genes, several relevant genes involved in the anthocyanin biosynthetic pathway, such as *chalcone synthase* (*CHS*), *chalcone isomerase* (*CHI*), *dihydroflavonol reductase* (*DFR*) and *anthocyanidin synthase* (*ANS*), were isolated from a subtractive EST library of flower buds. After partial cDNA sequences of these

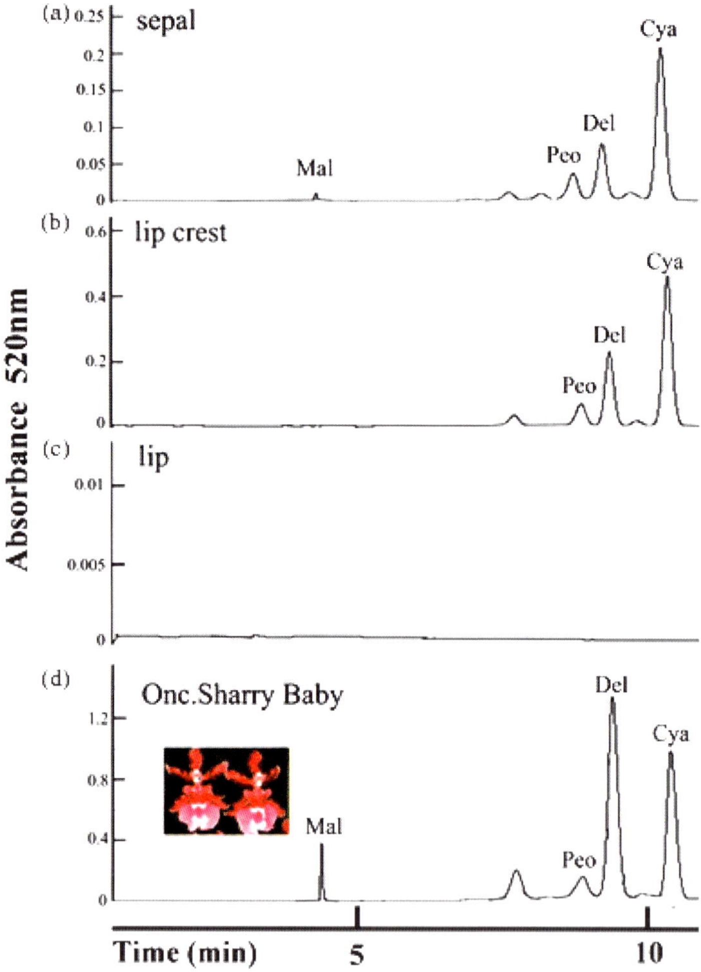

Fig. 6.2. HPLC profiles from anthocyanin analysis of *Oncidium* "Gower Ramsey" and *Oncidium* "Sharry Baby". Anthocyanins were extracted and analyzed from (a) the sepal, (b) the lip crest and (c) the lip of "Gower Ramsey" and (d) the flower bud of Sharry Baby. Peaks were identified with commercial standards and traced at 520 nm. Mal: malvidin-3-*O*-galactoside; Peo: peonidin-3-*O*-glucoside; Del: delphinidin-3-*O*-glucoside; Cya: cyanidin-3-*O*-glucoside. [From Chiou and Yeh (2008). The authors wish to thank *Plant Molecule Biology* for providing the figures cited in the text.]

genes were identified through blast search on the NCBI website, 5′-RACE and 3′-RACE were employed to get full-length cDNAs. A 1,488-bp *OgCHS* encoding 394 deduced amino acid residues, a 835-bp *OgCHI* encoding 219 deduced amino acid residues, a 1,241-bp *OgDFR* encoding 282 deduced amino acid residues, and a 968-bp *OgANS* encoding 229 deduced amino acid residues were successfully amplified The protein sequences of OgCHS, OgCHI, OgDFR, OgANS exhibited 78 percent and 75 percent, 62 percent and 57 percent, 67 percent and 55 percent, 61 percent and 51 percent identity to protein sequences of CHS, CHI, DFR, ANS from *Zea mays* and *Arabidopsis thaliana*, respectively.

Northern blot analysis was carried out to examine gene expression pattern in roots, leaves, flower buds at five developmental stages (S1–S5), floral lip tissues and the flower bud of "Sharry Baby". The results revealed that *OgCHS, OgCHI, OgDFR,* and O*gANS* were all actively expressed gradually from S1 to S4 and reached maximal expression at S5. In contrast, both *OgCHI* and *OgDFR* genes were not expressed in floral lip tissues, but *OgCHS* and *OgANS* remained actively expressed (Fig. 6.3). The correlation between the activation of *OgCHI* and *OgDFR* genes and the HPLC profile in floral lip tissue suggested that anthocyanins could not be synthesized because *OgCHI* and *OgDFR* were not expressed. Interestingly, strong expression of all four genes, *OgCHS, OgCHI, OgDFR,* and O*gANS,* in the flower bud

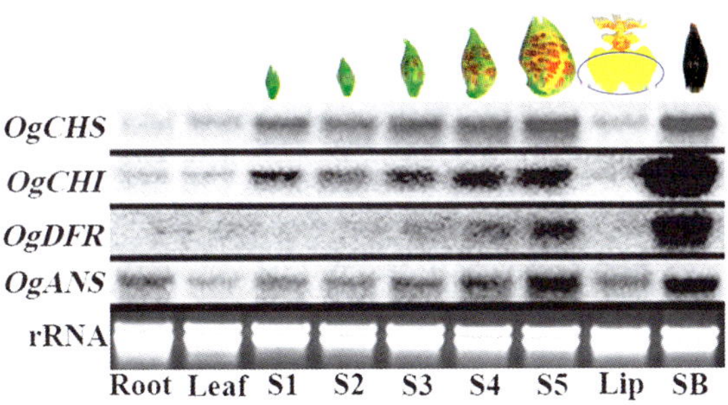

Fig. 6.3. Northern blot analysis of the four anthocyanin biosynthetic genes during flower development. [From Chiou and Yeh (2008). The authors wish to thank *Plant Molecule Biology* for providing the figures cited in the text.]

of Sharry Baby resulted in a large accumulation of anthocyanins based on HPLC data (Fig. 6.2d).

As the Northern blot analysis in Fig. 6.3 shows, anthocyanin deficiency in the floral lip might result from the expression deficiency of both *OgCHI* and*OgDFR* genes. We summarized that the recovery of these two genes could complete the anthocyanin pathway and produce anthocyanin compounds displaying the same red pigment as the lip crest tissue. Hence, *OgCHI* and *OgDFR* cDNA genes were together subcloned into a pGEM-T Easy vector backbone driven individually by the Pchrc promoter, which is a 1.7-kb 5′-upstream region of the gene encoding chromoplast-specific carotenoid-associated protein that is specifically expressed in flowers (unpublished data). The yellow lip tissue was bombarded with this construct. After incubation on an MS agar medium for two days, many red color spots were clearly observed in the lip tissue (Fig. 6.4). To identify the components in these spots, the spotted tissue was extracted in methanol solvent and analyzed by HPLC. Our analysis demonstrated that the red spots contained the

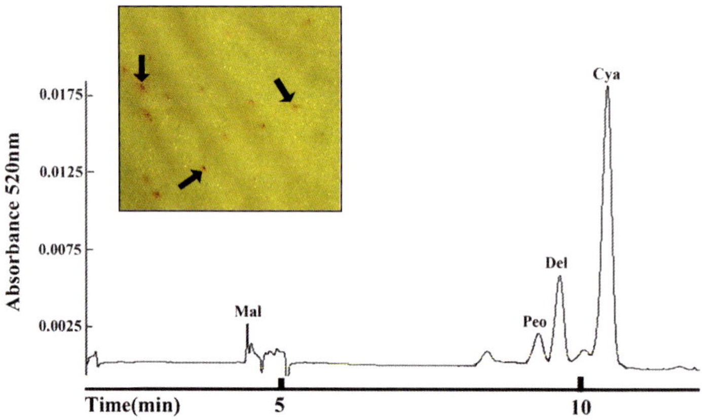

Fig. 6.4. Transient expression of *OgCHI* and *OgDFR* genes in yellow lip resulting in the development of red spots. The figure shows a digital image of lip tissue bombarded with *OgCHI* and *OgDFR* under the control of the Pchrc promoter (arrows indicate the red spots) and the HPLC profile. [From Chiou and Yeh (2008). The authors wish to thank *Plant Molecule Biology* for providing the figures cited in the text.]

same anthocyanin compounds as the sepal tissue, compounds such as peonidin-3-*O*-glucoside, delphinidin-3-*O*-glucoside and cyanidin-3-*O*-glucoside (Fig. 6.4). These results indicated that the expression of *OgCHI* and *OgDFR* together could fully counter the absence of anthocyanin synthesis in lip tissue, whereas the red spots could not be observed when lip tissue was bombarded with either *OgCHI* or *OgDFR* alone (data not shown).

6.2.3. *Regulatory analysis of anthocyanin biosynthesis*

Thus far, it has been demonstrated that transcription factors such as MYB and bHLH are required in anthocyanin biosynthesis to regulate the expression of structural genes. To test whether MYB proteins also regulate structural gene expression in our biological system, we designed degenerated primers corresponding to the most conserved regions of *MYB* genes regulating anthocyanin synthesis in different plant species. RT-PCR was performed to amplify a 214-bp cDNA fragment, which encodes the R2R3 domain (Fig. 6.5b). This cDNA fragment was then used as a probe to screen a cDNA library generated from the poly (A^+) RNA isolated from flower buds. With this approach, two *MYB* related candidate genes, named *OgMYB1* and *OgMYB33*, were isolated and characterized. The 937-bp full-length *OgMYB1* contains a coding region of 255 deduced amino acid residues, whereas the other 1,428-bp full-length *OgMYB33* encodes 301 deduced amino acid residues.

A phylogenetic analysis of the deduced amino acid sequences of *MYBs* related to anthocyanin regulation or other physiological processes indicated that *OgMYB1* placed within the group of anthocyanin regulators including *ZmPL, ZmC1, VvMYB5a*. In contrast, *OgMYB33* was not included in this group (Fig. 6.5a). This result suggests that *OgMYB33* may be involved in other physiological functions in *Oncidium* Gower Ramsey.

Alignment of the deduced amino acid sequences of *MYBs* related to anthocyanin regulation showed high sequence homology within the R2R3 domain at the N-terminus (Fig. 6.5b). The motif $[D/E]Lx_2[R/K]x_3Lx_6Lx_3R$ (from amino acid 76 to 95) that interacts with

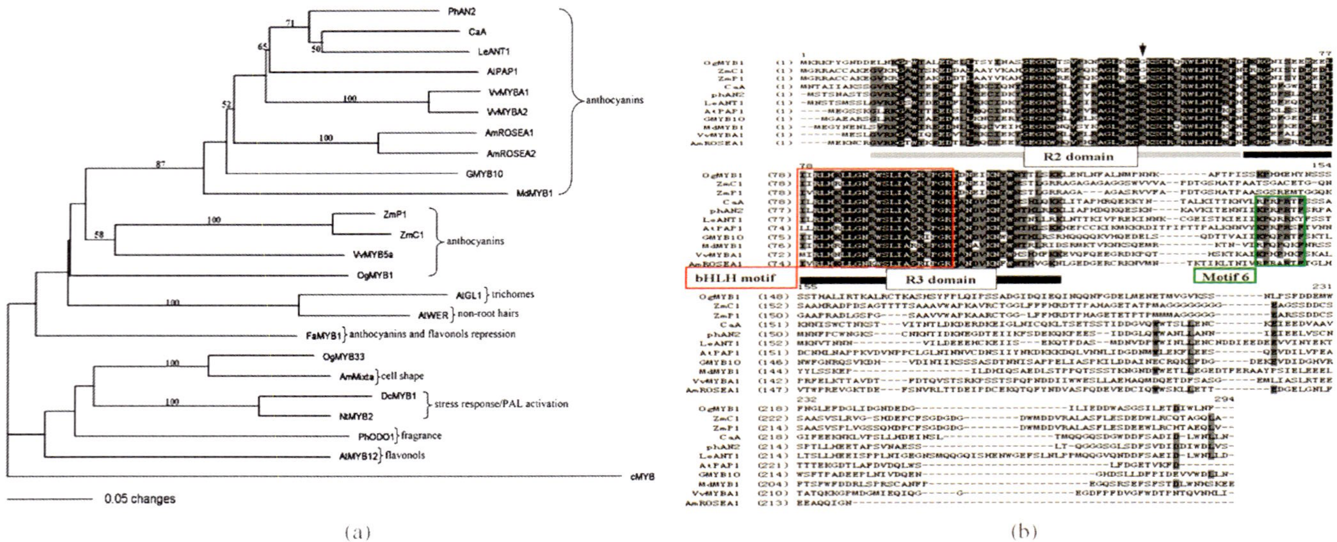

Fig. 6.5. Comparison of the deduced amino acid sequence of *OgMYB1* with other MYB proteins. (a) Phylogenetic analysis indicates the similarity of OgMYB1 with other MYBs that have been characterized as regulators of the anthocyanin pathway or other physiological processes. (b) Protein sequence alignment of OgMYB1 with other MYBs involved in the anthocyanin pathway. The R2 and R3 domains are underlined in black and gray, respectively. The bHLH binding motif is boxed in **purple** in the R3 domain. Motif 6 identified in C-terminal is boxed in **green**. A specific residue of interest that is common to anthocyanin regulators is indicated with a black arrow within the R2 domain. [From Chiou and Yeh (2008). The authors wish to thank *Plant Molecule Biology* for providing the figures cited in the text.]

the R-like bHLH protein is located in the R3 domain (Zimmermann *et al.* 2004). Takos *et al.* (2006) suggested that the R2 domain of the anthocyanin regulator has an Arg at position 39 and a conserved region in the motif 6 in dicots. By contrast, in monocots, Arg is replaced by Gly and the motif 6 is absent. This suggests that the regulatory mechanism of the anthocyanin MYBs is distinct between dicots and monocots.

The transcription level of *OgMYB1* was monitored by RT-PCR using gene specific primers designed to 3′ UTR. The results demonstrated that *OgMYB1* is highly expressed from S1 to S5, and in lip crest and Sharry Baby. However, *OgMYB1* mRNA accumulation was absent in floral lip tissue (Fig. 6.6a). This suggests that *OgMYB1* expression is

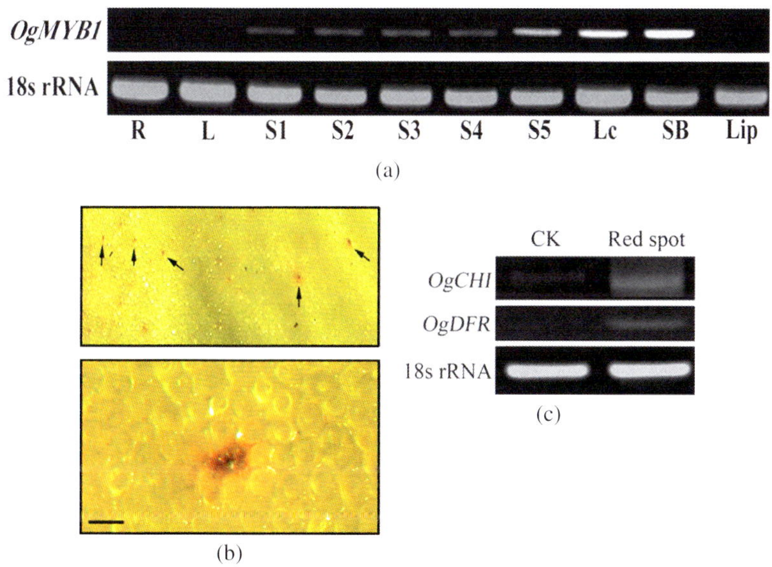

Fig. 6.6. Expression analysis of the transcripts of *OgMYB1* and transient expression of *OgMYB1* bombarded into lip tissue. (a) Transcript detection of *OgMYB1* by RT-PCR with gene-specific primers. (b) Microscope images of the red spots in lip tissue which were induced to synthesize anthocyanin by transient expression of *OgMYB1* driven by the Pchrc promoter. (c) RT-PCR analysis of *OgCHI* and *OgDFR* transcripts in the red spots of floral lip tissues, after bombarding with *OgMYB1* or empty vector (CK). [From Chiou and Yeh (2008). The authors wish to thank *Plant Molecule Biology* for providing the figures cited in the text.]

responsible for anthocyanin synthesis in floral tissues, and *OgCHI* and *OgDFR* genes may be specifically regulated by *OgMYB1.*

We therefore investigated whether activation of *OgMYB1* expression could induce anthocyanin biosynthesis in yellow lip tissue. In a further step, *OgMYB1* was bombarded into yellow lip tissue. As expected, transient expression of *OgMYB1* resulted in red spots on the yellow lip tissue. This demonstrated that OgMYB1 could regulate anthocyanin synthesis in lip tissues (Fig. 6.6b). HPLC analysis of the red spots identified the same anthocyanin compounds as those found in petal and sepal tissues (data not shown).

To examine whether *OgCHI* and *OgDFR* expression was regulated by *OgMYB1,* RNA samples from the red spots of transformed tissue were extracted and *OgCHI* and *OgDFR* expression levels were identified by RT-PCR. *OgCHI* and *OgDFR* were found to be more actively expressed in the bombarded tissue compared to the control tissue bombarded with empty vector only (Fig. 6.6c). These results suggest that the blockage of the anthocyanin synthesis pathway and the absence of anthocyanin compounds in lip tissue are probably caused by the absence of transcriptional activation of *OgMYB1*

6.3. Conclusion

HPLC analysis revealed that the red coloration in floral tissues such as sepal, petal and lip crest was composed of malvidin-3-*O*-galactoside, peonidin-3-*O*-glucoside, delphinidin-3-*O*-glucoside and cyanidin-3-*O*-glucoside compounds. In contrast, these pigments were not detected in yellow lip tissue. Expression analysis of four anthocyanin biosynthetic pathways, i.e. *chalcone synthase* (*OgCHS*), *chalcone isomerase* (*OgCHI*), *dihydroflavonol 4-reductase* (*OgDFR*) *and anthocyanidin synthase* (*OgANS*), showed that they are all active during floral development. However, *OgCHI* and *OgDFR* genes are specifically down-regulated in yellow lip tissue. Bombardment of lip tissue with *OgCHI* and *OgDFR* genes driven by a flower-specific promoter, Pchrc (chromoplast-specific carotenoid-associated gene), demonstrated that the transient expression of these two genes resulted in anthocyanin production in yellow lip tissue. Further analysis of a

R2R3 MYB transcription factor, *OgMYB1*, revealed that although it is actively expressed during floral development, it is not expressed in yellow lip tissue. Transient expression of *OgMYB1* in lip tissues by bombardment can also induce formation of red pigments through the activation of *OgCHI* and *OgDFR* transcription. These results demonstrate that differential expression of *OgMYB1* is critical to determining the color pattern of the floral organ in *Oncidium* Gower Ramsey.

References

Adams KL, Percifield R, Wendel JF. (2004) Organ-specific silencing of duplicated genes in a newly synthesized cotton allotetraploid. *Genetics* **168**: 2217–2226.

Aharoni A, De Vos CH, Wein M *et al.* (2001) The strawberry FaMYB1 transcription factor suppresses anthocyanin and flavonol accumulation in transgenic tobacco. *Plant J* **28**: 319–332.

Borevitz JO, Xia Y, Blount J *et al.* (2000) Activation tagging identifies a conserved MYB regulator of phenylpropanoid biosynthesis. *Plant Cell* **12**: 2383–2394.

Chang S, Puryean J, Cairney J. (1993) A simple and efficient method for isolating RNA from pine tree. *Plant Mol Biol Rep* **11**: 113–116.

Chiou CY, Yeh KW. (2008) Differential expression of MYB gene (OgMYB1) determines color patterning in floral tissue of *Oncidium* Gower Ramsey. *Plant Mol Biol* **66**: 379–388.

Chiou CY, Wu KQ, Yeh KW. (2008) Characterization and promoter activity of chromoplast specific carotenoid associated gene (CHRC) from *Oncidium* Gower Ramsey. *Biotechnol Lett* **30**: 1861–1866.

Davies KM, Schwinn KE. (2003) Transcriptional regulation of secondary metabolism. *Funct Plant Biol* **30**: 913–925.

De Jong WS, Eannetta NT, De Jong DM *et al.* (2004) Candidate gene analysis of anthocyanin pigmentation loci in the Solanaceae. *Theor Appl Genet* **108**: 423–432.

Deluc L, Barrieu F, Marchive C *et al.* (2006) Characterization of a grapevine R2R3-MYB transcription factor that regulates the phenylpropanoid pathway. *Plant Physiol* **140**: 499–511.

Dooner HK, Robbins TP, Jorgensen RA. (1991) Genetic and developmental control of anthocyanin biosynthesis. *Annu Rev Genet* **25**: 173–199.

Durbin ML, Lundy KE, Morrell PL *et al.* (2003) Genes that determine flower color: The role of regulatory changes in the evolution of phenotypic adaptations. *Mol Phylogenet Evol* **29**: 507–518.

Feldman M, Levy AA. (2005) Allopolyploidy — A shaping force in the evolution of wheat genomes. *Cytogenet Genome Res* **109**: 250–258.

Forkmann G, Martens S. (2001) Metabolic engineering and applications of flavonoids. *Curr Opin Biotechnol* **12**: 155–160.

Gombart AF, Grewal J, Koeffler HP. (2007) ATF4 differentially regulates transcriptional activation of myeloid-specific genes by C/EBPε and C/EBPα. *J Leukoc Biol* **81**: 1617–1625.

Goodman CD, Casati P, Walbot V. (2004) A multidrug resistance-associated protein involved in anthocyanin transport in *Zea mays*. *Plant Cell* **16**: 1812–1826.

Grotewold E. (2006) The genetics and biochemistry of floral pigments. *Annu Rev Plant Biol* **57**: 761–780.

Hieber AD, Mudalige-Jayawickrama RG, Kuehnle AR. (2006) Color genes in the orchid *Oncidium* Gower Ramsey: Identification, expression, and potential genetic instability in an interspecific cross. *Planta* **223**: 521–531.

Holton TA, Cornish EC. (1995) Genetics and biochemistry of anthocyanin biosynthesis. *Plant Cell* **7**: 1071–1083.

Kobayashi S, Ishimaru M, Hiraoka K *et al.* (2002) Myb-related genes of the Kyoho grape (*Vitis labruscana*) regulate anthocyanin biosynthesis. *Planta* **215**: 924–933.

Koes R, Verweij W, Quattrocchio F. (2005) Flavonoids: A colorful model for the regulation and evolution of biochemical pathways. *Trends Plant Sci* **10**: 236–242.

Liu B, Wendel JF. (2003) Epigenetic phenomena and the evolution of plant allopolyploids. *Mol Phylogenet Evol* **29**: 365–379.

Matsui S. (1994) Floral carotenoids in species and hybrids of the *Laeliinae*. *Lindleyana* **9**: 213–217.

Martin C, Paz-Ares J. (1997) MYB transcription factors in plants. *Trends Genet* **13**: 67–73.

Matsui S, Nakamura M. (1988) Distribution of flower pigments in perianth of *Cattleya* and allied species. *J Japan Soc Hort Sci* **57**: 222–232.

Nakano I, Paucar AA, Bajpai R *et al.* (2005) Maternal embryonic leucine zipper kinase (MELK) regulates multipotent neural progenitor proliferation. *J Cell Biol* **170**: 413–427.

Osborn TC, Pires JC, Birchler JA *et al.* (2003) Understanding mechanisms of novel gene expression in polyploids. *Trends Genet* **19**: 141–147.

Quattrocchio F, Wing J, van der Woude K *et al.* (1999) Molecular analysis of the *anthocyanin2* gene of petunia and its role in the evolution of flower color. *Plant Cell* **11**: 1433–1444.

Quattrocchio F, Wing JF, van der Woude K *et al.* (1998) Analysis of bHLH and MYB domain proteins: Species-specific regulatory differences are caused by divergent evolution of target anthocyanin genes. *Plant J* **13**: 475–488.

Schaefer HM, Wilkinson DM. (2004) Red leaves, insects and coevolution: A red herring? *Trends Ecol Evol* **19**: 616–618.

Schwinn K, Venail J, Shang Y *et al.* (2006) A small family of MYB-regulatory genes controls floral pigmentation intensity and patterning in the genus *Antirrhinum*. *Plant Cell* **18**: 831–851.

Swofford DL. (2001) *PAUP*: Phylogenetic Analysis Using Parsimony Version 4.0*. Sinauer Associates, Sunderland, MA.

Takos AM, Jaffe FW, Jacob SR *et al.* (2006) Light-induced expression of a MYB gene regulates anthocyanin biosynthesis in red apples. *Plant Physiol* **142**: 1216–1232.

Tan J, Wang HL, Yeh KW. (2005) Analysis of organ-specific, expressed genes in *Oncidium* orchid by subtractive expressed sequence tags library. *Biotechnol Lett* **27**: 1517–1528.

Thammasiri K, Tang CS, Yamamoto HY *et al.* (1986) Carotenoids and chlorophyll in yellow-flowered *Dendrobium* species. *Lindleyana* **1**: 215–218.

Tohge T, Nishiyama Y, Hirai MY *et al.* (2005) Functional genomics by integrated analysis of metabolome and transcriptome of *Arabidopsis* plants over-expressing an MYB transcription factor. *Plant J* **42**: 218–235.

Winkel-Shirley B. (2001) Flavonoid biosynthesis. A colorful model for genetics, biochemistry, cell biology, and biotechnology. *Plant Physiol* **126**: 485–493.

Wu XM, Lim SH, Yang WC. (2003) Characterization, expression and phylogenetic study of R2R3-MYB genes in orchid. *Plant Mol Biol* **51**: 959–972.

Zhang F, Gonzalez A, Zhao M *et al.* (2003) A network of redundant bHLH proteins functions in all TTG1-dependent pathways of *Arabidopsis*. *Development* **130**: 4859–4869.

Zimmermann IM, Heim MA, Weisshaar B *et al.* (2004) Comprehensive identification of *Arabidopsis thaliana* MYB transcription factors interacting with R/B-like BHLH proteins. *Plant J* **40**: 22–34.

CHAPTER 7

Qualitative and Quantitative Analysis of Pigments in *Phalaenopsis* Species

Ping-Chung Kuo* and Tian-Shung Wu†,‡

In Taiwan, *Phalaenopsis* hybrids are some of the most significant ornamental flowers for export worldwide. The *Phalaenopsis* hybrids produced by Taiwan's agricultural industry are available in various patterns and colors, including white, yellow, and red-purple cultivars. In this chapter we illustrate the compounds characterized in the *Phalaenopsis* species. The major components in the native *Phalaenopsis* species and cultivars are also analyzed with the assistance of high-performance liquid chromatography. The bioactivity of the pigments from the *Phalaenopsis* species and the color variations at different pH values are discussed.

7.1. Introduction

Phalaenopsis cultivars are some of the most economically important orchids in the market. The species is native to Asia and Australia and has white and yellow flowers. In the United States, Japan, and many European countries, both the domestic production and the import of this orchid is increasing rapidly. In Taiwan, *Phalaenopsis* hybrids are some of the most significant ornamental flowers for export worldwide, and the government has emphasized the importance of developing these hybrids as a dedicated agricultural industry. The *Phalaenopsis* hybrids recently produced by Taiwan are available in various patterns and colors, including white, yellow, and red-purple cultivars. Among these new varieties, dark red flowers, which some orchid growers describe

*Department of Biotechnology, National Farmosa University, Yunlin, Taiwan.
E-mail: pcckuoo@sunws.nfu.edu.tw
†Department of Chemistry, National Cheng Kung University, Tainan 701, Taiwan.
‡Corresponding author. E-mail: tswu@mail.ncku.edu.tw

as "black," are attracting the attention of both customers and scientists. Exactly how these pigments accumulated in the flowers has motivated studies by numerous research groups. However, no systematic and detailed analyses of the chemical constituents in the flowers of *Phalaenopsis* cultivars have been undertaken, and the components in the flowers and their biosynthetic pathway has remained unclear. In this chapter we will illustrate the compounds isolated and identified in the *Phalaenopsis* species previously and those accomplished by our group. These results are allowing better interpretations of the biosynthetic enzymes and genes involved and could provide the basis for the development of new varieties with even higher economic value.

7.2. Constituents of the *Phalaenopsis* Species

7.2.1. *Anthocyanins*

The secondary metabolites of the Orchidaceae plants have been intensively investigated and various constituents such as bibenzyl (Majumder and Basak, 1991a; Majumder and Pal, 1993; Majumder *et al.*, 1997; Leong *et al.*, 1997), stilbenoid (Majumder and Basak, 1991b; Majumder and Ghosal, 1993; Majumder and Ghosal, 1994; Majumder *et al.*, 1996), phenanthrene (Leong *et al.*, 1997; Majumder and Sen, 1991; Majumder and Maiti, 1989), and phenanthropyran derivatives (Majumder and Sabzabadi, 1988; Manako *et al.*, 2001), have been reported. However, due to the rarity and cost of plant materials, few reports on the constituents of the *Phalaenopsis* speices have been published, and only pyrrolizidines **1–4** (Brandänge and Lüning, 1969; Brandänge *et al.*, 1971; Brandänge *et al.*, 1972), anthocyanins **5–8** (Griesbach, 1990; Tatsuzawa *et al.*, 1997), and phenanthropyrans **9–10** (Manako *et al.*, 2001) were identified in the literature. The chemical structures of compounds **1–10** are shown in Fig. 7.1. Among these, the major pigments in *Phalaenopsis* hybrids are the anthocyanins, which have already been studied and extensively elucidated in the literature (Griesbach, 1990; Tatsuzawa *et al.*, 1997). Griesbach purified a sinapyl cyaniding 3,7,3′-triglucoside from the flowers of *Phalaenopsis schilleriana* (Griesbach, 1990), but its structure was not elucidated in detail. In a paper by Tatsuzawa *et al.* (1997), four acylated anthocyanins

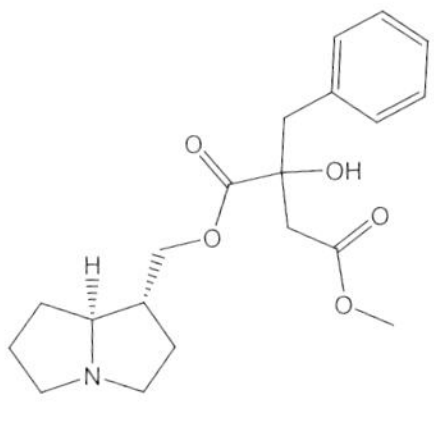

phalaenopsin (**1**)

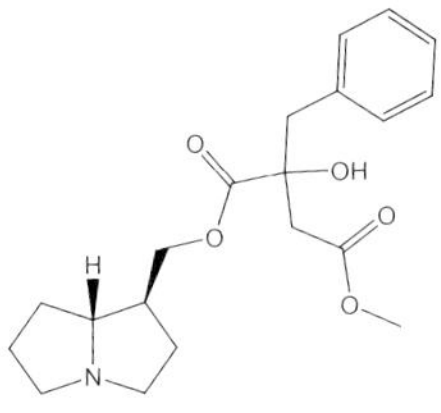

phalaenopsin L (**2**)

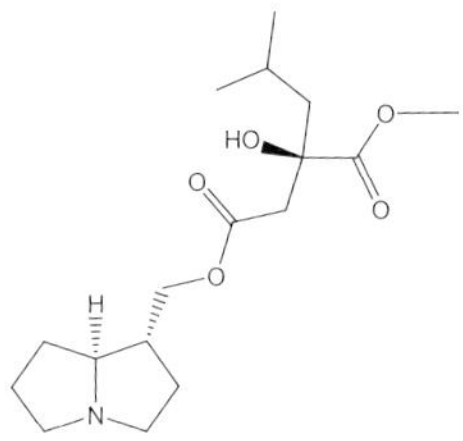

2-hydroxy-2-isobutyl-succinic acid 4-(hexahydro-pyrrolizin-1-ylmethyl) ester 1-methyl ester (**3**)

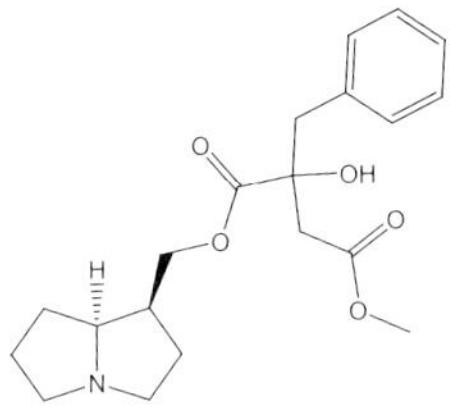

isoretronecanol (**4**)

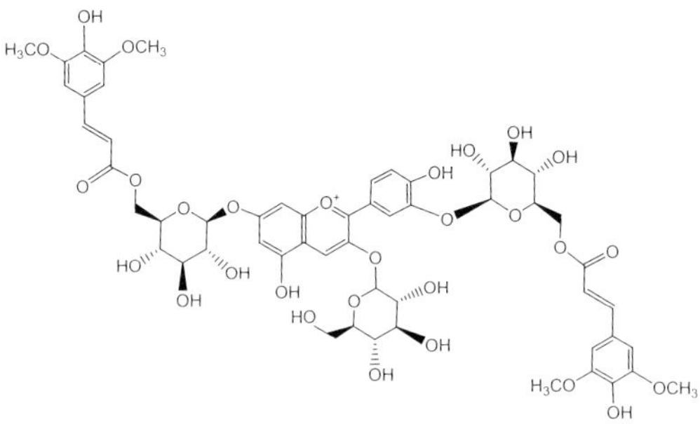

cyanidin 3-*O*-(*β*-*D*-glucopyranoside)-7,3'-di-*O*-[6-*O*-(sinapyl)-*β*-*D*-glucopyranoside] (**5**)

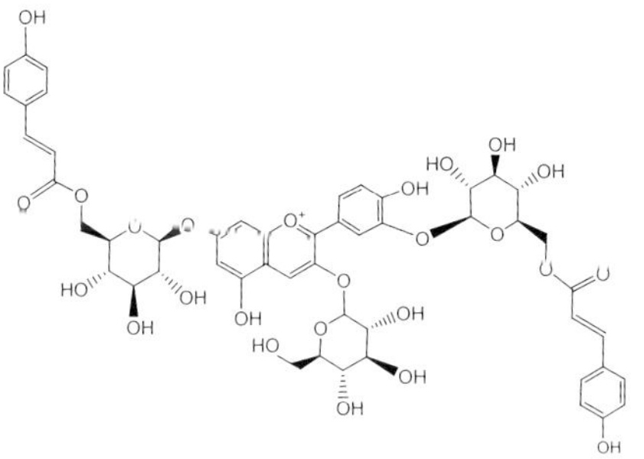

cyanidin 3-*O*-(*β*-*D*-glucopyranoside)-7,3'-di-*O*-[6-*O*-(hydroxycinnamyl)-*β*-*D*-glucopyranoside] (**6**)

Fig. 7.1. Chemical structures of compounds **1–12** purified from *Phalaenopsis* species.

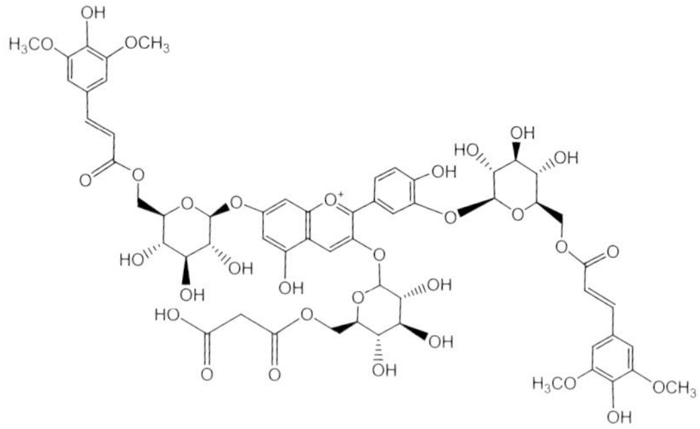

cyanidin 3-*O*-[6-*O*-(malonyl)-*β*-*D*-glucopyranoside]-7,3'-di-*O*-[6-*O*-(sinapyl)-*β*-*D*-glucopyranoside] (**7**)

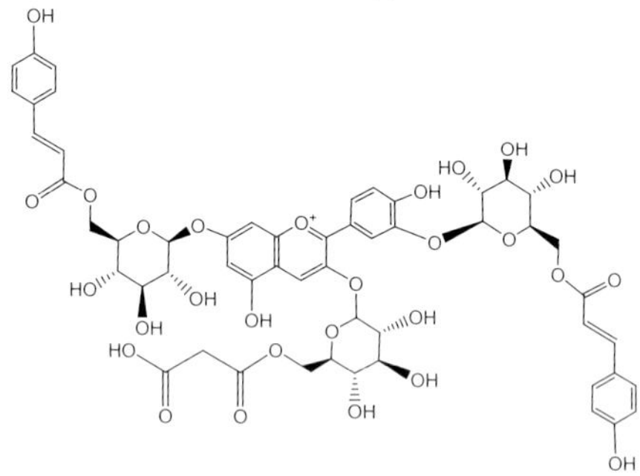

cyanidin 3-*O*-[6-*O*-(malonyl)-*β*-*D*-glucopyranoside]-7,3'-di-*O*-[6-*O*-(hydroxycinnamyl)-*β*-*D*-glucopyranoside] (**8**)

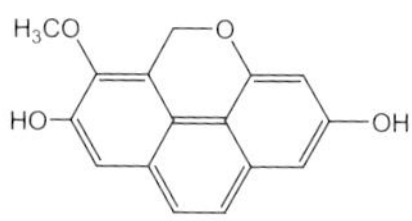

3-methoxy-2,7-dihydroxy-5*H*-phenanthro[4,5-bcd]pyran (**9**)

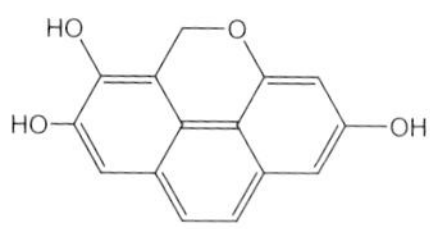

2,3,7-trihydroxy-5*H*-phenanthro[4,5-bcd]pyran (**10**)

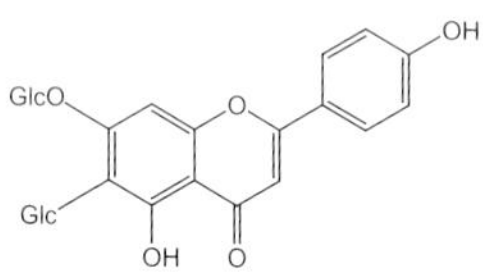

saponarin (**11**)

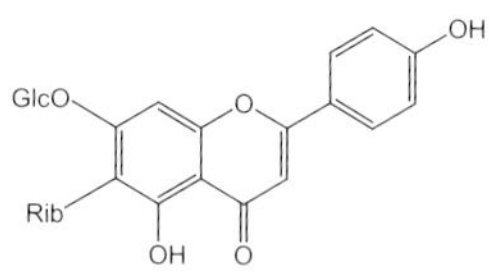

6-C-*β*-D-ribofuranosyl-7-O-*β*-D-glycopyranosyl apigenin (**12**)

Fig. 7.1. *(Continued)*

were isolated from the red-purple flowers of *Phalaenopsis* hybrid cultivars. In addition Tatsuzawa *et al.*, also investigated the distribution of these pigments in the flowers of *P. equestris, P. intermedia, P. leucorrhoda, P. sanderiana*, and *P. schilleriana* cultivar "Pink Butterfly" and proved that anthocyanins **5** and **7** were major components in these species.

In our study, we also utilized spectrometric and mass data to characterize the anthocyanins **5** and **7** as the major pigments in the fresh flowers of *Phalaenopsis* hybrids, *Doritaenopsis* East Sun Beauty (Chen, 2009). According to these results, cyanidin 3-*O*-[β-D-glucopyranoside]-7,3′-[6-*O*-(sinapyl)-β-D-glucopyranoside] (**5**) and its malonyl derivative (**7**) are responsible for the red to purple colors in the flowers of *Phalaenopsis* cultivars.

Recently, Yoshida *et al.* (2009) studied metalloanthocyanin, a supramolecular self-assembled metal-complex pigment composed of anthocyanins, flavones, and metal ions in a stoichiometric ratio, and the color development by intramolecular stacking. Most of the available ionization methods in mass spectrometry, including EI, CI, FAB, and MALDI, destroyed the supramolecules and yielded only the mass of the monomeric molecular ions. Only electrospray ionizaiotn mass spectrometry (ESI-MS) obtained the exact molecular weight of the supramolecules (Kondo *et al.* 1994). Although no reports on the anthocyanins in *Phalaenopsis* species have indicated the presence of metalloanthocyanin supramolecules, the major pigments accumulated in a similar manner to metalloanthocyanins in other species (Fig. 7.2).

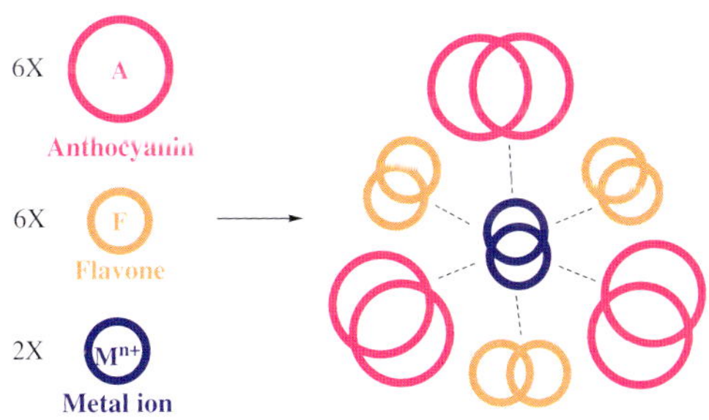

Fig. 7.2. Formation of a metalloanthocyanin by self-assembly.

7.2.2. *Flavonoids*

Flavonoids continue to attract scientific interest due to their structural diversity, biological and ecological importance. They affect plant interactions with microsymbionts (Romeo *et al.*, 1998), insect predators and pollinators (Harborne, 1986; Stafford, 1997), and also function in pigmentation and act as protectants against UV irradiation (Ylstra *et al.*, 1992; Brouillard, 1988; Harborne, 1994). Almost all higher plants produce flavonoids and some of these constituents are fairly unique, with many specific compounds being accumulated during plant growth and development. Consequently, plants with different flavonoid accumulation patterns may be used to study the regulation of the flavanone metabolic pathway and to elucidate potential biosynthetic mechanisms. However, there have been no published reports related to the observation of flavonoids in the *Phalaenopsis* species.

In our study (Chen, 2009) two flavonoids, saponarin (**11**) and 6-*C*-β-D-ribofuranosyl-7-*O*-β-D-glycopyranosyl apigenin (**12**), were characterized and their structures are given in Fig. 7.1. According to our quantitative analytic results, these flavonoids were the major constituents in the flowers of *Phalaenopsis* cultivars, instead of anthocyanins. This is reasonable since the flavonoids are the precursors of anthocyanins in the biosynthetic pathway.

7.2.3. *Others*

In addition to the pyrrolizidines **1–4** (Brandänge and Lüning, 1969; Brandänge *et al.*, 1971, 1972) and phenanthropyrans **9–10** (Manako *et al.*, 2001), which were identified in the *Phalaenopsis* species in the previous literature, in our research (Chen, 2009) the mixtures of β-sitosteryl acetate (**13**) and stigmasteryl acetate (**14**) (Zhang *et al.*, 2005), mixtures of β-sitosterol (**15**) and stigmasterol (**16**) (Kuo *et al.*, 2004), cyclotirucanenol (**17**) (Khan *et al.*, 1988), vanillic acid (**18**) (Kuo *et al.*, 2004), *p*-methoxybenzaldehyde (**19**) (Chang *et al.*, 2003), canthin-6-one (**20**) (Kuo *et al.*, 2004), and methyl 4-β-D-glucopyranosyloxy-*trans*-cinnamate (**21**) (Yamauchi *et al.*, 1981) were also characterized with the aid of 1D and 2D NMR spectral analyses

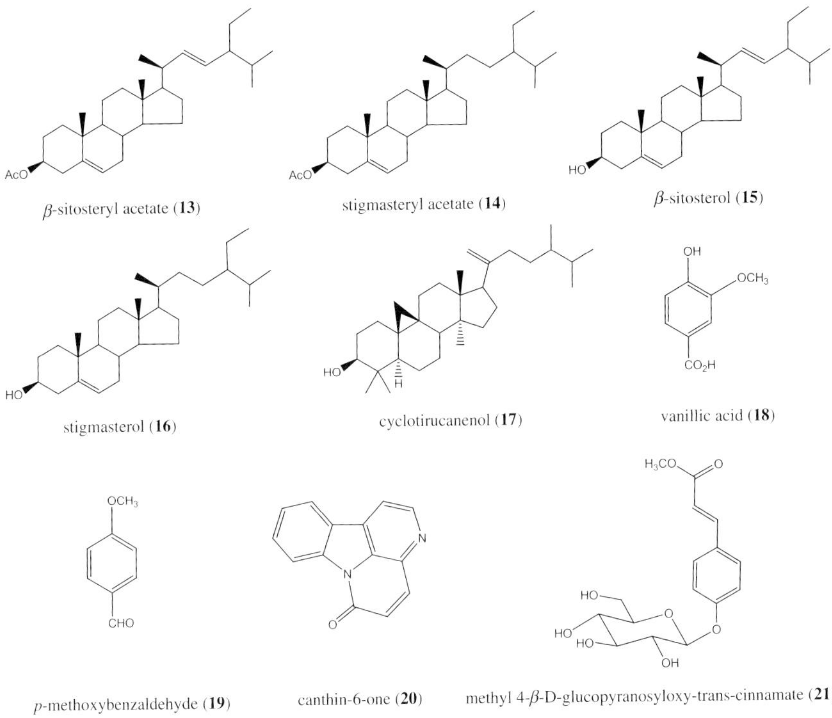

Fig. 7.3. Chemical structures of minor constituents **13–21** purified from *Phalaenopsis* species.

along with mass spectrometry. The chemical structures of these minor constituents **13–21** are given in Fig. 7.3. Among these compounds, methyl 4-β-D-glucopyranosyloxy-*trans*-cinnamate (**21**) possesses the C_6-C_3 basic skeleton and is related to the key precursors in the biosynthetic pathway of flavonoids and anthocyanins.

7.2.4. *The bioactivity of pigments from Phalaenopsis species and their application*

Globally, consumers are paying more and more attention to cosmetics made from natural products. In response, the development of skin-care cosmetics consisting of herbal preparations has been the focus of many research groups and companies.

The anti-aging and whitening effects of chemical constituents from the flowers of *Phalaenopsis* hybrids have been evaluated using α,α-diphenyl-β-picrylhydrazyl free radical (DPPH) scavenging and inhibition of tyrosinase assays (Kuo *et al.*, 2005). Anthocyanin **5** and flavonoid **11** exhibited moderate scavenging activity, with IC_{50} values of 27.3 and 307.1 μM, respectively, compared to the reference compound, vitamin E (IC_{50} value of 12.5 μM). In addition, flavonoid **12** displayed significant anti-tyrosinase activities, with an IC_{50} value of 41.6 μM, compared to the reference compound, kojic acid (IC_{50} value of 30.0 μM). These findings indicate that the crude extract and partial purified fractions of the flowers of *Phalaenopsis* cultivars may play significant roles in preparing new anti-aging and whitening cosmetic products.

7.3. Quantification of Pigments in *Phalaenopsis* Species

7.3.1. *Analysis of native Phalaenopsis species*

In the research furnished by our group, high-performance liquid chromatography (HPLC) was utilized to analyze the contents of pigments in native Formosan *Phalaenopsis* species (flower materials were kindly identified and made available by Dr. C. C. Tsai, Kaohsiung District Agriculture Research and Extension Station, Council of Agriculture, Taiwan) (Chen, 2009). In total, six species, *Phalaenopsis schilleriana, P. equestris, P. X intermedia, P. aphrodite* subsp. *formosana, P. aphrodite,* and *P. stuartiana,* were quantitatively analyzed to determine the constituent quantities of flavonoids and anthocyanins. Among these native species, *P. schilleriana* and *P. equestris* have pink-red flowers while other species have white flowers. Only trace amounts of anthocyanins could be observed in the red *Phalaenopsis* species. The major constituents determined by HPLC analysis in the native Formosan *Phalaenopsis* species are flavonoids **11** and **12**.

7.3.2. *Analysis of Phalaenopsis cultivars*

In addition, our analysis included 12 cultivars of *Phalaenopsis* hybrids (Fig. 7.4): (a) *Phalaenopsis aphrodite* var. *formosana,* (b) *Doritaenopsis*

Fig. 7.4. Pictures of flowers of 12 *Phalaenopsis* cultivars (a)–(l).

King Car 1111, (c) *D*. V3 Sogo Yukidian, (d) *D*. I-Hsin Black Jack, (e) *P*. Little Mary, (f) *D*. Plum Rose, (g) *P*.Taida Salu, (h) *D*. King Hsiang Rose *X P. equestris*, (i) *D*. Jubau Rose, (j) *D*. East Sun Beauty, (k) *P*. Little Richard, and (l) *D*. I-Hsin Sesame. These greenhouse-grown cultivars, kindly offered by Ox Orchids (Tainan, Taiwan) and authenticated by Prof. W. H. Chen (Department of Life Science, National University of

Kaohsiung), were also subjected to quantitative analysis of their pigments with the aid of HPLC and the standards purified and characterized by our group (Chen, 2009).

All the cultivars contained anthocyanin **5** and flavonoids **11** and **12** as the major pigments except for the white species, *P. aphrodite* var. *formosana*, *D.* King Car 1111, and *D.* V3 Sogo Yukidian, which did not show the presence of anthocyanins. Although the flowers of other examined cultivars are red, only *D.* I-Hsin Black Jack ("black" flowers) was shown to comprise a relatively larger quantity of anthocyanin **5**. In addition, *D.* Plum Rose and *P.* Taida Salu contained larger quantities of anthocyanin among the screening cultivars, but only one-third to one-fifth of the total flavonoid (**11** and **12**) content.

7.4. Conclusions and Future Prospects

Although several minor constituents isolated from the *Phalaenopsis* species were reviewed in this chapter, the major components in *Phalaenopsis* flowers are the anthocyanins **5** and **7** and flavonoids **11** and **12**. In order to study the color variations at different pH values, the UV-visible spectra of flower extracts and mixtures constituting the major pigments were recorded and compared by our group (Chen, 2009). The visible spectra of crude extract of *Phalaenopsis* hybrids were composed of two fractions: the anthocyanin pigment fraction, which was blue, and the flavonoid co-pigment fraction, which was yellow, at high pH values. Consequently, the color of the petals was mainly determined by the pH values and the contents of co-pigments. Thus, a suitable strategy to develop blue hues in future *Phalaenopsis* hybrids should be controlling the intracellular pH genes in flower petals and suitable adjustment of the pH values.

Acknowledgments

The authors are grateful for partial financial support from the Council of Agriculture and the National Science Council, Taiwan. The authors would also like to thank the National Center for High-Performance Computing, Taiwan, for the use of its computing facilities.

References

Brandänge S, Lüning B. (1969) Studies on Orchidaceae alkaloids. XII. Pyrrolizidine alkaloids from *Phalaenopsis amabilis* B1. and *P. mannii* Rchb. f. *Acta Chem Scand* **23**: 1151–1154.

Brandänge S, Lüning B, Moberg C, Sjöstrand E. (1971) Studies on Orchidaceae alkaloids. XXIV. A pyrrolizidine alkaloid from *Phalaenopsis cornu-cervi* Rchb. f. *Acta Chem Scand* **25**: 349–350.

Brandänge S, Lüning B, Moberg C, Sjöstrand E. (1972) Studies on Orchidaceae alkaloids. XXX. Investigation of fourteen *Phalaenopsis* Species. A new pyrrolizidine alkaloid from *Phalaenopsis equestris* Rchb. f. *Acta Chem Scand* **26**: 2558–2560.

Brouillard R. (1988) In: *The Flavonoids — Advances in Research Since 1980*, p. 525. Chapman and Hall, New York.

Chang FR, Lee YH, Yang YL *et al.* (2003) Secoiridoid glycoside and alkaloid constituents of *Hydrangea chinensis. J Nat Prod* **66**: 1245–1248.

Chen GF. (2009) Flower pigment constituents and their quantitative analysis of *Phalaenopsis* species. Master's Thesis. Graduate Institute of Electro-Optical and Materials Science, National Formosa University.

Griesbach, RJ. (1990) Flavonoid copigments and anthocyanin of *Phalaenopsis schilleriana. Lindleyana* **5**: 231–234.

Harborne JB. (1986) In: Cody B, Middleton Jr E, Harborne JB (eds.), *Plant Flavonoids in Biology and Medicine: Biochemical, Pharmacological, and Structure-Activity Relationships*, p. 15. Alan R Liss, New York.

Harborne JB. (1994) In: *The Flavonoids — Advances in Research Since 1986*, pp. 1–22. Chapman and Hall, London.

Khan AQ, Ahmed Z, Kazmi NH, Malik A. (1988) The structure and absolute configuration of cyclotirucanenol, a new triterpene from *Euphorbia tirucalli. Z Naturforsch B* **43**: 1059–1062.

Kondo T, Ueda M, Yoshida K *et al.* (1994) Direct observation of a small-molecule associated supramolecular pigment, commelinin, by electrospray ionization mass spectroscopy. *J Am Chem Soc* **116**: 7457–7458.

Kuo PC, Damu AG, Lee KH, Wu TS. (2004) Cytotoxic and antimalarial constituents from the roots of *Eurycoma longifolia. Bioorg Med Chem* **12**: 537–544.

Kuo PC, Damu AG, Cherng CY, Jeng JF. (2005). Isolation of a natural antioxidant, dehydrozingerone from *Zingiber officinale* and synthesis of its analogues for recognition of effective antioxidant and antityrosinase agents. *Arch Pharm Res* **28:** 518–528.

Leong YW, Kang CC, Harrison LJ, Powell AD. (1997) Phenanthrenes, dihydrophenanthrenes and bibenzyls from the orchid *Bulbophyllum vaginatum. Phytochemistry* **44**: 157–165.

Majumder PL, Sabzabadi E. (1988) Agrostophyllin, a naturally occurring phenanthropyran derivative from *Agrostophyllum khasiyanum. Phytochemistry* **27**: 1899–1901.

Majumder PL, Maiti DC. (1989) Flaccidinin and oxoflaccidin, two phenanthrene derivatives of the orchid *Coelogyne flaccida. Phytochemistry* **28**: 887–890.

Majumder PL, Basak M. (1991a) Two bibenzyl derivatives from the orchid *Cirrhopetalum andersonii. Phytochemistry* **30**: 321–324.

Majumder PL, Basak M. (1991b) Two stilbenoids from the orchid *Cirrhopetalum andersonii. Phytochemistry* **30**: 3429–3432.

Majumder PL, Sen RC. (1991) Pendulin, a polyoxygenated phenanthrene derivative from the orchid *Cymbidium pendulum. Phytochemistry* **30**: 2432–2434.

Majumder PL, Pal S. (1993) Cumulatin and tristin, two bibenzyl derivatives from the orchids *Dendrobium cumulatum* and *Bulbophyllum triste. Phytochemistry* **32**: 1561–1565.

Majumder PL, Ghosal S. (1993) Two stilbenoids from the orchid *Arundina bambusifolia. Phytochemistry* **32**: 439–444.

Majumder PL, Ghosal S. (1994) Two stilbenoids from the orchid *Arundina bambusifolia. Phytochemistry* **35**: 205–208.

Majumder PL, Lahiri S, Mukhoti N. (1996) Four stilbenoids from the orchid *Agrostophyllum khasiyanum. Phytochemistry* **42**: 1157–1161.

Majumder PL, Roychowdhury M, Chakraborty S. (1997) Bibenzyl derivatives from the orchid *Bulbophyllum protractum. Phytochemistry* **44**: 167–172.

Manako Y, Wake H, Tanaka T *et al.* (2001) Phenanthropyran derivative from *Phalaenopsis equestris. Phytochemistry* **58**: 603–605.

Romeo J, Downum K, Verpoorte R. (1998) Phytochemistry of plant-microorganism interactions. In: *Recent Advances in Phytochemistry*, p. 32. Plenum Publishing, New York.

Stafford HA. (1997) Roles of flavonoids in symbiotic and defense functions in *Legume* roots. *Bot Rev* **63**: 27–39.

Tatsuzawa F, Saito N, Seki H *et al.* (1997) Acylated cyanidin glycosides in the red-purple flowers of *Phalaenopsis. Phytochemistry* **45**: 173–178.

Yamauchi T, Abe F, Taki M. (1981) Protoplumericin, an iridoid bis-glucoside in *Allamanda neriifolia. Chem Pharm Bull* **29**: 3051–3055.

Ylstra B, Touraev A, Moreno RMB *et al.* (1992) Flavonols stimulate development, germination, and tube growth of tobacco pollen. *Plant Physiol* **100**: 902–907.

Yoshida K, Mori M, Kondo T. (2009) Blue flower color development by anthocyanins: From chemical structure to cell physiology. *Nat Prod Rep* **26**: 884–915.

Zhang X, Geoffroy P, Miesch M *et al.* (2005) Gram-scale chromatographic purification of beta-sitosterol. Synthesis and characterization of beta-sitosterol oxides. *Steroids* **70**: 886–895.

CHAPTER 8

Biosynthetic Pathway of Pigments in *Phalaenopsis* Species

Ping-Chung Kuo* and Tian-Shung Wu†,‡

In the present chapter, the common biosynthetic pathway of flavonoids and anthocyanins in the *Phalaenopsis* species and those in other plants are discussed and compared. The flavonoid pathway is part of the larger phenylpropanopid metabolism, which produces a range of other secondary metabolites, such as phenolic acids, lignins, lignans, stilbenes, and chalcones. Although the accumulation of anthocyanins in plants is a sophisticated process involving several enzymes, the key enzymes in the *Phalaenopsis* species that determine the colors of flowers are UDP-glucose:flavonoid glucosyltransferases (UFGTs) related to the glucosylation of the pigments. According to the results from enzymatic activity assays, the biosynthetic pathway of the major pigments in the *Phalaenopsis* species was revised.

8.1. Introduction

There are three types of chemically distinct pigments — betalains, carotenoids and flavonoids — responsible for the colors of flowers (Grotewold, 2006). Among them, flavonoids are one important class of the secondary plant metabolites which have attracted a great deal of attention over the decades as they have been found to play key roles as major pigments in red, blue and purple colors. In addition, they also affect plant interactions with microsymbionts (Romeo *et al.*, 1998), insect predators and pollinators (Harborne, 1986; Stafford, 1997), and

*Department of Biotechnology, National Farmosa University, Yunlin, Taiwan.
E-mail: pcckuoo@sunws.nfu.edu.tw
†Department of Chemistry, National Cheng Kung University, Tainan 701, Taiwan.
‡Corresponding author. E-mail: tswu@mail.ncku.edu.tw

act as protectants against UV irradiation (Ylstra *et al.*, 1992; Brouillard, 1988; Harborne, 1994). Virtually all higher plants produce flavonoids. However, some of them are rather unique and contain many specific compounds that are accumulated during plant growth and development. Consequently, plants with different flavonoid accumulation patterns can be used to study the regulation of the flavonoid metabolic pathway and to elucidate potential biosynthetic mechanisms.

Anthocyanins are also important secondary plant products and constitute a major subgroup of flavonoids. They play the main roles as pigments in flowers, and the biosynthetic pathway of anthocyanins is one of the most extensively studied pathways of secondary plant products, reflecting their broad distribution among the angiosperms (Grotewold, 2006; Suzuki *et al.*, 2000). Anthocyanins are water soluble pigments that exist in almost vascular plants, and they usually accumulate in vacuoles or anthocyanoplasts (Harborne, 1994; Suzuki *et al.*, 2000). In respect of their chemical structures, anthocyanins are glycosides of anthocyanidins, which have differences in the number and position of substituted hydroxy and methoxy groups. Moreover, the substitution pattern of hydroxy and methoxy groups is the determining factors for the color of anthocyanins (Harborne, 1994; Delgado-Vargas *et al.*, 2000). As the research results obtained by our group show, the anthocyanins act as the major pigments in the *Phalaenopsis* species. The role of flavones and anthocyanins in providing stable flower colors in the angiosperms has more recently been outlined by Harborne and Williams, (1998, 2001). In the present chapter, the common biosynthetic pathway of flavonoids and anthocyanins in the *Phalaenopsis* species and that in other plants will be discussed and compared.

8.2. Common Biosynthetic Pathway of Anthocyanins

8.2.1. *Phenylpropanoid metabolism leading to flavonoids and anthocyanins*

Flavonoid biosynthesis is probably the best identified of all the secondary metabolic pathways (Fig. 8.1) (Holton and Cornish, 1995; Winkel-Shirley, 2001; Su and Hsu, 2003; Springob *et al.*, 2003). Since 1994 significant progress has been made in the characterization and

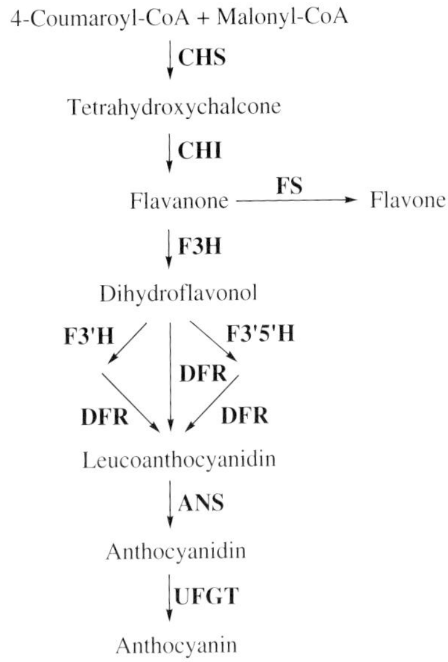

Fig. 8.1. The common biosynthetic pathway of anthocyanins.

analysis of genes and cDNAs for flavonoid biosynthetic enzymes and regulatory factors. The flavonoid pathway is part of the larger phenylpropanopid metabolism, which produces a range of other secondary metabolites, such as phenolic acids, lignins, lignans, stilbenes, and chalcones. The key precursor in the biosynthesis of flavonoids is phenylalanine, which results from the shikimate and arogenate pathways, and malonyl-CoA, derived from citrate produced by the tricarboxylic acid (TCA) cycle. Various branches of the flavonoid biosynthetic pathways have already been extensively studied, including well-characterized ones leading to the colored anthocyanins and proanthocyanidins, and the generally colorless flavones, flavonols and isoflavonoids.

The flavonoid and anthocyanin pathway starts with the formation of the C_{15} backbone by the first enzyme, chalcone synthase (CHS), which utilizes 4-coumaroyl-CoA and malonyl-CoA to produce tetrahydroxychalcone. Chalcone isomerase (CHI) then catalyzes the stereospecific isomerization of the yellow-colored tetrahydroxychalcone to the colorless naringenin. With the assistance of flavanone 3-hydroxylase

(F3H), naringenin is transformed into dihydrokaempferol and further hydroxylated to result in the colorless dihydroflavonol by flavonoid 3′-hydroxylase (F3′H) or flavonoid 3′,5′-hydroxylase (F3′5′H). The dihydroflavonol is successively converted to anthocyanidin by the catalytic function of dihydroflavonol 4-reductase (DFR) and anthocyanidin synthase (ANS). At the last step, the unstable anthocyanidin is stabilized by glucosylation at the 3-*O* position by UDP-glucose:flavonoid 3-*O*-glucosyltransferase (3GT or UFGT) in neutral cytosol, and the anthocyanidin 3-*O*-glucoside is then transported to acidic vacuoles (Holton and Cornish, 1995; Koes *et al.*, 2005). The role of ANS in the biosynthetic pathway is to catalyze reduction of the leucoanthocyanidins to the corresponding colorful anthocyanidins. For this reason, sometimes ANS is put forward as the determining factor in flower colors. However, the anthocyanidin pseudobases are thought to be too unstable to accumulate in the cell, and should be further converted into the stable anthocyanins through the 3-*O*-glycosylation. The 3-*O*-glycosylation is a prerequisite for further modifications, such as the second glycosylation, acylation, methylation or prenylation, in anthocyanin molecules. These modifications contribute to increased solubility and stability of the anthocyanins, and may influence the color variation of anthocyanins by inter- and intramolecular stacking (Springob *et al.*, 2003). Several studies on the biosynthesis of anthocyanins have already been published and the detailed mechanism has been described in the above-cited literature. Not all the enzymes related to anthocyanin biosynthesis will be covered in the present chapter. Only the determining enzyme, such as glucosyltransferase, will be discussed.

8.2.2. *Colors of different anthocyanins*

There are more than 500 different anthocyanins reported to be purified from plant sources. Generally, the anthocyanins consist of an aglycone (anthocyanidin), sugar(s), and usually acyl group(s). In respect of their structures, the anthocyanidins are derivatives of 2-phenylbenzopyrylium (flavylium cation), and more than 30 monomeric anthocyanidins have been properly identified.

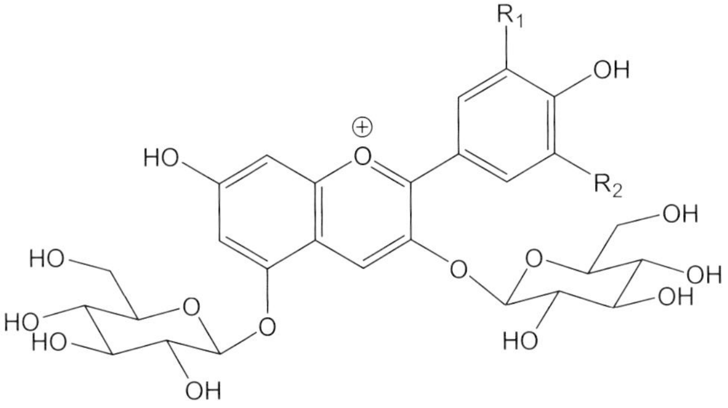

Pelargonidin	(**1**)	$R_1 = H, R_2 = H$
Cyanidin	(**2**)	$R_1 = OH, R_2 = H$
Delphinidin	(**3**)	$R_1 = OH, R_2 = OH$
Peonidin	(**4**)	$R_1 = OCH_3, R_2 = H$
Petunidin	(**5**)	$R_1 = OCH_3, R_2 = OH$
Malvidin	(**6**)	$R_1 = OCH_3, R_2 = OCH_3$

Fig. 8.2. Chemical structures of basic anthocyanins pelargonidin (**1**), cyanidin (**2**), delphinidin (**3**), peonidin (**4**), petunidin (**5**), and malvidin (**6**).

However, around 90 percent of the anthocyanins are based on only six basic structures: pelargonidin (**1**), cyanidin (**2**), delphinidin (**3**), peonidin (**4**), petunidin (**5**) and malvidin (**6**) (Fig. 8.2) (Harborne, 1986).

Anthocyanins exist in various tissues of higher plants, including flowers, fruits, seed coats, leaves, stems, tubers and roots (Grotewold, 2006; Harborne, 1994; Goto and Kondo, 1991). They are responsible for most pink, red, mauve, and blue colors and the degree of blueness is influenced by several factors. The first is the hydroxylation/methoxylation status of the anthocyanin B-ring. Correspondingly, the derivatives of pelargonidin (**1**), cyanidin (**2**) and delphinidin (**3**) (Fig. 8.2) display red-orange, red-purple, and purple-blue colors, respectively (Holton and Cornish, 1995; Winkel-Shirley, 2001). However, in its monomeric state, anthocyanin never exhibits a blue color in weakly acidic or neutral media; instead, a purple color is observed, and under such conditions the molecule rapidly loses its color. The pH-dependent structural changes in the anthocyanidin chromophore and the resulting color changes are generally investigated using kinetic studies and shown as in Fig. 8.3 (Brouillard, 1988, 1982; Brouillard and Delaporte, 1977).

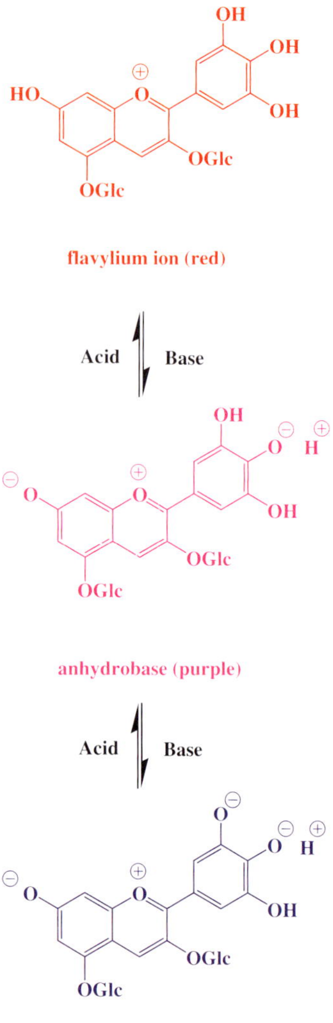

Fig. 8.3. Structural changes in the anthocyanidin chromophore and their pH-dependent color changes in aqueous solution.

Another factor is co-pigmentation, which was suggested as early as 1931 (Robinson and Robinson, 1931). Here, colorless compounds, such as amino acids, sugars and flavonoids stabilize the anthocyanins, resulting in a bathochromic shift in the absorption spectrum. Hayashi *et al.* reported that the major pigment in *Commelina communis* was composed not only of anthocyanin but also of flavone and Mg^{2+} ions

(Hayashi *et al.*, 1958; Mitsui *et al.*, 1959). This means the colors present in tissues may be different from the color of anthocyanin itself.

8.2.3. *Stabilization of anthocyanins*

Polyacylated anthocyanins, which are anthocyanins with two or more aromatic acyl residues, such as *p*-coumaroyl, caffeoyl, feruloyl, sinapyl, and *p*-hydroxyl benzyl moieties, are found mainly in bluish flower petals (Brouillard, 1988; Harborne, 1994; Goto and Kondo, 1991; Goto *et al.*, 1986; Goto, 1987). Between 1983 and 1992, Goto and Kondo published various structures of polyacylated anthocyanins, proposing intramolecular sandwich-type stacking of aromatic acyl residues onto the anthocyanin chromophore (Goto *et al.*, 1983; Goto *et al.*, 1984; Kondo *et al.*, 1987; Kondo *et al.*, 1989; Kondo *et al.*, 1990a; Kondo *et al.*, 1990b; Kondo *et al.*, 1991). Yoshida *et al.* also reported evidence for intramolecular stacking in gentiodelphin (Yoshida *et al.*, 1992). However, until now no X-ray structures have been solved and reported due to the difficulties in crystallizing these molecules. Thus, the mechanism of stabilization of anthocyanins has been proposed as follows. Simple anthocyanins are stabilized mainly by intermolecular aggregate formation, either homomolecular or heteromolecular with flavones. The more complicated anthocyanins possessing two or more aromatic acyl groups are stabilized by sandwich-type intramolecular stacking onto the anthocyanidin chromophore and two acyl groups due to the hydrophobic interaction among aromatic nuclei and hydrophilic assistance resulted from the sugar moieties. The most complicated metalloanthocyanins are composed of metal-chelated anthocyanins, which are further stabilized by self-association and co-pigmentation with flavone molecules (Goto, 1989).

8.3. Biosynthesis of Anthocyanins in *Phalaenopsis* Species

8.3.1. *Role of glucosyltransferase (GT) in anthocyanin formation*

Most anthocyanins contain one, two or three saccharide units. The sugar moieties are connected to the anthocyanidins through *O*-linkages.

However, Saito *et al.* (2003) reported a 8-*C*-glucosylcyanidin 3-[6-(malonyl)glucoside] from the purple flowers of *Tricyrtis formosana* cultivars. This was the first report of a natural *C*-glycosylanthocyanin. Usually the anthocyanin monosaccharides are represented by glucose, galactose, rhamnose, arabinose, xylose and glucuronic acid. Among them, the glucosyl moieties have been identified in as much as 90 percent of anthocyanins (Andersen and Jordheim, 2006).

Glucosylation at the 3-*O* position by UDP-glucose:flavonoid 3-*O*-glucosyltransferase (3GT) is indicated to be the key step in stabilizing the biosynthetic anthocyanin, as mentioned above. In addition, UDP-glucose:flavonoid 5-*O*-glucosyltransferase (5GT) modifies the anthocyanin molecule to allow for more stable complexes in co-pigmentation, resulting in flowers with a bright reddish-purple color from dull violet or pure red (Yamazaki *et al.*, 1999). Glucosylation at other positions is furnished with the aid of several glucosyltransferases, including UDP-glucose:flavonoid 7-*O*-glucosyltransferase (7GT) and UDP-glucose:flavonoid *C*-glucosyltransferase (CGT). Moreover, the UDP-glucose:anthocyanidin 3-*O*-glucoside-2″-*O*-glucosyltransferase (3GGT) identified in *Ipomoea nil* mediates the glucosylation of anthocyanidin 3-*O*-glucosides to yield anthocyanidin 3-*O*-sophorosides (Morita *et al.*, 2005). Although the function of these flavonoid and anthocyanin glycosides in plant cells remains unclear, they are assumed to represent the defense compounds against potential pathogens (Hirotani *et al.*, 2000) and exhibit antioxidant bioactivity (Marinova *et al.*, 2007). In theory, glucosyltransferase (GT) only applies the glucose to the chromophore, such as flavonoid and anthocyanidin, and does not affect the colors of anthocyanins. However, molecules with colors from yellow to red would fade quickly if the GT were absent. Thus, it is clear that GT is the key enzyme in pigment formation.

8.3.2. *Genes involved in color formation in plants*

According to the biosynthetic pathway of anthocyanins, several enzymes including chalcone synthase (CHS), chalcone isomerase (CHI), flavanone 3-hydroxylase (F3H), flavonoid 3′-hydroxylase (F3′H) or flavonoid 3′, 5′-hydroxylase (F3′5′H), dihydroflavonol 4-reductase

(DFR), anthocyanidin synthase (ANS), and UDP-glucose:flavonoid 3-*O*-glucosyltransferase (3GT or UFGT) are involved, and their activities may influence color formation in plants. The gene regulation of plant pigmentation has been performed in several studies, especially in *Zea mays* (maize), *Anfirrhinum majus* (snapdragon), *Arabidopsis thaliana, Petunia hybrida,* and *Vitis vinifera* (grape) (Grotewold, 2006; Aradhya *et al.*, 2003). Two full-length cDNA clones, Den-CHS-4 and Den-DFR-1, respectively encoding CHS and DFR, were reported from the flower bud of the *Dendrobium* species (Mudalige-Jayawickrama *et al.*, 2005). Northern analysis indicated that both genes are expressed in all development stages of buds, with the strongest expression in medium-sized buds. Anthocyanin accumulation in the *Oncidium* species is dependent on DFR activity. In contrast, the *Oncidium DFR* gene expression levels were up-regulated in unopened flower buds, as well as in the sepal, petals, and the lip crest of fully opened flowers (Lowry and Keong, 1973; Hieber *et al.*, 2006). In the *Bromheadia* orchid, *DFR* expression by Northern analysis was only detected in anthocyanin-rich young leaves, and only low levels of *DFR* expression were detectable in the colored flower parts. Glycosylation and subsequent acylation play significant roles not only in modifying flower color but also in increasing solubility and stability of hydrophobic flavonoids by external hydrogen bonding of sugar residues with surrounding water molecules (Goto *et al.*, 1982; Yoshida, 2000). These should be furnished with the assistance of UFGT enzymes. It has been shown that *UFGT* is expressed in the berry skins of red grape cultivars but not in the white ones, and expression of the other anthocyanin biosynthetic genes were detected in both white and red cultivars (Boss *et al.*, 1996). However, there were no differences in either coding or promoter sequences of UFGT between white and red cultivars, so the phenotypic change from white to red is considered to be the result of a mutation in a regulatory gene controlling the expression of *UFGT* (Kobayashi *et al.*, 2001). In addition, delivery of *MybA*, a gene of a transcription factor related to coloring, into the somatic embryos of grapes resulted in the same coloration and *UFGT* gene expression (Kobayashi *et al.*, 2002). Recently, mutants with the *anthocyaninless1* gene (*ANL1*) encoding a *UFGT* gene were isolated from *Arabidopsis thaliana*. The mutants accumulated less anthocyanin in plant tissues and were unable to accumulate any visible anthocyanin in the cotyledons and the upper part of the hypocotyls (Kubo, 2007). In our preliminary studies on the characterization of the biosynthetic enzymes in *Phalaenopsis* hybrids, in addition to the common biosynthetic enzymes, three glycosyltransferases (UFGT1-3) have been found. Both in the white and red *Phalaenopsis* hybrids, the northern blotting of all the enzymes did not yield any substantial differences except for UFGT3. In white cultivars of *Phalaenopsis* hybrids, there was no occurrence of UFGT3 and subsequent glycosylation and stabilization of anthocyanin may not be feasible. Thus, there were no anthocyanins in the white *Phalaenopsis* hybrids, and as a result, the colors of these flowers were white to light yellow. In contrast, in the red series, UFGT3 was present, which yielded stabilized anthocyanins giving the red color in the flowers. According to these results, UFGT3 in *Phalaenopsis* hybrids could act as a key enzyme in determining the colors of the flowers and thus *UFGT* might be a determining gene in color formation.

8.3.3. *Expression of UFGT genes in Phalaenopsis species*

Three *UFGT* genes, *PeUFGT 1–3*, were discovered in the *Phalaenopsis* species, and the alignment showed that the amino acid sequences of the three proteins were not similar. In a northern blot analysis, *PeUFGT3* was significantly expressed in the red flowers but there was no expression in the white flowers. In contrast, both the *PeUFGT1* and *PeUFGT2* genes were strongly expressed in the white and red *Phalaenopsis* hybrids and there were no differences between the two cultivars. Similar results were observed in different flower organs of *P.* "Luchia Lady" where *PeUFGT3* was more strongly expressed in the red lip (Cheng, 2005). In order to analyze the function and enzyme activity of *PeUFGT* gene products, our group expressed the recombinant PeUFGT proteins in *E. coli* and carried out the functional analysis of three PeUFGT enzymes using corresponding flavonoids as substrates. Our results established that PeUFGT1 was a 7GT and PeUFGT2 was a 6CGT as these enzymes transformed the apigenin to form saponarin, apigenin 6-*C*-glucosido-7-*O*-glucopyranoside, which is the major co-pigment in *Phalaenopsis* cultivars. In addition, PeUFGT3 was evidenced as a 3GT which catalyzed kaempferol to form kaempferol-3-*O*-glucopyranoside (Wu, 2009).

In addition, the effect of *PeUFGT3* on anthocyanin biosynthesis was investigated using RNA interference (RNAi) technology. The *PeUFGT3* gene-silencing experiment demonstrated different levels of flower color bleaching and a reduced level of *PeUFGT3* transcription activity closely correlated with the flower phenotype. Furthermore, the content of anthocyanin also significantly decreased in flowers with *PeUFGT3* gene silencing (Hsu, 2007). Our gene-silencing results also illustrated that the proper accumulation of anthocyanin is blocked by the knock-down expression of *PeUFGT3*.

8.3.4. *Proposed pathway for the pigments in Phalaenopsis species*

According to the putative pathway for the biosynthesis of anthocyanin and flavone derivatives and the experimental results of the enzymatic activity assays, the biosynthetic pathway of the major pigments in *Phalaenopsis* species was revised as shown in Fig. 8.4. The flavonone naringenin accumulated through the common pathway with the assistance of CHS and CHI catalysis. The naringenin substrate is

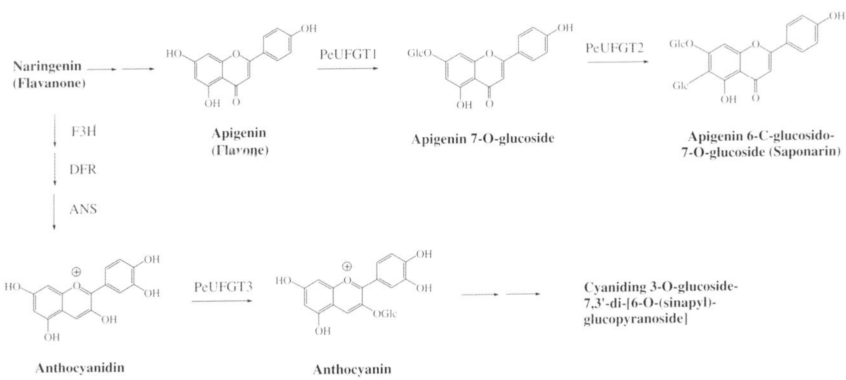

Fig. 8.4. Revised biosynthetic pathway of anthocyanin in the *Phalaenopsis* species.

not only transformed into dihydroflavonol; in contrast, the catalytic function of flavone synthase changes part of the flavonone to form apigenin, and two glucoses are successively inserted into the molecule under the catalysis of PeUFGT1 and PeUFGT2 to yield saponarin. In previous studies, saponarin is synthesized from its precursor isovitexin (apigenin 6-*C*-glucoside) through 7-*O*-glucosidation (Marinova *et al.*, 2007; Brazier-Hicks *et al.*, 2009). However, in the *Phalaenopsis* species, saponarin results from apigenin 7-*O*-glucoside via 6-*C*-glucosidation. This indicates that the GTs in the *Phalaenopsis* species may have different substrate specificity from those in other species. In addition, another fraction of naringenin could also be transformed into anthocyanidin by the catalytic function of F3H, F3′H, F3′5′H, DFR, and ANS. Although in our experiment we cannot confirm PeUFGT3 as UDP-glucose:anthocyanidin 3-*O*-glucosyltransferase, we propose that PeUFGT3 would transport glucose to the OH-3 position of anthocyanidin, resulting in stabilized anthocyanidin. This would be further modified by other enzymes to yield cyanidin 3-*O*-glucoside-7,3′-di-[6-*O*-(sinapyl)-glucopyranoside], which is the major pigment in the *Phalaenopsis* species.

8.4. Conclusions and Future Prospects

Although the accumulation of anthocyanins in plants is a sophisticated process involving several enzymes, the key enzymes in the *Phalaenopsis* species that determine the colors of flowers are UFGTs related to the glucosylation of the pigments. It would be interesting to apply this information to modulate the contents of flavonoid glycosides and anthocyanins in recombinant plants. With appropriate control of the biosynthetic enzymes, it may be possible to produce various economically valuable *Phalaenopsis* hybrids exhibiting new colors.

Acknowledgments

The authors are grateful for partial financial support from the Council of Agriculture and the National Science Council, Taiwan. The authors also wish to thank the National Center for High-Performance Computing, Taiwan, for the use of its computing facilities.

References

Andersen ØM, Jordheim M. (2006) In: Andersen ØM, Markham KR (eds.), *Flavonoids, Chemistry, Biochemistry and Applications*, pp. 475–495. CRC Press, Boca Raton.

Aradhya MK, Dangl GS, Prins BH *et al.* (2003) Genetic structure and differentiation in cultivated grape, *Vitis vinifera* L. *Gene Res* **81**: 179–192.

Boss PK, Davies C, Robinson SP. (1996) Expression of anthocyanin biosynthesis pathway genes in red and white grapes. *Plant Mol Biol* **32**: 565–569.

Brazier-Hicks M, Evans KM, Gershater MC *et al.* (2009) The *C*-glycosylation of flavonoids in cereals. *J Biol Chem* **284**: 17926–17934.

Brouillard R, Delaporte B. (1977) Chemistry of anthocyanin pigments. II. Kinetic and thermodynamic study of proton transfer, hydration and tautomeric reactions of malvidin 3-glucodise. *J Am Chem Soc* **99**: 8461–8468.

Brouillard R. (1982) In: Markakis P (ed.), *Anthocyanins as Food Colors*, pp. 1–40. Academic Press, New York.

Brouillard R. (1988) In: *The Flavonoids — Advances in Research Since 1980*, p. 525. Chapman and Hall, New York.

Cheng HY. (2005) Cloning and characterization of a flower color related gene (UDP-glucose:flavonoid 3-*O*-glucosyltransferase) in *Phalaenopsis*. Master's Thesis. Institute of Biotechnology, National University of Kaohsiung.

Delgado-Vargas F, Jiménez AR, Paredes-López O. (2000) Natural pigments: Carotenoids, anthocyanins, and betalains — Characteristics, biosynthesis, processing, and stability. *Cri Rev Food Sci Nutr* **40**: 173–289.

Goto T, Kondo T, Tamura H *et al.* (1982) Structure of gentiodelphin, an acylated anthocyanin isolated from *Gentiana makinori*, that is stable in dilute aqueous solution. *Tetrahedron Lett* **23**: 3695–3698.

Goto T, Kondo T, Tamura H *et al.* (1983) Structure of platyconin, a diacylated anthocyanin isolated from the chinese bell-flower *Platycodon grandiflorum*. *Tetrahedron Lett* **24**: 2181–2184.

Goto T, Kondo T, Kawai T, Tamura H. (1984) Structure of cinerarin, a tetra-acylated anthocyanin isolated from the blue garden cineraria, *Senecio cruentus*. *Tetrahedron Lett* **25**: 6021–6024.

Goto T, Tamura H, Kawai T *et al.* (1986) Chemistry of metalloanthocyanins. *Ann NY Acad Sci* **471**: 155–173.

Goto T. (1987) Structure, stability and color variation of natural anthocyanins. *Prog Chem Org Nat Prod* **52**: 113–158.

Goto T. (1989) Structure of complex natural anthocyanins. *J Chin Chem Soc* **36**: 1–9.

Goto T, Kondo T. (1991) Structure and molecular stacking of anthocyanins. Flower color variation. *Angew Chem Int Ed Engl* **30**: 17–33.

Grotewold E. (2006) The genetics and biochemistry of floral pigments. *Annu Rev Plant Biol* **57**: 761–780.

Harborne JB. (1986) In: Cody B, Middleton Jr E, Harborne JB (eds.), *Plant Flavonoids in Biology and Medicine: Biochemical, Pharmacological, and Structure-Activity Relationships*, p. 15. Alan R Liss, New York.

Harborne JB. (1994) In: *The Flavonoids — Advances in Research Since 1986*, pp. 1–22. Chapman and Hall, London.

Harborne JB, Williams CA. (1998). Anthocyanins and other flavonoids. *Nat Prod Rep* **15**: 631–652.

Harborne JB, Williams CA. (2001) Anthocyanins and other flavonoids. *Nat Prod Rep* **18**: 310–333.

Hayashi K, Abe Y, Mitsui S. (1958) Anthocyanins. XXX. Blue anthocyanin from flowers of *Commelina*, the crystallization and some properties. *Proc Japan Acad* **34**: 373–378.

Hieber AD, Mudalige-Jayawickrama RG, Kuehnle AR. (2006) Color genes in the orchid *Oncidium* Gower Ramsey: Identification, expression, and potential genetic instability in an interspecific cross. *Planta* **223**: 521–531.

Hirotani M, Kuroda R, Suzuki H, Yoshikawa T. (2000) Cloning and expression of UDP-glucose:flavonoid 7-*O*-glucosyltransferase from hairy root cultures of *Scutellaria baicalensis. Planta* **210**: 1006–1013.

Holton TA, Cornish EC. (1995) Genetics and biochemistry of anthocyanin biosynthesis. *Plant Cell* **7**: 1071–1083.

Hsu CY. (2007) Flower color modification via downregulation of putative UDP-glucose:flavonoid 3-*O*-glucosyltransferase gene with RNA interference in *Phalaenopsis*. Master's Thesis. Institute of Biotechnology, National University of Kaohsiung.

Kobayashi S, Ishimaru M, Ding CK *et al.* (2001) Comparison of UDP-glucose:flavonoid 3-*O*-glucosyltransferase (UFGT) gene sequence between white grapes (*Vitis inifera*) and their sports with red skin. *Plant Sci* **160**: 543–550.

Kobayashi S, Ishimaru M, Hiraoka K, Honda C. (2002) *Myb*-related genes of the Kyoho grape (*Vitis labruscana*) regulate anthocyanin biosynthesis. *Planta* **215**: 924–933.

Koes R, Verweij W, Quattrocchio F. (2005) Flavonoids: A colorful model for the regulation and evolution of biochemical pathways. *Trend Plant Sci* **10**: 236–242.

Kondo T, Kawai T, Tamura H, Goto T. (1987) Structure determination of heavenly blue anthocyanin, a complex monomeric anthocyanin from the morning glory *Ipomea tricolor*, by means of the negative NOE method. *Tetrahedron Lett* **28**: 2273–2276.

Kondo T, Yamashiki J, Kawahori K, Goto T. (1989) Structure of lobelinin A and B, novel anthocyanins acylated with three and four different organic acids, respectively. *Tetrahedron Lett* **30**: 6055–6058.

Kondo T, Oki K, Yoshida K, Goto T. (1990a) Structure of violdelphin, an anthocyanin from violet flower of *Delphinium hybridum. Chem Lett* **19**: 137–138.

Kondo T, Ueda M, Goto T. (1990b) Structure of ternatin B1, a pentaacylated anthocyanin substituted on the B-ring asymmetrically with two long chains. *Tetrahedron* **46**: 4749–4756.

Kondo T, Suzuki K, Yoshida K *et al.* (1991) Structure of cynanodelphin, a tetra-*p*-hydroxybenzoated anthocyanin from blue flower of *Delphinium hybridum. Tetrahedron Lett* **32**: 6375–6378.

Kubo H, Nawa N, Lupsea SA. (2007) Anthocyaninless1 gene of *Arabidopsis thaliana* encodes a UDP-glucose:flavonoid-3-*O*-glucosyltransferase. *J Plant Res* **120**: 445–449.

Lowry JB, Keong SC. (1973) A preliminary study of Malaysian orchid pigments. *Malays J Sci* **2**: 115–121.

Marinova K, Kleinschmidt K, Weissenbock G, Klein M. (2007) Flavonoid biosynthesis in barley primary leaves requires the presence of the vacuole and controls the activity of vacuolar flavonoid transport. *Plant Physiol* **144**: 432–444.

Mitsui S, Hayashi K, Hattori S. (1959) Further studies on commelinin, a crystalline blue metallc-anthocyanin from the flowers of commelinin. Studies on anthocyanins XXXI. *Proc Japan Acad* **35**: 169–174.

Morita Y, Hoshino A, Kikuchi Y *et al.* (2005) Japanese morning glory *dusky* mutants displaying reddish-brown or purplish-gray flowers are deficient in a novel glycosylation enzyme for anthocyanin biosynthesis, UDP-glucose:anthocyanidin 3-*O*-glucoside-2″-*O*-glucosyltransferase, due to 4-bp insertions in the gene. *Plant J* **42**: 353–363.

Mudalige-Jayawickrama RG, Champagne MM, Hieber AD, Kuehnle AR. (2005) Cloning and characterization of two anthocyanin biosynthetic genes from *Dendrobium* orchid. *J Am Soc Hort Sci* **130**: 611–618.

Robinson GM, Robinson R. (1931) Survey of anthocyanins. I. *Biochem J* **25**: 1687–1705.

Romeo J, Downum K, Verpoorte R. (1998) Phytochemistry of plant-microorganism interactions. In: *Recent Advances in Phytochemistry*, p. 32. Plenum Publishing, New York.

Saito N, Tatsuzawa F, Miyoshi K *et al.* (2003) The first isolation of C-glycosylanthocyanin from the flowers of *Tricyrtis formosana*. *Tetrahedron Lett* **44**: 6821–6823.

Springob K, Nakajima J, Yamazaki M, Saito K. (2003) Recent advances in the biosynthesis and accumulation of anthocyanins. *Nat Prod Rep* **20**: 288–303.

Stafford HA. (1997) Roles of flavonoids in symbiotic and defense functions in *Legume* roots. *Bot Rev* **63**: 27–39.

Su V, Hsu BD. (2003) Cloning and expression of a putative cytochrome P450 gene that influences the colour of *Phalaenopsis* flowers. *Biotech Lett* **25**: 1933–1939.

Suzuki K, Xue H, Tanaka Y *et al.* (2000) Flower color modification of *Torenia hybrida* by cosuppresion of anthocyanin biosynthesis genes. *Mol Breeding* **6**: 239–246.

Winkel-Shirley B. (2001) Flavonoid biosynthesis. A colorful model for genetics, biochemistry, cell biology, and biotechnology. *Plant Physiol* **126**: 485–493.

Wu CF. (2009) Identification and functional analysis of UDP-glucosyltransferase for pigments biosynthesis in *Phalaenopsis* orchids. Master's Thesis. Institute of Life Science, National Cheng Kung University.

Yamazaki M, Gong Z, Fukuchi-Mizutani M *et al.* (1999) Molecular cloning and biochemical characterization of a novel anthocyanin 5-*O*-glucosyltransferase by mRNA differential display for plant forms regarding anthocyanin. *J Biol Chem* **274**: 7405–7411.

Yoshida K, Kondo T, Goto T. (1992) Intramolecular stacking conformation of gentiodelphin, a diacylated anthocyanin from *Gentiana makinoi*. *Tetrahedron* **48**: 4313–4326.

Yoshida K, Toyama Y, Kameda K, Kondo T. (2000) Contribution of each cafferoyl residue of the pigment molecules of gentiodelphin to blue color development. *Phytochemistry* **54**: 85–92.

Ylstra B, Touraev A, Moreno RMB *et al.* (1992) Flavonols stimulate development, germination, and tube growth of tobacco pollen. *Plant Physiol* **100**: 902–907.

CHAPTER 9

Biosynthetic Regulation of Floral Scent in *Phalaenopsis*

Yu-Yun Hsiao*, Wen-Chieh Tsai†, Wen-Huei Chen‡ and Hong-Hwa Chen*,†,‡,§

Floral scent is a key modulating factor in plant–insect interactions and thus plays a central role in successful pollination. Orchidaceae, containing more than 25,000 species, is one of the largest monocotyledon families. Pollinators have played an important role in driving floral diversification in orchids, which has obviously been advantageous for the evolution of this highly successful family. We investigate how flower fragrance is generated in certain *Phalaenopsis* orchids by identifying floral expressed-sequence-tags (ESTs) and deducing the floral scent biosynthesis pathway of *Phalaneopsis bellina* using bioinformatics analysis. The major chemical components in the *P. bellina* flower are found to be monoterpene of geraniol, linalool, and their derivatives, and maximally emitted from day 5 post-anthesis to day 7 post-anthesis. The encoded enzymes collection in the biosynthetic pathway from glycerolaldehyde-3-phosphate (G3P) to geraniol and linalool are recognized through data mining of the *P. bellina* floral EST database (dbEST). Transcripts preferentially expressed in *P. bellina* include those encoding lipoxygenase, epimerase, diacylglycerol kinase and geranyl diphosphate synthase (GDPS). EST filtering results showed that transcripts encoding signal transduction, Myb transcription factors and methyltransferase, in addition to those for scent biosynthesis, are detected by *in silico* hybridization of the *P. bellina* unigene database against those of the scentless species, rice and *Arabidopsis*. Altogether, we pinpoint 66 percent of the biosynthetic steps from G3P to geraniol, linalool and their derivatives. We then focused on the key enzymes in this pathway, the geranyl diphosphate synthase (GDPS) which catalyzes the formation of GDP for further synthesis of geraniol, linalool and their

*Department of Life Sciences, National Cheng Kung University, Tainan 701, Taiwan.
†Institute of Tropical Plant Sciences, National Cheng Kung University, Tainan 701, Taiwan.
‡Orchid Research Center, National Cheng Kung University, Tainan 701, Taiwan.
§Corresponding author. E-mail: hhchen@mail.ncku.edu.tw

derivatives. The novel PbGDPS enzyme lacks the conserved aspartate-rich motif $DD(X)_{2-4}D$, possesses a dual prenyltransferase activity, producing both GDP and farnesyl diphosphate (FDP), and functions as a homodimer. Homology modeling and site-direct mutagenesis of PbGDPS indicate that the Glu-rich motif may provide a binding site for Mg^{2+} and catalyze the formation of prenyl products. Expression pattern results show that the *PbGDPS* was flower specific and maximal PbGDPS expression is concomitant with maximal emission of monoterpenes on day 5 post-anthesis, thereby suggesting that PbGDPS plays a crucial role in scent production/emission in orchid flowers. In conclusion, these approaches demonstrate the opportunity to study floral scent in the non-model plants such as orchids, and moreover to accelerate molecular breeding in the orchid industry.

9.1. Introduction

9.1.1. *Significance of the floral/orchid scent*

Floral scent is a key modulating factor in plant–insect interactions and thus plays a central role in successful pollination. Closely related plant species that rely on different insects for pollination produce different odors (Raguso and Pichersky, 1995; Pichersky and Gershenzon, 2002). The floral scent is of paramount importance to plant reproduction and evolution (Reinhard *et al.*, 2004). Orchidaceae is one of the largest monocotyledon families, containing more than 25,000 species. In orchids, large quantities of pollen are formed in masses spread by animals (bees, moths, flies and birds) and floral scents serve as attractants for species-specific pollinators (Van der Pijl and Dodson, 1969). These pollinators have played a major role in orchid evolution. The range of odors produced by orchids is enormous, providing an inexhaustible basis for specificity. Such diversity is advantageous in the evolution of an obviously successful family.

Orchidaceae have sophisticated flower morphologies, including two whorls of perianth segments, three sepals and three petals, one of which is highly evolved as a labellum (lip) and serves as a landing platform for pollinators. The Orchidaceae species pollinated by bees, wasps and bumble bees cover a wide range of scents, from rosy- and ionone-floral to aromatic- and spicy-floral. Many volatile components have been identified in Orchidaceae flowers from the

American tropics, African tropics, Australian tropics and parts of Europe (Kaiser, 1993).

9.1.2. *Floral scent of Phalaenopsis*

Phalaenopsis Alliance is undoubtedly the most widely grown orchid in the world. According to Van der Pijl and Dodson, the genus contains rosy-floral, aromatic-floral, sweeted-floral to spicy-floral scents. Examples include *P. fasciata, P. hieroglyphica* and *P. schilleriana* (Van der Pijl and Dodson, 1969); *P. bellina,* classified in subgenus *Polychilos,* is native to Malaysia and numerous commercial varieties have been bred because of its pleasant fragrance.

Qualitative and quantitative analyses of volatile compounds from *P. bellina* flowers were performed using GC-MS. Monoterpenoids, phenylpropanoids, benzenoids and fatty acid derivatives were detected. Monoterpenoids, including geraniol, linalool and their derivatives (Table 9.1), accounted for more than 80 percent of the

Table 9.1 Volatiles Emitted by *P. bellina* and *P. equestris* Flowers

	P. bellina (Scented)		P. equestris (Scentless)	
Compounds	ng/flower/hrs	SE	ng/flower/hrs	SE
Monoterpene				
Myrcene	12.0	7.1	—	—
β-pinene oxide	3.7	1.2	—	—
Epoxy-terpenyl acetate	14.0	9.1	—	—
Linalool	105.4	15.3	—	—
3,7-dimethyl-6-octen-1-ol	2.3	1.4	—	—
3,7 dimethyl-1,5,7-octatrien-3-ol	4.3	2.6	—	—
2,6-dimethyl-3,7-octadiene-2,6-diol	18.3	6.0	—	—
2,6-dimethyl-1,7-octadiene-3,6-diol	20.3	9.1	—	—
Linalool oxide	3.4	2.4	—	—
trans-geraniol	163.4	1.6	—	—
Nerol	6.0	1.9	—	—
3,7-dimethyl-2,6-octadienal	3.1	0.1	—	—
Geranic acid	8.6	5.3	—	—
2,6-dimethyl-2,6-octadiene-1,8-diol	18.0	12.7	—	—
Total	**382.8**	16.1	—	—

(*Continued*)

Table 9.1 (*Continued*)

Compounds	P. bellina (Scented)		P. equestris (Scentless)	
	ng/flower/hrs	SE	ng/flower/hrs	SE
Phenylpranoid				
3-phenyl-2-propen-1-ol	3.7	2.6	—	—
2,4-bis(1,1-dimethylethyl)-phenol	4.0	1.7	—	—
2,6-bis(1,1-dimethylethyl)-4-methyl-phenol	18.9	11.1	104.4	13.0
2,6-dimethoxy-4-(2-propenyl)-phenol	12.0	8.5	4.6	4.0
Total	**38.6**	17.1	**109.0**	14.0
Benzenoid				
1,2,3-trimethoxy-5-(2-propenyl)-benzene	27.1	12.7	—	—
3,5-di-*tert*-butyl-4-hydroxybenzaldehyde	7.1	0.8	—	—
3,5-di-*tert*-butyl-4-hydroxybenzyl alcohol	6.0	4.2	6.5	2.1
3-methylphenyl butanoic acid ester	—	—	10.0	1.3
2-methylphenyl butanoic acid ester	—	—	10.5	2.0
Naphthalene	—	—	6.2	0.6
Total	**40.2**	20.8	**33.2**	7.0
Fatty Acid Derivatives				
2-hexanol	0.7	0.5	40.2	2.0
4-hydroxy-4-methyl-2-pentanone	2.6	1.5	8.6	1.1
3-hexanone	—	—	17.3	1.9
5-hexen-2-one	—	—	7.6	0.5
2-nonanol	—	—	22.2	2.0
1-methyl-cyclopentanol	—	—	18.5	5.1
4,6-dimethyl-dodecane	—	—	19.8	4.6
Hexadecane	—	—	37.5	15.0
4-methyl-dodecane	—	—	25.3	0.5
4-methyl-2-propyl-1-pentanol	—	—	10.1	0.3
n-nonylaldehyde	—	—	3.6	0.1
2-butoxy-ethanol	—	—	3.3	0.1
2-ethyl-1-hexanol	—	—	85.8	0.8
Heptadecanoic acid	—	—	30.7	1.0
Total	**3.3**	1.8	**330.5**	35.0

SE: Standard error.

total volatiles collected from the *P. bellina* flowers. They included geraniol, nerol, 2,6-dimethyl-octa-3,7-diene-2,6-diol, 2,6-dimethyl-octa-1,7-diene-3,6-diol, 3,7-dimethyl-2,6-octadienal, geranic acid and 2,6-dimethyl-octa-2,6-diene-1,8-diol (Table 9.1). In contrast, no monoterpenoid derivatives were emitted in the scentless *P. equestris* flowers; fatty acid derivatives, phenylpropanoids, and benzenoids were the major volatiles. These compounds are barely detectable by the human nose.

9.1.3. *Difficulty of orchid scent research*

The global flower industry thrives on novelty. Domestication of wild species in conjunction with traditional breeding has long been the principle path for the generation of novelty in the industry. Traits such as flower color, shape and fragrance are primary novelty markers as they are key determinants in consumer choice. Floral scent plays not only an important role in attracting pollinators, but is also important to consumer choice in flower purchase due to its sensual associations. However, many modern floricultural varieties have lost their scent during classic breeding programs. Breeders in cut-flower and ornamental markets have focused on producing plants with improved vase life, shipping characteristics and visual aesthetic values (i.e., color and shape). Owing to the lack of direct selection, or perhaps because of a negative correlation with any of these traits, many cultivated flowers have lost their scents.

Breeding and development of new varieties is one of the most important factors that keep the *Phalaenopsis* orchid as a leading product among potted flowers in the world market (Griesbach, 2002). Through tremendous breeding efforts, various types of *Phalaenopsis* varieties with attractive colors and graceful appearance have been developed. The breeding of *Phalaenopsis* is usually divided into two types: the standard big flower type and the novel type. The standard big flower type includes the white, semi-alba, pink and strips, which are usually derivatives of the white *P. amabilis* and the pink *P. schilleriana*. Varieties of the novel type are usually smaller flowers with special coloration when *P. equestris, P. gigantea, P. stuartiana,* and so on are involved in the pedigree (Chen *et al.*, 2004; Harper, 1991). In a pedigree analysis of the 12 popular big white Taisuco hybrids, it was found that the average genetic contribution of *P. Doris* to these large white *Phalaenopsis* hybrids is 50.47 percent (Jia *et al.*, 1999). Through the genetic flow from *P. Doris* species to the modern big flower hybrids, it is predicted that the superior hybrids were developed firstly by improving the genetic characteristics of *P. Doris,* a tetraploid with a large genomic capacity to accumulate additive alleles. Then, it was backcrossed and hybridized with its relatives to recombine and to accumulate desirable additive alleles for the large flower size and other favorable traits. Wild species with scent traits possess diploid genomes, but commercial big flower varieties have tetraploid genomes. It is impossible for the interbreeding between the scented *Phalaenopsis* species and the big flower parent to introduce scent traits from wild species into commercial varieties because the triploid progeny are sterile.

9.1.4. *Reasons for studying P. bellina*

To date, no simple, efficient, and reliable culture methods for scented orchids have been developed. Although the sympatric speciation of orchids is linked to differences in their floral odors, large genome size, long life cycle and regeneration time of these plants, together with inefficient transformation systems, the investigation of orchid scent biology is difficult. Furthermore, several scented and scentless species are cross-incompatible, which restricts the production of scented offspring by traditional breeding techniques. In some successful cases of cross-breeding, the progenies have diluted scent or have lost the ability to produce scent. Prior to this study, the biosynthetic pathways of orchid flower fragrance were not well understood. Little was known about the enzymes and genes controlling scent production in monocotyledons such as orchids.

To our knowledge, the growing cycles of *Phalaenopsis* orchids are two to three years. Using traditional hybridization to transmit useful traits into commercial varieties is a long process that will take years to achieve (Arditti, 1992). In addition, intraspecific and/or interspecific incompatibility limits the work of variety improvement. All five subgenera of *Phalaenopsis* have the same chromosome number ($2n = 2x = 38$) that can be divided into small, medium and large chromosome groups, according to their chromosome sizes and nuclear DNA contents (Kao *et al.*, 2001; Lin *et al.*, 2001; Shindo and Kamemmoto, 1963). Most commercial cultivars are derived from species with small chromosomes such as *P. amabilis, P. aphrodite, P. stuartiana, P. schilleriana, P. equestris,* and so on. Species with strong scents have large chromosomes, such as *P. amboinnensis, P. bellina, P. venosa, P. violacea,* and so on. It is difficult to make a cross successfully between species with small and large chromosomes due to interspecific incompatibility.

Most of the F1 plants are sterile even if a few of these interspecific hybrids are obtained. The study of chromosome associations (Arends, 1970) has shown that there is a high frequency of bivalents in hybrids between species both with small chromosomes and between species both with large chromosomes, but a low frequency of bivalents in hybrids between one species with small and one with large chromosomes. There is an urgent need to double the chromosomes of diploid scent species to form tetraploids to accelerate the improvement of commercial fragrant *Phalaenopsis* orchids.

P. bellina, which belongs to the large flower family, has a strong and enticing scent for humans. In addition, there are some native tetraploid species used to breed commercial fragrant *Phalaenopsis* orchids and it is thus an important parent for scented cultivar breeding. Floral scent is a composite characteristic that is determined by a complex mixture of low-molecular-mass volatile molecules, and dominated by monoterpenoid, sesquiterpenoid, phenylpropanoid, benzenoid compounds, and fatty acid derivatives. If we hope to identify the scent biosynthesis pathway and isolate the control genes, we need to establish what the major scent component is. Previously, we demonstrated that floral scents in *P. bellina* (Orchidaceae, monocot) are rich in the monoterpenes, geraniol and linalool and their derivatives. We can use *P. bellina* to study monoterpenoids in *Phalaenopsis* orchids — how the various scent compounds are made, where they are made, how they are emitted from the flower, the nature of the floral scent biosynthesis pathway, and finally, how researchers are using the newly identified scent genes to genetically engineer flowers that will produce new scents in *Phalaenopsis*.

9.2. Scent Biosynthesis Pathway

9.2.1. *Strategy to isolate scent biosynthesis pathway in P. bellina*

Floral scent is one of the important strategies for ensuring fertilization and for determining seed or fruit set. Research on plant scents has been hampered mainly by the invisibility of this character, its dynamic nature, and complex mixtures of components that are present in very small quantities. Most progress in scent research, as in other areas of plant biology, has come from the use of molecular and biochemical techniques. We investigate how flower fragrance is generated in certain *Phalaenopsis* orchids by determining the chemical components of the floral scent, identifying floral expressed-sequence-tags (ESTs), and deducing the pathways of floral scent biosynthesis in *Phalaneopsis bellina* using bioinformatics analysis.

In this chapter, we describe a strategy that combines chemical analysis, genomics and bioinformatics to uncover the scent biosynthesis pathway and the relevant genes (Fig. 9.1). First, volatile analysis showed that monoterpenoids, including geraniol, linalool and their derivatives, are major compounds of the *P. bellina* fragrance (Table 9.1). It is reasonable to assume that monoterpenoids are biosynthesized in these flowers. Second, we found that the scent-emitting structures of *P. bellina* were mostly in the perianth but not in the column (Hsiao *et al.*, 2008a). The column of an orchid flower is a specialized organ comprising a fused stamen and carpel, and many genes are involved in regulating its development. To reduce the complexity, we constructed a cDNA library from *P. bellina* flower buds with the columns removed immediately before blooming.

A total of 2,359 individual 5′ ESTs were retained from the *P. bellina* flower bud with an average length of 916 bp, and the confidence level

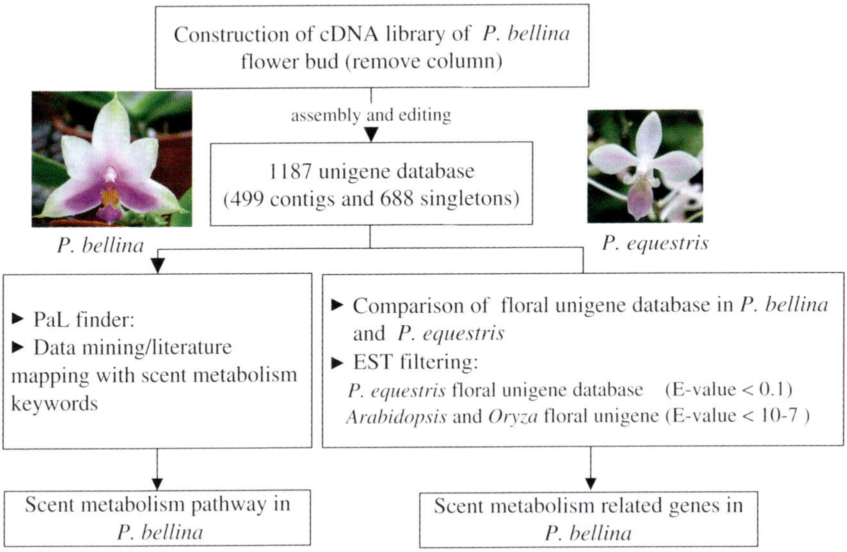

Fig. 9.1. Strategy of identification of *Phalaenopsis* scent metabolism pathway and scent candidate genes involved in the DXP-geraniol-linalool pathway.

of the sequences called using PHRED was 22.8 per base. From 247 cDNA clones that were analyzed by restriction enzyme digestion, the average insert size was 1.2 ± 0.32 kb. For better quality annotation, the ESTs were first refined and then the related genes were assembled as unigenes. This process generated 1,187 unigenes: 499 contigs and 688 singletons. In the *P. bellina* floral dbEST, 1,748 (74.1 percent) of the 2,359 ESTs showed similarities to known sequences in the Uniprot database. The sequences in the flower bud dbEST were functionally characterized using a Gene Ontology scheme (Ashburner *et al.*, 2000; Kaiser, 1993). Details of the gene species included in each group are given in Fig. 9.2. GO allowed 45.6 percent of the total ESTs to be placed in the molecular function category, 18.5 percent in the biological process category, and 1.8 percent in the cellular component category. The remaining 34 percent either showed insufficient similarities to any proteins (no hits, 25.9 percent) or hit proteins without a GO identifier (unclassified, 8.1 percent). Among the molecular

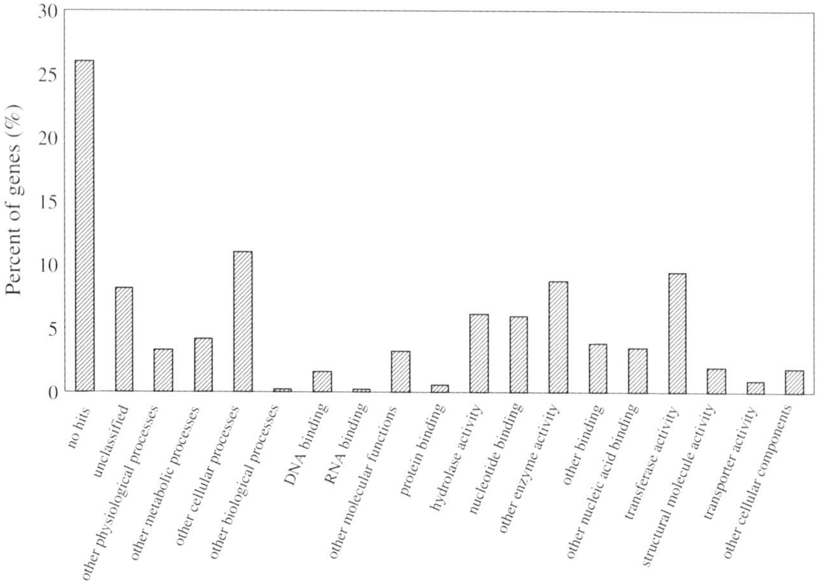

Fig. 9.2. Functional classification of *P. bellina* flower bud dbEST.

functions, the categories most highly represented were the transferases (9.4 percent), other enzymes excluding hydrolases, kinases and transferases (8.7 percent), and hydrolases (5.9 percent). Among the biological processes, the largest proportion (10.9 percent) of functionally assigned ESTs fell into cellular processes excluding signal transduction, cell organization and biogenesis and transport; other metabolic processes (excluding protein metabolism, DNA metabolism, RNA metabolism, electron transport, energy pathways and transcription) accounted for 4.1 percent. Together, these two classes of molecular functions and biological processes accounted for 64.1 percent of the assignable ESTs (Fig. 9.2). Third, to identify candidate genes in the DXP-geraniol-linalool pathway of *Phalaenopsis*, we developed a Pathway and Literature finder system (PaL), to map ESTs to metabolic pathways, and correlated the EST annotations and metabolic components to the related literature (Hsiao *et al.*, 2006).

9.2.2. *Scent biosynthesis pathway and related pathway*

9.2.2.1. *Scent biosynthesis pathway*

We find deoxyxylulose-5-phosphate synthase (DXPS), deoxyxylulose-5-phosphate reductase (DXPR), 4-diphosphocytidyl-2-C-methyl-D-erythritol 2-phosphate cyclase (DMEC), and geranyl diphosphate synthase (GDPS) related ESTs are mapped to the methylerythritol phosphate (MEP) pathway, which is included in the steroid biosynthesis pathway in the KEGG database by using PaL finder. We also elucidated the floral scent biosynthesis pathway in *P. bellina* further in accordance with the GC-MS data on geraniol, 2,6-dimethyl-octa-2,6-dien-1,8-diol and citronellol. The major steps in the scent biosynthesis pathway from pyruvate and G3P to geraniol, linalool and their derivatives were identified. These include genes encoding DXPS, DXPR, DMEC, epimerase (EPI), GDPS and cytochrome P450. The biosynthetic pathway for linalool and its derivatives, including linalool oxide, were also deduced (Fig. 9.3).

9.2.2.2. *Scent related gene*

We hope to identify the gene involved in the production, transformation and emission of scent materials. Comparisons of the most abundant ESTs by calculating the enrichment factor (defined below) for different transcripts in the floral dbESTs of *P. bellina* and *P. equestris* (Hsiao *et al.*, 2006; Tsai *et al.*, 2004) facilitate the provisional identification of scent metabolism genes. The enrichment factor is obtained by dividing the proportion of a certain transcript in the scented species by that in the scentless species. Statistical significance was determined using the simple Monte Carlo Test ($\alpha \leq 0.05$). GDPS, EPI, LOX and diacylglycerol kinase (DAK) showed 16.6-, 10.0-, 9.9- and 9.5-fold enrichments, respectively (Table 9.2). GDPS was significantly differentially expressed in the scented species (0.30 percent in *P. bellina* vs. 0.02 percent in *P. equestris*; see Table 9.2). GDPS participates in the biosynthesis of monoterpenes in plastids (Sommer *et al.*, 1995) primarily by supplying the essential precursor, suggesting that *P. bellina* may

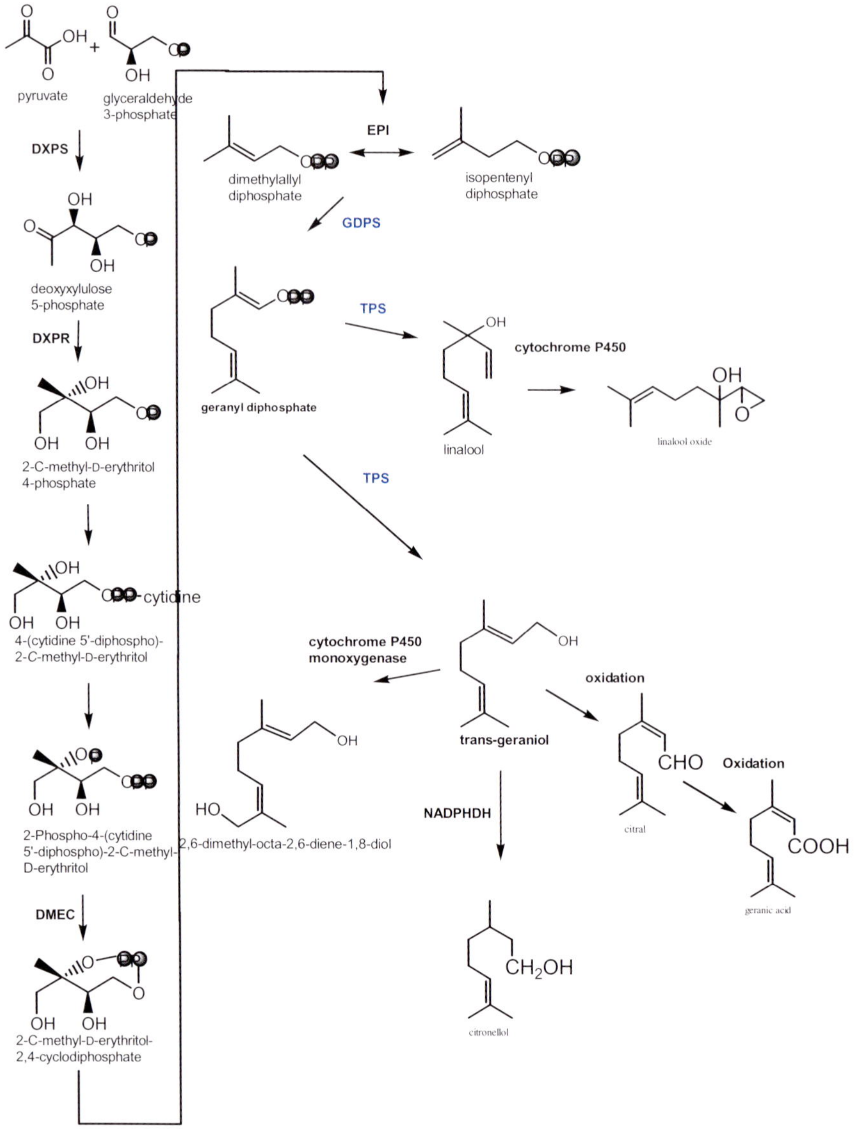

Fig. 9.3. Putative metabolic pathway from pyruvate and glyceraldehyde-3-phosphate to scent synthesis and related enzymes in *P. bellina*. DXPS: deoxyxylulose-5-phosphate synthase; DXPR: deoxyxylulose-5-phosphate reductoisomerase; DMEC: 4-diphosphocytidyl-2-C-methyl-D-erythritol 2-phosphate cyclase; EPI: epimerase; GDPS: geranyl diphosphate synthase; NADPHDH: NADPH dehydrogenase. Adapted from Hsiao *et al.* (2006).

Table 9.2 Differentially Expressed Transcripts Between *P. bellina* (Scented) and *P. equestris* (Scentless) Species in the Floral cDNA Libraries

	Abundance in Floral dbESTs		
Putative Functions	P. bellina (%)[a]	P. equestris (%)[b]	Enrichment Factor[c]
Geranyl diphosphate synthase	0.30	0.02	16.6
Epimerase/dehydratase	1.61	0.16	10.0
Lipoxygenase	2.12	0.21	9.9
Diacylglycerol kinase	0.34	0.04	9.5
Expansin	0.68	0.16	4.2
Peroxidase	1.36	0.39	3.5
Elongation factor	1.19	0.37	3.2
O-methyltransferase	0.78	0.30	2.6
Cytochrome P450	0.97	0.48	2.0
Heat shock protein	1.10	0.66	1.7
Class I chitinase	0.38	0.27	1.4
Molybdopterin biosynthesis cofactors	0.20	0.0	ND[d]

[a]Number of EST divided by 2,359 ESTs of *P. bellina* floral dbESTs.
[b]Number of EST divided by 5,593 ESTs of *P. equestris* floral dbESTs.
[c]Proportion (%) of ESTs in *P. bellina* and *P. equestris* floral bud libraries and the enrichment factor in the floral scent bud library (proportion of *P. bellina* flowers).
[d]ND: undetermined.

produce scent molecules in the plastids rather than the cytoplasm. Among the ESTs of the *P. bellina* flower bud, EPI (1.61 percent in *P. bellina* vs. 0.16 percent in *P. equestris*; see Table 9.2) is a multifunctional protein with hydratase, dehydrogenase, epimerase and isomerase activities involved in fatty acid beta-oxidation in plant peroxisomes and glyoxysomes (Engeland and Kindl, 1991; Guhnemann-Schafer *et al.*, 1994). DAK, enriched in the scented species (0.34 percent in *P. bellina* vs. 0.04 percent in *P. equestris*; see Table 9.2), converts diacylglycerol to phosphatidic acid, which has been shown to accumulate rapidly in plant cells in response to stimuli and may function as a signal molecule (Munnik *et al.*, 1998a; Munnik *et al.*, 1998b; Laxalt and Munnik, 2002). If so, DAK may serve as a signal molecule for day-fragrance in *P. bellina* rather than being involved in scent biosynthesis directly.

In addition, expansin showed a 4.2-fold higher abundance in the scented species (0.68 percent in *P. bellina* vs. 0.16 percent in *P. equestris*; see Table 9.2). Expansin loosens plant cell walls and thereby stimulates cell enlargement (Cosgrove *et al.*, 2002). In orchids, expansin may burst the cell walls in the epithelium to release scent components. Indeed, transmission electron microscopy shows that bulges extend from the outer epidermis cell walls during flowering of *P. bellina* (unpublished results). Interestingly, our results show that peroxidase (1.36 percent in *P. bellina* vs. 0.39 percent in *P. equestris*, a 3.5-fold enrichment) might be related to scent metabolism.

More than two-fold increases of O-methyltransferase transcripts (0.78 percent in *P. bellina* vs. 0.30 percent in *P. equestris*, a 2.6-fold enrichment) and cytochrome P450 transcripts (0.97 percent in *P. bellina* vs. 0.48 percent in *P. equestris*, a 2-fold enrichment) were also observed in *P. bellina* flowers. Previous research has shown that O-methyltransferase is involved in many reactions in volatile biosynthesis (Lavid *et al.*, 2002). O-methyltransferase methylates a wide range of substrates such as catechols, phenylpropanoids, orcinol and isoquinoline, all of which are involved in modifying scent molecules upon induction (Tholl *et al.*, 2004). Cytochrome P450 can act as a hydroperoxide lyase and catalyze the cleavage of lipoxygenase products (fatty acid hydroperoxides), forming omega-oxoacids and volatile C6- and C9-aldehydes and alcohols (Noordermeer *et al.*, 2001). However, in some cases, fewer enzymes than expected are required to synthesize the various hydroxylated structures, such as cytochrome P450 mono-oxygenase in *Mentha* species (Bertea *et al.*, 2001) and geraniol 10-hydroxylase in *Catharanthus roseus* (Burlat *et al.*, 2004). Molybdopterin biosynthesis cofactors (0.21 percent in *P. bellina* vs. 0.0 percent in *P. equestris*; see Table 9.2) are involved in various types of oxidative metabolism, including nitrogen metabolism and phytohormone biosynthesis (Schwarz *et al.*, 1997).

To detect still more scent-related genes, we performed EST filtering using less stringent conditions to remove genes homologous between *P. bellina* and *P. equestris*. The EST filtering results revealed genes involved in the biosynthetic pathway from pyruvate and G3P to linalool, geraniol and their derivatives in floral unigenes of *P. bellina*, consistent with the chemical analysis results, showing that no linalool and geraniol were emitted from the floral organs of *P. equestris*. Further EST filtering using TBLASTX against the collected floral unigenes of rice and *Arabidopsis*, which both lack fragrance, removed the genes involved in morphogenesis (Hsiao *et al.*, 2006). A total of 330 unigenes common to the above two EST filtering results constituted a more refined set of scent-related genes in *P. bellina* (Table 9.3). These transcripts included genes for NADPH dehydrogenase (NADPHDH), EPI, molybdopterin biosynthesis cofactors and GDPS. Concomitantly, EPI and GDPS were among the highly expressed transcripts, and molybdopterin biosynthesis cofactors were only expressed in the *P. bellina* flower buds (Table 9.2). Our results also suggest that NADPHDH and cytochrome P450 are required for the formation of linalool and geraniol derivatives (Fig. 9.3).

Table 9.3 EST Filtering Results Using Unigenes of *P. bellina* Against Those of *P. equestris, Arabidopsis* and Rice

Annotation	Scent_Scentless (820)[a]	Scent_At + Os (365)[b]	Scent_Scentless_At + Os (330)[c]
NADPH dehydrogenase	6	6	6
Molybdopterin biosynthesis cofactors	3	3	3
Membrane protein	6	5	3
Sensor protein	2	2	2
Geranyl diphosphate synthase	2	2	2
Ferrochelatase	1	1	1
Epimerase/dehydratase	1	1	1
Mitogen-activated protein kinases	1	1	1
Photosystem protein	1	1	1
Myb family protein	3	1	1
O-methyltransferase	1	1	1
No hits found	120	87	69
Hypothetical protein	81	49	43
Unknown protein	93	35	25

[a]The number of unigenes obtained in the EST filtering result at an E-value $< e^{-5}$.
[b]The number of unigenes obtained in the EST filtering result at an E-value $< e^{-7}$.
[c]The number of unigenes that were simultaneously present in previous two ESTs filtering results.

Meanwhile, transcripts encoding signal transduction factors such as sensor proteins, membrane proteins and mitogen-activated protein kinase were also identified (Table 9.3), suggesting that scent emission may be related to stimuli that elicit a series of signal transduction processes leading to gene expression and scent production. We also detected ferrochelatase, as well as the Myb family of proteins (Table 9.3), which have been shown to regulate the biosynthesis of the petunia flower fragrance (Verdonk *et al.*, 2005).

9.2.3. *Critical enzyme for scent biosynthesis*

From our chemical and bioinformatics analyses, we deduced a monoterpene biosynthesis pathway of 15 steps in the *P. bellina* flower, from G3P to geraniol, linalool and their derivatives. Expression of the scent-related genes was confirmed using RNA blot hybridization (Hsiao *et al.*, 2006). These results were consistent with the bioinformatics analysis of floral ESTs preferentially expressed in the scent species. Comparison of the most abundant transcripts in floral libraries of scent and scentless species showed the transcripts of LOX, EPI, DAK and GDPS to be highly enriched in the scent species. Interestingly, genes encoding for scent biosynthesis, signal transduction, Myb and methyltransferase were also identified.

9.3. Characterization and Significance of Geranyl Diphosphate Synthase Gene in *Phalaenopsis*

Terpenoids belong to a large family of plant secondary metabolites, and their corresponding alcohols possess useful properties such as fragrance, flavor, insecticidal properties, and characteristics that make them useful as pharmaceutical agents. Terpene metabolites are not only essential for plant growth and development (e.g., gibberellin phytohormones) but are also important intermediaries in the various interactions of plants with the environment (Tholl *et al.*, 2006). All monoterpenes are derived from the same substrate, geranyl diphosphate (GDP, C_{10}), which is catalyzed by GDP synthase (GDPS), a member of the short-chain trans-prenyltransferase family, via the condensation of dimethylallyl diphosphate (DMADP) with isopentenyl diphosphate (IDP). The short-chain prenyltransferases also include farnesyl diphosphate synthase (FDPS) and geranylgeranyl diphosphate synthase (GGDPS), which synthesize farnesyl diphosphate (FDP, C_{15}) and geranylgeranyl diphosphate (GGDP, C_{20}), respectively (Reiling *et al.*, 2004; Ogura and Koyama, 1998). These enzymes provide the acyclic branch-point intermediates for isoprenoid biosynthesis and control the flux into various terpenoid products (Liang *et al.*, 2002) (Fig. 9.4). The alignment of different GDPS sequences has revealed several conserved regions. These include two prominent aspartate-rich (Asp-rich) motifs, designated as the first aspartate-rich motif (FARM) and the second aspartate-rich motif (SARM), that share the consensus sequence DD(X)2-4D (where X represents any amino acid) and are critical for catalytic function and substrate binding in the trans-prenyltransferase family (Wang and Ohnuma, 2000; Takaya *et al.*, 2003).

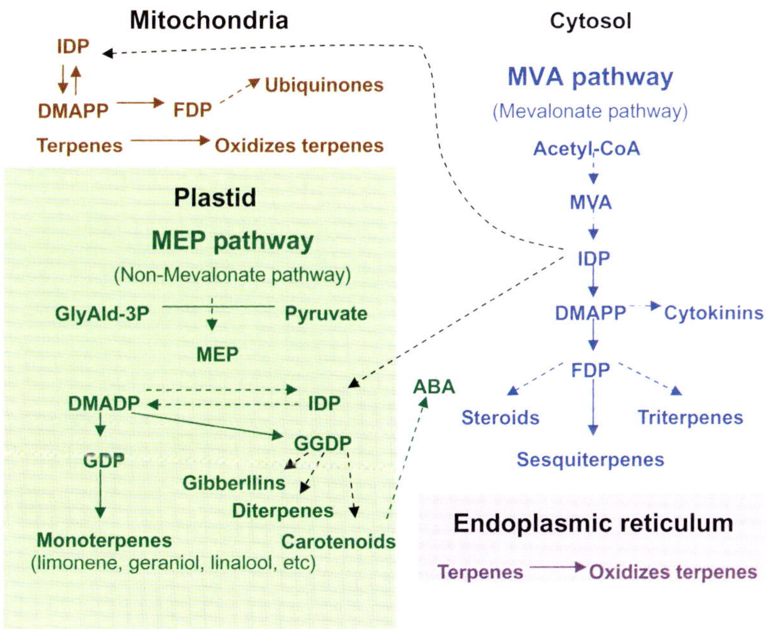

Fig. 9.4. Generalized schematic of isoprenoid biosynthesis in plants.

The distribution of GDPS appears to be limited within nature. It has been described in the plants *Mentha piperita* (Burke *et al.*, 1999), *Arabidopsis thaliana* (Bouvier *et al.*, 2000), *Abies grandis* (Burke and Croteau, 2002), *Antirrhinum majus* and *Clarkia breweri*, and it has only been identified in the bark beetle (Gilg *et al.*, 2005). The *Arabidopsis* GDPS, which contains the Asp-rich motifs, is presumed to be a functional homodimer, as is the *A. grandis* GDPS (Bouvier *et al.*, 2000; Burke and Croteau, 2002). However, GDPS from *M. piperita, A. majus,* and *C. breweri* comprises heterodimers of small subunits (*Mentha*_SSU, *Antirrhinum*_SSU and *Clarkia*_SSU) and large subunits (*Antirrhinum*_LSU) (Tholl *et al.*, 2004; Burke *et al.*, 1999). The large subunit shares high amino acid sequence identity with GGDPS from plants (50–75 percent). However, the small subunits share only 22–38 percent identity with GGDPS. Notably, the small subunit from these plants lacks the Asp-rich motifs and is inactive *per se* (Tholl *et al.*, 2004). It appears to control the length of the chain synthesized by the catalytic large subunit, because interaction of the small subunit with GGDPS results in the conversion of functional GGDPS into GDPS.

9.3.1. *The role of PbGDPS*

9.3.1.1. *Molecular cloning and sequence analysis of PbGDPS*

The full-length cDNA of *P. bellina* GDPS (PbGDPS) was isolated from a *P. bellina* floral cDNA library (Hsiao *et al.*, 2006) and sequenced. It was 1,102 bp in length and contained an open reading frame of 693 bp that encoded a protein of 231 amino acids. The amino acid sequence of PbGDPS showed low amino acid identity (30–32 percent) with GDPS from *M. piperita, A. majus* and *C. breweri* (*Mentha*_SSU, *Antirrhinum*_SSU and *Clarkia*_SSU) (Fig. 9.5). In addition, PbGDPS lacked the conserved FARM and SARM motifs (red boxes) that comprise an essential substrate binding element in all other short-chain prenyltransferases identified thus far (Ogura and Koyama 1998). Instead, a glutamate-rich (Glu-rich) sequence (EAEVE) was identified in the SARM motif (Fig. 9.5). Furthermore, the first 30 amino acids at the N-terminus formed a plastid targeting sequence (Fig. 9.5, horizontal line).

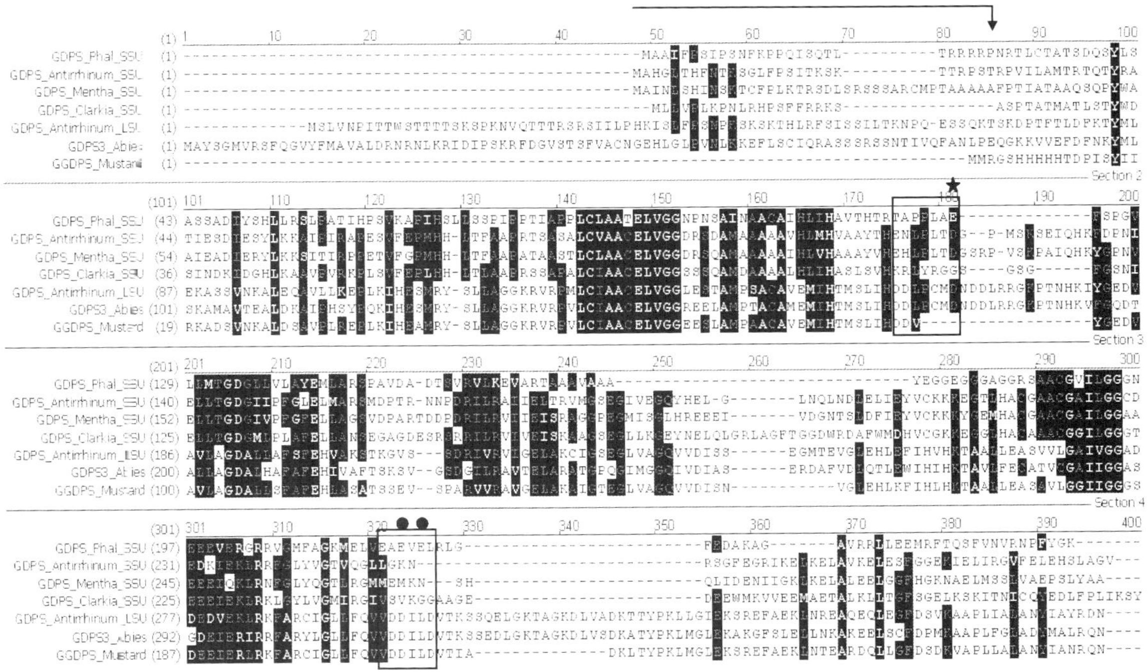

Fig. 9.5. Alignment of amino acid sequences of GDPS in *Abies grandis*, large subunit GDPS of *Antirrhinum majus* (GDPS_Antirrhinum_LSU), small subunit of *Antirrhinum majus* (GDPS_Antirrhinum_SSU), *Clarkia breweri* (GDPS_Clarkea_SSU), *Mentha piperita* (GDPS_Mentha_SSU), *Phalaenopsis bellina* (GDPS_Phal_SSU) and GGDPS of mustard (GGDPS_Mustard). The conserved aspartate-rich motif (DDXXD) and the putative active site (EAEVE) in GDPS_Phal_SSU are shown in the red box. Dashes indicate gaps inserted for optimal alignment. The horizontal line indicates the putative N-terminal transit peptide region. The arrowhead indicates the truncation site of GDPS_Phal_SSU. The blue circles indicate the amino acid residues mutated in this study. The star denotes the glutamate in the FARM region.

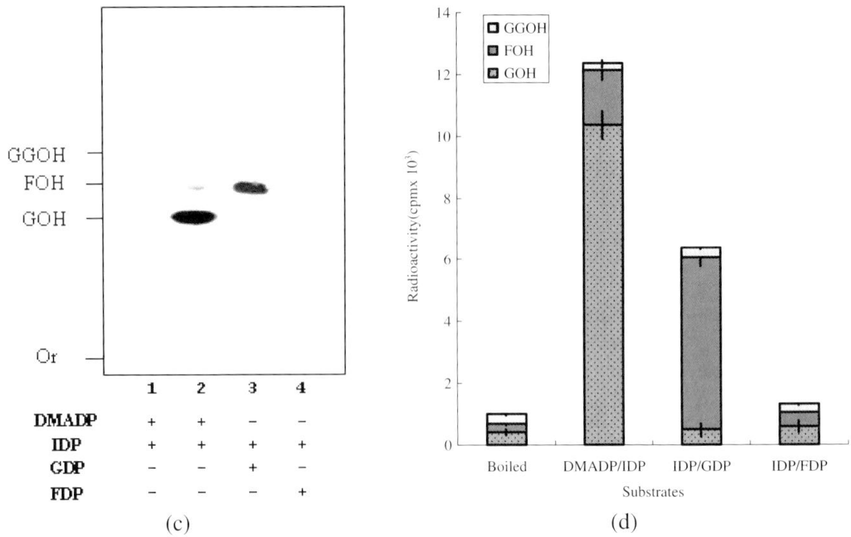

Fig. 9.6. (*Continued*)

occurred on day 5 post-anthesis with a 7-fold increase relative to the basal level, followed by a 2.5-fold decrease on day 7 post-anthesis, and it returned to the basal level on day 10 post-anthesis (Fig. 9.8a). The emission of linalool and geraniol during flower development was analyzed by time-course headspace collections of the volatile scent compounds emitted by a single living flower during the daytime in summer over a period of 9 h (from 9 am to 6 pm). Headspace collections began on the day of flowering and continued until flower senescence (day 14). Emission of linalool and geraniol increased strongly on day 2 post-anthesis, peaked on day 5 to 7 post-anthesis, and then declined gradually thereafter (Fig. 9.8b). According to these results, the highest expression of *PbGDPS* during flower development was concomitant with the maximal floral emission of monoterpenes (Fig. 9.8b), which suggests that PbGDPS plays a crucial role in scent production/emission in orchid flowers.

Further evidence of the involvement of PbGDPS in floral scent emission came from the analysis of PbGDPS expression in various orchid species with different scent-producing abilities (Fig. 9.9a). The major

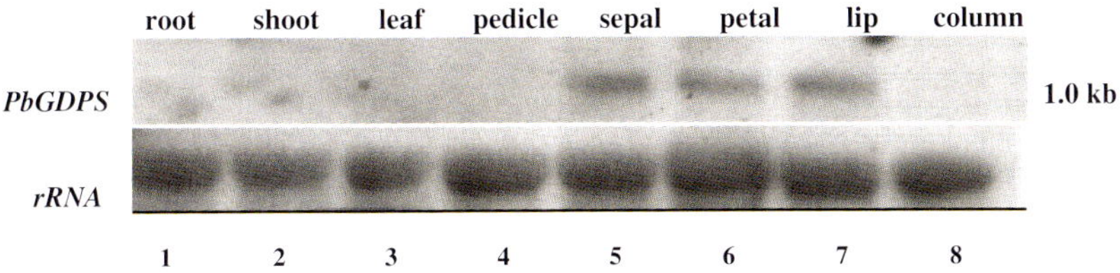

(a)

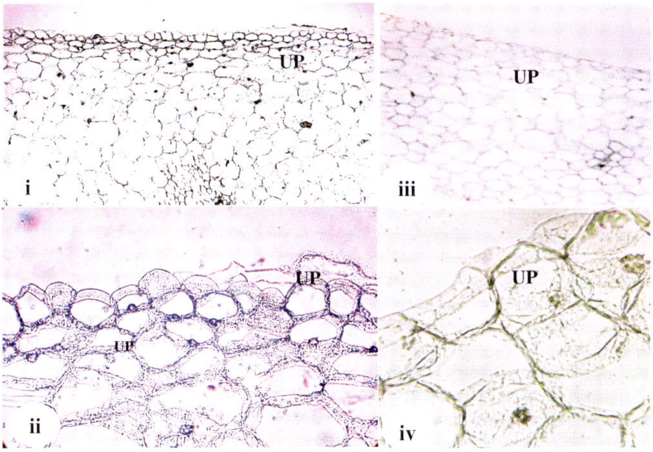

(b)

Fig. 9.7. (a) Spatial expression (lanes 1–8) of *PbGDPS* analyzed by northern blot hybridization. The ribosomal RNA indicated that the same amount of total RNA was loaded in each lane. (b) *In situ* localization of *PbGDPS* on the petal epidermis on day 5 post-anthesis (D + 5). Sections were hybridized with antisense 3′-specific RNA probe (i, ii) or sense 3′-specific RNA probe (iii, iv). UP = Upper epidermis.

volatiles emitted from the flowers of *P. bellina* are linalool and geraniol. The flowers of *Doritaenopsis* Kenneth Schubert "five", the progeny of a cross between *D. pulcherrima* and *P. bellina*, emitted only linalool, and the level of linalool was much lower than that emitted from the parental scented species *P. bellina* (Fig. 9.9c). In contrast, the progeny of another cross between *D. pulcherrima* and *P. bellina, D.* Kenneth Schubert, did not produce any scent that can be detected by humans. We also analyzed two varieties of *P. equestris*. The scented flowers of *P. equestris* "W-72" emit volatiles that are not monoterpenes (Table 9.1), as do

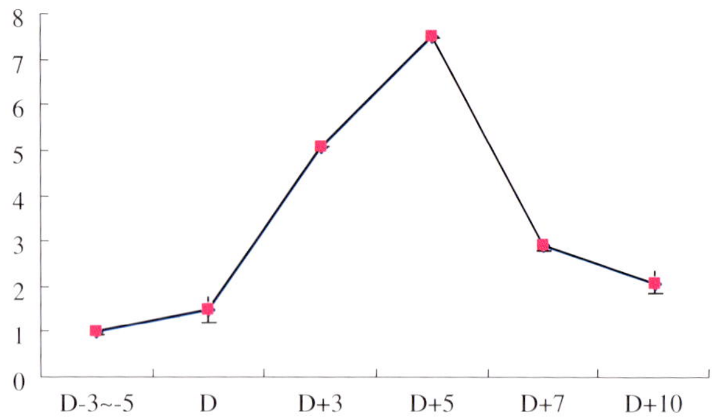

(a)

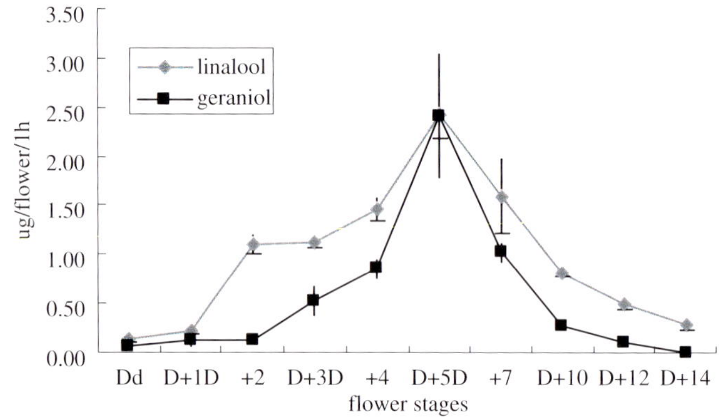

(b)

Fig. 9.8. (a) Quantitative real-time RT-PCR analysis with *PbGDPS*-specific primers. Results presented were normalized to the amplification of the ubiquitin reference gene. The calculation was based on three repeats and two individual experiments for the *P. bellina* flower, respectively. (b) Quantification of linalool and geraniol emitted from flowering day (Dd) to day 14 post-anthesis (D + 14) of *P. bellina*.

those of the scentless species *P. equestris*. The results of expression analysis of PbGDPS in these plants revealed that it was predominantly expressed in the scented species *P. bellina* and its scented offspring *D*. Kenneth Schubert "five" (Fig. 9.9b, lanes 1 and 4). In addition, the expression levels of PbGDPS in the scented species correlated with the amount of scent emitted (Fig. 9.9c). PbGDPS was expressed at lower

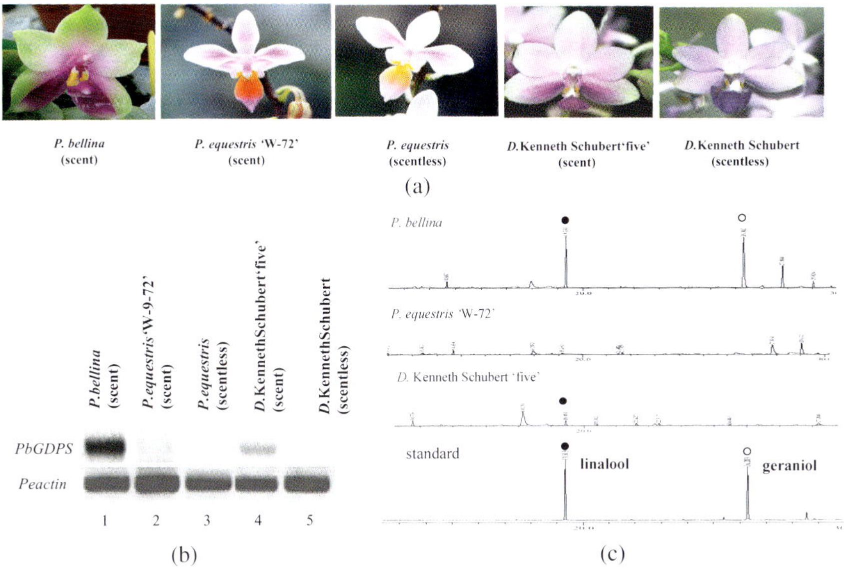

Fig. 9.9. Scented species of *P. bellina*, *P. equestris* "W9-72" and *D.* Kenneth Schubert "five" and scentless species of *P. equestris* and *D.*Kenneth Schubert. (a) Scented and scentless *Phalaenopsis* species. (b) Detection of *PbGDPS* transcripts by RT-PCR with *PbGDPS* gene-specific primers and RNA isolated from various flowers on day 5 post-anthesis. (c) The volatile components were collected from three flowers during each sampling for 9 h (from 9 am to 6 pm) from day 5 post-anthesis and analyzed by GC-MS. The resulting products were separated by gas chromatography and identified by comparison of mass spectra and retention times with those of authentic standards (bottom panel). The black circle is linalool and the white circle is geraniol.

levels in the scented offspring than in the parental line, which correlates with the lower production of linalool in the former. In contrast, no PbGDPS expression was detected in *P. equestris* or the scentless offspring *D.* Kenneth Schubert (Fig. 9.9b).

9.3.2. *PbGDPS protein behaviors*

9.3.2.1. *Homology modeling of the PbGDPS structure*

To understand how PbGDPS can act as a prenyltransferase with no Asp-rich motifs, we performed molecular modeling. We found no

GDPS structures within the Protein Data Bank (PDB). SWISS-MODEL (Schwede *et al.*, 2003; Guex and Peitsch, 1997; Arnold *et al.*, 2006) chose the X-ray crystal structure of *Sinapis alba* (mustard) GGDPS (SaGGDPS; PDB ID: 2J1O and 2J1P) as the best template (Fig. 9.10a). Sequence alignment of PbGDPS and SaGGDPS revealed 49 percent structural similarity. The modeled structure of the PbGDPS monomer

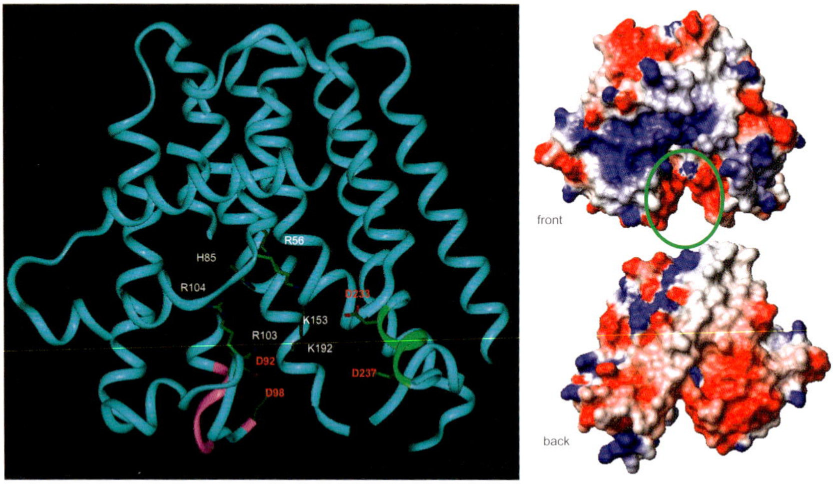

(a) (b)

Fig. 9.10. Comparison of three-dimensional PbGDPS modeling structure and SaGGDPS crystal structure (2J1P_A). (a) SaGGDPS crystal structure showed residues contacting with diphosphates in the two substrate-binding aspartate-rich motifs directly or via Mg^{2+}. The first and second aspartate-rich motifs are in purple and green, respectively. (b) The surface electrostatic potential image of SaGGDPS displays positive, negative, and neutral charges shown in blue, red and gray, respectively. The green circle indicates the binding pocket. (c) Modeling structure of PbGDPS. The positions corresponding to two common substrate-binding aspartate-rich motifs in SaGGDPS are in purple and green. The residue (E192) in the glutamate-rich region corresponding to D237 of SaGGDPS and the residue E194 corresponding to D98 of SaGGDPS, which was far away from the binding site in the modeling structure, are labeled red. (d) The surface electrostatic potential image of PbGDPS displays positive, negative, and neutral charges shown in blue, red and gray, respectively. The green circle indicates the binding pocket.

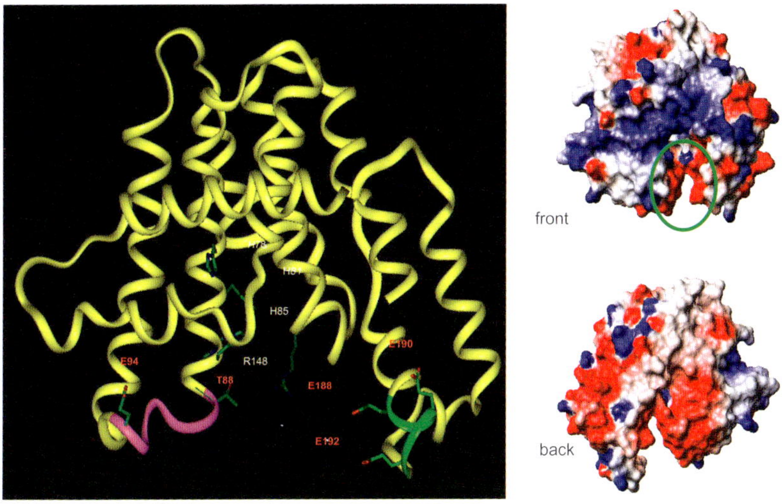

(c) (d)

Fig. 9.10. (*Continued*)

showed a single central core (Figs. 9.10b and 9.10d) surrounded by 14 α-helices, which formed a reaction cavity in the centre of the enzyme (Fig. 9.10b). PbGDPS lacks the common consensus FARM and SARM sequences, but the model of the PbGDPS structure allowed us to infer that the Glu-rich motif in PbGDPS and the SARM in SaGGDPS occupied equivalent positions within these structures (Fig. 9.10a). In addition, a Glu residue in PbGDPS corresponded to the third Asp in the FARM of SaGGDPS (Fig. 9.5, blue star). The Glu-rich motif probably provides a chelating site for Mg^{2+} ions, which are required for the enzyme activity, similarly to the more common SARM.

To establish that the proposed PbGDPS structure would be functionally and catalytically competent, we compared the locations and positional orientations of its reactive residues with those of SaGGDPS. The surface electrostatic potential of PbGDPS at the active site was similar to that of SaGGDPS, with a positive and a negative cluster located at the front and back, respectively (Fig. 9.10). The substrate-binding pocket of the PbGDPS model structure was 15 ± 2 Å, shallower than that of SaGGDPS ($\sim$20 Å; Fig. 9.10, green circle). The pocket showed a basic-amino-acid cluster that corresponded to the IDP-binding site in SaGGDPS, which contains lysine (K53), arginine (R56) and histidine (H85). The corresponding basic-amino-acid cluster identified in PbGDPS contained three histidine residues, at positions 78, 81, and 85 (Fig. 9.10). Furthermore, the modeled structure revealed a hydrophobic surface around helices 6 and 7, which could represent a potential dimerization site (Figs. 9.10c and 9.10d).

9.3.2.2. *Site-directed mutagenesis of the Glu-rich motif*

To test the hypothesis that the function of the Glu-rich motif resembles that of the Asp-rich motif, we replaced the two glutamate residues (E192 and E194; Fig. 9.10) with two nonpolar alanine residues (Ala, A) or two polar aspartate residues (Asp, D), and named the mutants PbGDPS-Ala (EAAVA) and PbGDPS-Asp (EADVD), respectively. The site-directed mutagenized proteins were expressed in *E. coli* and purified in the same way as the wild-type protein. The functional activity of the purified proteins and the chain lengths of their products were analyzed. PbGDPS-Ala showed virtually no prenyltransferase activity, and PbGDPS-Asp (Fig. 9.11a). PbGDPS-Asp generated GDP as the major product and FDP as a minor product (Fig. 9.11b). However, the amounts of GDP and FDP produced by the PbGDPS-Asp mutant were lower than those produced by the wild-type protein (Fig. 9.11b). Although PbGDPS-Ala also produced GDP, the formation of FDP was not detected (Fig. 9.11b). These results suggested that the glutamate residues in the motif were crucial for PbGDPS catalytic efficiency. This is a novel GDPS with dual-functional homodimer but without Asp-rich motifs in plants.

9.3.3. *Phylogenetic relationships of PbGDPS*

To determine the phylogenetic relationships between PbGDPS and other plant GDPS proteins, we constructed a phylogenetic tree of the known short-chain prenyltransferase genes using the derived amino acid sequences of the respective genes as input data. The tree showed that GDPSs are related more closely evolutionarily to GGDPSs than to

FDPSs (Fig. 9.12). In addition, the GDPS proteins scattered into four clades. These results agreed with the hypothesis that GDPSs might have multiple origins during plant evolution (Tholl *et al.*, 2004). Among these clades, the first, GGDPS/GDPS-a, included plant GGDPSs and gymnosperm GDPSs, which homodimerize to form the active enzyme (Burke and Croteau, 2002). The second clade, GDPS-b, consists of *Arabidopsis* and tomato GDPSs, which also form active GDPS by homodimerization (Bouvier *et al.*, 2000). Members of the first two clades

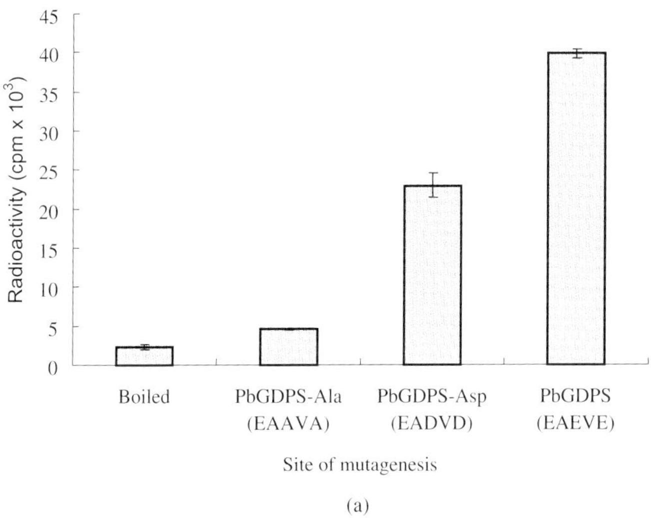

Fig. 9.11. Site-directed mutagenesis of the Glu-rich motif in the corresponding SARM motif of PbGDPS. (a) Prenyltransferase activity detected for PbGDPS and PbGDPS-Ala (EAAVA), PbGDPS-Asp (EADVD) mutated proteins (100 μg) with Mg^{2+}. Boiled protein was included as a negative control in the same assay. Each column represents the average of values from three independent experiments. Vertical bars indicate standard errors. (b) TLC autoradiograms of prenyl alcohols derived from products with wild-type, PbGDPS-Ala (EAAVA) and PbGDPS-Asp (EADVD) mutant proteins by using various prenyldiphosphates with different chain lengths as substrates. "Or" refers to the origin site and authentic standards including geraniol (GOH, C_{10}), farnesol (FOH, C_{15}), and geranylgeraniol (GGOH, C_{20}). Gieraniol (GOH), R_f = 7.5 cm; farnesol (FOH), R_f = 9 cm; and geranylgeraniol (GGOH), R_f = 9.6 cm.

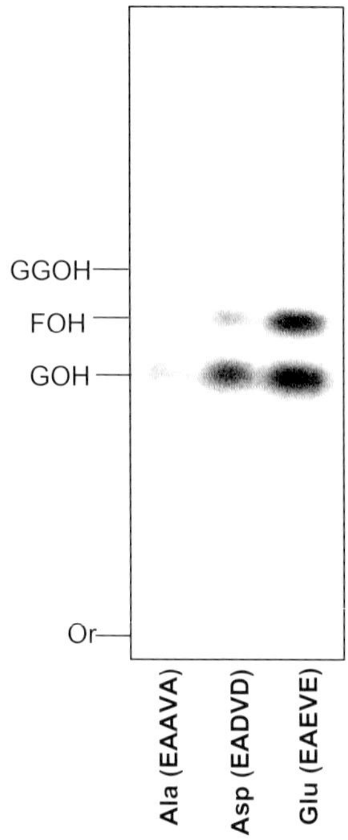

(b)

Fig. 9.11. (*Continued*)

all contained the Asp-rich motifs. In contrast, members of both the third and fourth clades, GDPS-c and GDPS-d, did not contain these motifs. At present, the GDPS-c clade contains GDPSs from *C. breweri, A. majus, M. piperita,* and *V. vinifera* (Tholl *et al.*, 2004; Burke *et al.*, 1999, 2004), and the GDPS-d clade contains GDPS from the monocot orchids. In the GDPS-c clade, the GDPSs from the eudicots comprise the small subunits of the heterodimeric enzymes and the catalytic activity subunit is provided by the large subunit. In contrast, PbGDPS, in the GDPS-d clade, formed a homodimer and was itself a functional dual-prenyltransferase, although it contained a Glu-rich, rather than an Asp-rich, motif. In addition, the topology was consistent with the last common ancestor of dicots and monocots containing a prototype of the GDPS-c/-d-type enzyme.

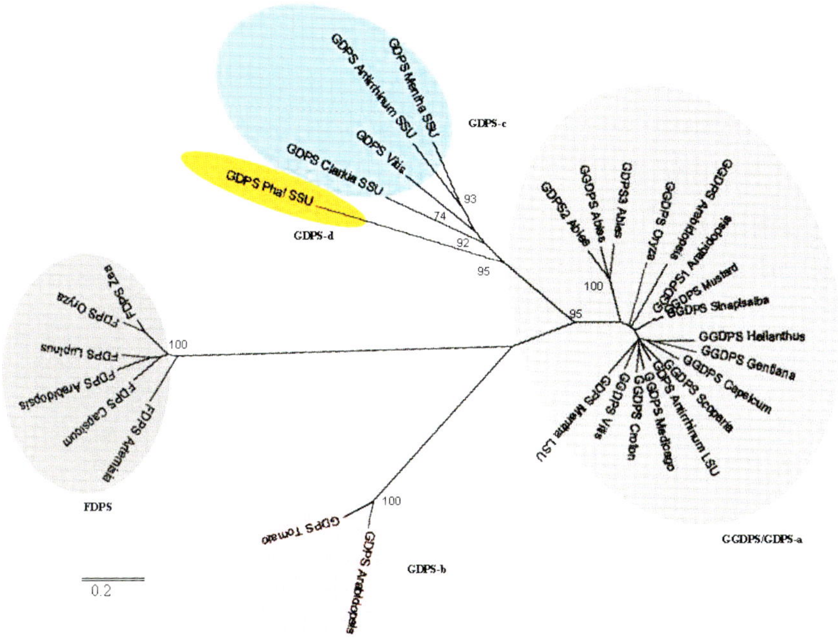

Fig. 9.12. Neighbor-joining tree based on degree of sequence similarities among plant GDPS-(a–d), FDPS and GGDPS.

9.4. Conclusions

Flower scent is a key modulating factor in plant-insect interactions and thus plays a central role in successful pollination. Closely related plant species that rely on different insects for pollination produce different odors. Orchid floral scents serve as attractants for species-specific pollinators that have played a major role in orchid evolution. The range of odors produced by orchids is enormous, providing an inexhaustible basis for specificity. Such diversity is advantageous in the evolution of an obviously successful family. However, it contains great mysteries surrounding orchid volatile production, biosynthesis, and their molecular mechanisms. Here, we have tried to understand the scent components, biosynthesis pathway and major synthetic enzymes involved in the mechanism of producing scent in the *P. bellina* flower. Nonetheless, there are still numerous areas in which significant progress can be made.

Qualitative and quantitative analyses showed the major flower fragrance components were monoterpenoids, phenylpropanoids, benzenoids and fatty acid derivatives. Monoterpenoids including geraniol, linalool and their derivatives (Table 9.1) accounted for more than 80 percent of the total volatiles collected from the *P. bellina* flowers. We were able to identify genes, including those for DXPS, DXPR, DMEC, GDPS, EPI, NADPHDH and cytochrome P450, involved in the DXP-geraniol-linalool pathway, by combining EST data mining with metabolic profiling and analysis of volatiles and EST filtering. In addition, we also showed that transcripts encoding signal transduction, Myb transcription factors and methyltransferase were well correlated with scent metabolism. The components of *P. bellina* flower scents are not only monoterpenoids but also phenylpropanoids, benzenoids and fatty acid derivatives. These components may also be involved in the modification of flower scent. The strategies developed and adopted here can also be applied to the study of phenylpropanoid, benzenoid and fatty acid biosynthetic pathways in orchids.

Floral scent research has mainly concentrated on the isolation and characterization of enzymes and genes involved in the final steps (terpene synthase, TPS) of scent biosynthesis (Pichersky *et al.*, 2006). However, the level of scent biosynthetic enzymes is not the only limiting factor, and the direct precursor (GDP) could also contribute to the regulation of scent production (Tholl *et al.*, 2004; Aharoni *et al.*, 2004; Chen *et al.*, 2000; Guterman *et al.*, 2006; Nogues *et al.*, 2006). In addition, FDP biosynthesis and GGDP biosynthesis have been much more carefully investigated at the enzymatic and molecular level than GDP biosynthesis. Here, we identified a functional homodimeric GDPS that lacks an Asp-rich motif for scent production in the flower of *P. bellina* (Orchidaceae, monocot). Spatial and temporal expression analyses of *PbGDPS_SSU* revealed that it is closely related to monoterpene emission, which indicates that it may play a key role in the regulation of scent production in *P. bellina* flowers (Hsiao *et al.*, 2008b).

9.5. Perspectives

All of the prospective applications for altering volatile emission in orchids require a genetic engineering approach, which must be achieved with a comprehensive understanding of the regulation of the volatile biosynthetic mechanism. Predictable manipulation of volatile production, emission, and composition of the fragrance requires the availability of genes that are responsible for the biosynthesis of volatile secondary metabolites, along with an understanding of the networks that regulate the emission of volatiles in orchids.

Acknowledgments

The authors are indebted to Tian-Hsiang Huang and Hei-Chia Wang, who developed the PaL finder methods, and Lina Lai for the GC-MS analysis (National Science Council Instrument Center, University of National Cheng Kung, Tainan, Taiwan). They also thank Dr. Shaio Chun Tu (Department of Biology and Biochemistry, University of Houston, Texas) for the enzyme kinetics analysis and Tao-Hsin Chang (Institute of Biochemical Sciences, National Taiwan University, Taipei, Taiwan) for the suggestion of protein purification and gel filtration.

References

Aharoni A, Giri AP, Verstappen FW *et al.* (2004) Gain and loss of fruit flavor compounds produced by wild and cultivated strawberry species. *Plant Cell* **16**: 3110–3131.

Arditti J. (1992) *Fundamentals of Orchid Biology*. John Wiley & Sons, New York.

Arends JC. (1970) Cytological observations on genome homology in eight interspecific hybrids of *Phalaenopsis*. *Genetica* **41**: 88–100.

Arnold K, Bordoli L, Kopp J *et al.* (2006) The SWISS-MODEL workspace: A web-based environment for protein structure homology modelling. *Bioinformatics* **22**: 195–201.

Ashburner M, Ball CA, Blake JA *et al.* (2000) Gene ontology: Tool for the unification of biology. The Gene Ontology Consortium. *Nat Genet* **25**: 25–29.

Bertea CM, Schalk M, Karp F *et al.* (2001) Demonstration that menthofuran synthase of mint (*Mentha*) is a cytochrome P450 monooxygenase: Cloning, functional expression, and characterization of the responsible gene. *Arch Biochem Biophys* **390**: 279–286.

Bouvier F, Suire C, d'Harlingue A *et al.* (2000) Molecular cloning of geranyl diphosphate synthase and compartmentation of monoterpene synthesis in plant cells. *Plant J* **24**: 241–252.

Burke CC, Wildung MR, Croteau R. (1999) Geranyl diphosphate synthase: Cloning, expression, and characterization of this prenyltransferase as a heterodimer. *Proc Natl Acad Sci USA* **96**: 13062–13067.

Burke C, Croteau R. (2002) Geranyl diphosphate synthase from *Abies grandis*: cDNA isolation, functional expression, and characterization. *Arch Biochem Biophys* **405**: 130–136.

Burke C, Croteau R. (2002) Interaction with the small subunit of geranyl diphosphate synthase modifies the chain length specificity of geranylgeranyl diphosphate synthase to produce geranyl diphosphate. *J Biol Chem* **277**: 3141–3149.

Burke C, Klettke K, Croteau R. (2004) Heteromeric geranyl diphosphate synthase from mint: Construction of a functional fusion protein and inhibition by bisphosphonate substrate analogs. *Arch Biochem Biophys* **422**: 52–60.

Burlat V, Oudin A, Courtois M *et al.* (2004) Co-expression of three MEP pathway genes and geraniol 10-hydroxylase in internal phloem parenchyma of *Catharanthus roseus* implicates multicellular translocation of intermediates during the biosynthesis of monoterpene indole alkaloids and isoprenoid-derived primary metabolites. *Plant J* **38**: 131–141.

Chen D, Ye H, Li G. (2000) Expression of a chimeric farnesyl diphosphate synthase gene in *Artemisia annua* L. transgenic plants via *Agrobacterium tumefaciens*-mediated transformation. *Plant Science* **155**: 179–185.

Chen F, Ro DK, Petri J *et al.* (2004) Characterization of a root-specific *Arabidopsis* terpene synthase responsible for the formation of the volatile monoterpene 1,8-cineole. *Plant Physiol* **135**: 1956–1966.

Cosgrove DJ, Li LC, Cho HT *et al.* (2002) The growing world of expansins. *Plant Cell Physiol* **43**: 1436–1444.

Engeland K, Kindl H. (1991) Evidence for a peroxisomal fatty acid beta-oxidation involving D-3-hydroxyacyl-CoAs. Characterization of two forms of hydro-lyase that convert D-(-)-3-hydroxyacyl-CoA into 2-trans-enoyl-CoA. *Eur J Biochem* **200**: 171–178.

Gilg AB, Bearfield JC, Tittiger C *et al.* (2005) Isolation and functional expression of an animal geranyl diphosphate synthase and its role in bark beetle pheromone biosynthesis. *Proc Natl Acad Sci* **102**: 9760–9765.

Griesbach RJ. (2002) *Development of Phalaenopsis Orchids for the Mass-Market.* ASHS Press, Alexandra, VA.

Guex N, Peitsch MC. (1997) SWISS-MODEL and the Swiss-PdbViewer: An environment for comparative protein modelling. *Electrophoresis* **18**: 2714–2723.

Guhnemann-Schafer K, Engeland K, Linder D *et al.* (1994) Evidence for domain structures of the trifunctional protein and the tetrafunctional protein acting in glyoxysomal fatty acid beta-oxidation. *Eur J Biochem* **226**: 909–915.

Guterman I, Masci T, Chen X *et al.* (2006) Generation of phenylpropanoid pathway-derived volatiles in transgenic plants: Rose alcohol acetyltransferase produces phenylethyl acetate and benzyl acetate in petunia flowers. *Plant Mol Biol* **60**: 555–563.

Harper T. (1991) Mutliflora *Phalaenopsis*: The contribution of *Phal. equestris* in breeding multifloras. *Amer Orch Soci Bull* **60**: 106–114.

Hsiao YY, Tsai WC, Kuoh CS *et al.* (2006) Comparison of transcripts in *Phalaenopsis bellina* and *Phalaenopsis equestris* (Orchidaceae) flowers to deduce monoterpene biosynthesis pathway. *BMC Plant Biol* **6**: 14.

Hsiao YY, Tsai WC, Chen WH *et al.* (2008a) *Floral Scent in Phalaenopsis*. Global Science Books, London.

Hsiao YY, Jeng MF, Tsai WC *et al.* (2008b) A novel homodimeric geranyl diphosphate synthase from the orchids *Phalaenopsis bellina* lacking a DD(X)2-4D motif. *Plant J* **55**: 719–733.

Jia JW, Crock J, Lu S *et al.* (1999) (3R)-Linalool synthase from *Artemisia annua* L.: cDNA isolation, characterization, and wound induction. *Arch Biochem Biophys* **372**: 143–149.

Kaiser R. (1993) *The Scent of Orchids*. Elsevier Press, Amsterdam.

Kao YY, Chang SB, Lin TY *et al.* (2001) Differential accumulation of heterochromatin as a cause for karyotype variation in *Phalaenopsis* orchids. *Ann Bot* **87**: 387–395.

Lavid N, Wang J, Shalit M *et al.* (2002) O-methyltransferases involved in the biosynthesis of volatile phenolic derivatives in rose petals. *Plant Physiol* **129**: 1899–1907.

Laxalt AM, Munnik T. (2002) Phospholipid signalling in plant defence. *Curr Opin Plant Biol* **5**: 332–338.

Liang PH, Ko TP, Wang AH. (2002) Structure, mechanism and function of prenyltransferases. *Eur J Biochem* **269**: 3339–3354.

Lin S, Lee HC, Chen WH *et al.* (2001) Nuclear DNA contents of *Phalaenopsis* species and *Doritis pulcherrima*. *J Amer Soc Horti Sci* **126**: 195–199.

Munnik T, Irvine RF, Musgrave A. (1998a) Phospholipid signaling in plants. *Biochim Biophys Acta* **1389**: 222–272.

Munnik T, van Himbergan JAJ, ter Riet B *et al.* (1998b) Detailed analysis of the turnover of polyphosphoinositides and phosphatidic acid upon activation of phospholipase C and D in *Chlamydomonas* cells treated with non-permeablizing concentrations of mastropara. *Planta* **207**: 133–145.

Nogues I, Brilli F, Loreto F. (2006) Dimethylallyl diphosphate and geranyl diphosphate pools of plant species characterized by different isoprenoid emissions. *Plant Physiol* **141**: 721–730.

Noordermeer MA, Veldink GA, Vliegenthart JF. (2001) Fatty acid hydroperoxide lyase: A plant cytochrome p450 enzyme involved in wound healing and pest resistance. *Chembiochem* **2**: 494–504.

Ogura K, Koyama T. (1998) Enzymatic aspects of isoprenoid chain elongation. *Chem Rev* **98**: 1263–1276.

Pichersky E, Gershenzon J. (2002) The formation and function of plant volatiles: Perfumes for pollinator attraction and defense. *Curr Opin Plant Biol* **5**: 237–243.

Pichersky E, Noel JP, Dudareva N. (2006) Biosynthesis of plant volatiles: Nature's diversity and ingenuity. *Science* **311**: 808–811.

Raguso RA, Pichersky E. (1995) Floral volatiles from *Clarkia breweri* and *C. concinna* (Onagraceae): Recent evolution of floral scent and moth pollination. *Plant Syst Evol* **194**: 55–67.

Reiling KK, Yoshikuni Y, Martin VJ *et al.* (2004) Mono and diterpene production in *Escherichia coli. Biotechnol Bioeng* **87**: 200–212.

Reinhard J, Srivivasan MV, Zhang S. (2004) Olfaction: Scent-triggered navigation in honeybees. *Nature* **427**: 411.

Schwarz G, Boxer DH, Mendel RR. (1997) Molybdenum cofactor biosynthesis: The plant protein Cnx1 binds molybdopterin with high affinity. *J Biol Chem* **272**: 26811–26814.

Schwede T, Kopp J, Guex N *et al.* (2003) SWISS-MODEL: An automated protein homology-modeling server. *Nucl Acid Res* **31**: 3381–3385.

Shindo K, Kamemmoto H. (1963) Karyotype analysis of some species of *Phalaenopsis. Cytologia* **28**: 390–398.

Sommer S, Severin K, Camara B *et al.* (1995) Intracellular localization of geranylpyrophosphate synthase from cell cultures of *Lithospermum erythrochizon. Phytochemistry* **38**: 623–627.

Takaya A, Zhang YW, Asawatreratanakul K *et al.* (2003) Cloning, expression and characterization of a functional cDNA clone encoding geranylgeranyl diphosphate synthase of *Hevea brasiliensis. Biochim Biophys Acta* **1625**: 214–220.

Tholl D, Kish CM, Orlova I *et al.* (2004) Formation of monoterpenes in *Antirrhinum majus* and *Clarkia breweri* flowers involves heterodimeric geranyl diphosphate synthases. *Plant Cell* **16**: 977–992.

Tholl D, Boland W, Hansel A *et al.* (2006) Practical approaches to plant volatile analysis. *Plant J* **45**: 540–560.

Tsai WC, Kuoh CS, Chuang MH *et al.* (2004) Four DEF-like MADS box genes displayed distinct floral morphogenetic roles in *Phalaenopsis* orchid. *Plant Cell Physiol* **45**: 831–844.

Van der Pijl L, Dodson CH. (1969) In: Fla CG (ed.), *Orchid Flowers.* University of Miami, Florida.

Verdonk JC, Haring MA, van Tunen AJ *et al.* (2005) ODORANT regulates fragrance biosynthesis in petunia flowers. *Plant Cell* **17**: 1612–1624.

Wang KC, Ohnuma S. (2000) Isoprenyl diphosphate synthases. *Biochim Biophys Acta* **1529**: 33–48.

CHAPTER 10

Warm- and Cool-Night Temperatures Alter Carbon Dioxide Fixation Pattern and Biochemical Metabolism in *Phalaenopsis aphrodite*

Ya-Chen Tseng[*], Yo-Ching Liu[†], Kai-Meng Tseng[†], Wen-Huei Chen[*,†] and Heng-Long Wang[*,†,‡]

After being acclimated to constant warm- (28°C day/28°C night) and cool-night (28°C day/20°C night) temperature regimes in growth chambers for two weeks, two groups of mature *Phalaenopsis aphrodite* subsp. *formosana* plants both clearly exhibited a diurnal oscillation in stomatal conductance, net CO_2 uptake rate, malate and starch levels, and phosphoenolpyruvate carboxylase (EC 4.1.1.31) and NAD^+-malic enzyme (EC 1.1.1.39) activities. Hence, *P. aphrodite* is an obligate crassulacean acid metabolism plant. Nevertheless, a different night temperature greatly affected both the stomatal conductance and the contribution of ambient and respiratory CO_2 to the nocturnal accumulation of malate. However, there appeared to be no significant difference between the two groups in the amounts of nocturnal accumulated malate and daily deposited starch. These results demonstrate that *P. aphrodite* is congruent with the characteristics of CAM plants in having great flexibility and plasticity in response to changes in environmental conditions. In addition, the formation of reproductive stem, viz. spike, was noticeably inhibited by a constant warm temperature, but induced by fluctuating warm-day and cool-night conditions. The relationship between the metabolic pool variation and spike induction in *Phalaenopsis* is also discussed.

[*]Institute of Biotechnology, National University of Kaohsiung, Kaohsiung 811, Taiwan.
[†]Department of Life Sciences, National University of Kaohsiung, Kaohsiung 811, Taiwan.
[‡]Corresponding author. E-mail: hlwang@nuk.edu.tw

10.1. Introduction

Phalaenopsis is an epiphytic orchid with thick, succulent leaves. Several authors have ascertained that *Phalaenopsis* is a crassulacean acid metabolism (CAM) plant (McWilliams, 1970; Endo and Ikusima, 1989; Ota *et al.*, 1991; Guo and Lee, 2006). Generally, the prominent characteristic of CAM plants is the fixation of atmospheric CO_2 via open stomata during the night (phase I) and early morning (phase II). The receptor of CO_2 is phosphoenolpyruvate (PEP) derived from the breakdown of stored carbohydrates, the product is malate that is stored in the vacuole and the reaction is catalyzed by PEP carboxylase (EC 4.1.1.31, PEPCase) (Osmond, 1978). The nocturnally accumulated malic acid is released from the vacuole at midday (phase III) and during the afternoon (phase IV), and is decarboxylated to form pyruvate and CO_2 through the mitochondrial NAD^+-linked malic enzyme (EC 1.1.1.39, NAD^+-ME) (Day, 1980) or the cytosolic $NADP^+$-linked malic enzyme (EC 1.1.1.40, $NADP^+$-ME) (Arron *et al.*, 1979). The released CO_2 goes *in situ* to photosynthetic assimilation via the Calvin cycle and provides precursors for subsequent nocturnal malic acid synthesis. Thus, CAM plants show a cyclical and reciprocal fluctuation of carbohydrates and organic acids according to a day-night rhythm (Osmond, 1978).

Temperature is one of the important environmental factors influencing the metabolism and development of CAM plants (Osmond, 1978; Lüttge, 2004). Although CAM characteristics can be well expressed under constant temperature (Lüttge and Beck, 1992), many studies have claimed that cool night temperatures (15–20°C) and temperature differences between day and night are both required for maximum nocturnal accumulation of malic acid in CAM plants (Lüttge, 2004; Neales, 1973; Haag-Kerwer *et al.*, 1992; Lootens and Heursel, 1998). Moreover, a different night temperature affects stomatal status and respiration activity (Lüttge, 1990) during the night. Hence, the contribution of ambient and respiratory CO_2 to the nocturnal accumulation of malate will be significantly different.

Phalaenopsis is a very popular orchid and the most valuable potted plant in the world. When *Phalaenopsis* was grown under a constant temperature higher than 28°C, spike initiation was significantly inhibited (Sakanishi *et al.*, 1980; Yoneda *et al.*, 1991). Conversely, a diurnal fluctuation of high day and low night temperature promoted spike induction in *Phalaenopsis* (Lee and Lin, 1984). These culturing conditions have been extensively adopted by orchid growers in Taiwan and Japan to schedule the flowering date of *Phalaenopsis*. Since the induction of spike emergence is the key step in controlling the flowering date, and the spiking mechanism is yet to be well understood, the aim of the present chapter is not merely to evaluate the effect of warm (28°C)- and cool (20°C)-night temperature on the photosynthetic responses and the diurnal variation of metabolic pools, including organic acids, starch and soluble carbohydrates, but also to decipher the relationship between the metabolites and spike induction in *Phalaenopsis*.

10.2. Different Night Temperature Affects Photosynthesis

10.2.1. *Chlorophyll a fluorescence*

Chlorophyll a fluorescence analysis is a rapid and non-invasive screening method for evaluating photosystem II (PSII) function, and it has been widely used as a sensitive probe of plants to reflect the interaction of plants with their environmental conditions (Röhacek and Bartak, 1999; Maxwell and Johnson, 2000; Calatayud *et al.*, 2006; Baker, 2008). Table 10.1 shows the maximum quantum yield (Φ_{P0}) is kept at around 0.8 and seemingly little affected by the difference in nighttime temperature. In general, the Φ_{P0} value of plants under normal conditions is close to 0.83, and it will decrease when stressed (Röhacek and Bartak, 1999). The effective quantum yield (Φ_P) also remained at similar levels, 0.52 ± 0.09 and 0.47 ± 0.04, respectively for warm- and cool-night temperature conditions. Clearly, the cool-night temperature did not affect the PSII photochemical efficiency of *P. aphrodite* (Table 10.1).

Table 10.1 The Effect of Cool-Night Temperature on the Maximum Quantum Yield (Φ_{P0}) and the Effective Quantum Yield (Φ_P) of *P. aphrodite*.

	28°C/28°C	28°C/20°C
F_0	119.50 ± 10.50	109.00 ± 0.00
F_M	627.50 ± 42.50	555.00 ± 10.00
Φ_{P0}	0.80 ± 0.01^a	0.80 ± 0.00^a
F'_0	109.50 ± 0.50	94.50 ± 4.50
F'_M	237.50 ± 42.50	180.00 ± 5.00
Φ_p	0.52 ± 0.09^a	0.47 ± 0.04^a

Data were expressed as means ± SE, $n = 3$ individuals plants. Means followed by the same letters are not significantly different at the 5% level.

10.2.2. *Gas exchange*

The stomatal opening of *P. aphrodite* exhibited a significant diurnal fluctuation: opening occurred essentially at night, remaining closed throughout most of the day, and reopened in the late afternoon (Fig. 10.1a). Similarly, the pattern of net CO_2 uptake rate (Fig. 10.1b) corresponded well to the status of stomatal opening (Fig. 10.1a). During the late afternoon, CO_2 fixation slightly increased, and the maximum was reached at the beginning of the night, with 5.1 and 2.3 μmole m^{-2}s^{-1} in cool- and warm-night conditions, respectively. It gradually decreased after 9 pm and became zero by the early morning (Fig. 10.1b). Under cool-night conditions, the stomatal conductance (Fig. 10.1a) and the net CO_2 uptake rate (Fig. 10.1b) were indeed significantly higher than those for the constant warm treatment (Lüttge, 2004; Neales, 1973; Haag-Kerwer *et al.*, 1992; Lootens and Heursel, 1998). Estimating the total nocturnal net carbon gain under the cool-night and constant-warm conditions by integrating the CO_2 uptake rate curves during the dark period (Fig. 10.1b), we obtained 143.3 and 58.8 mmole m^{-2}, respectively. Evidently, a cool-night favored ambient CO_2 uptake.

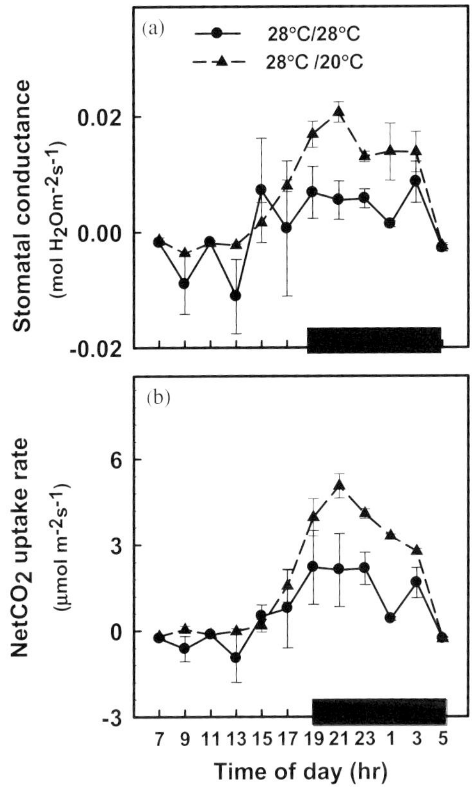

Fig. 10.1. The diurnal patterns of stomatal conductance (a) and net CO_2 uptake rate (b) of *P.* aphrodite under warm- (closed circles) and coolnight (closed triangles) conditions. Each point represents the average of two individual plants.

10.3. Different Night Temperature Affects Biochemical Metabolisms

10.3.1. *Organic acids*

CAM is primarily characterized by a large day–night change of malate. Figure 10.2a shows the diurnal oscillation of malate. Irrespective of

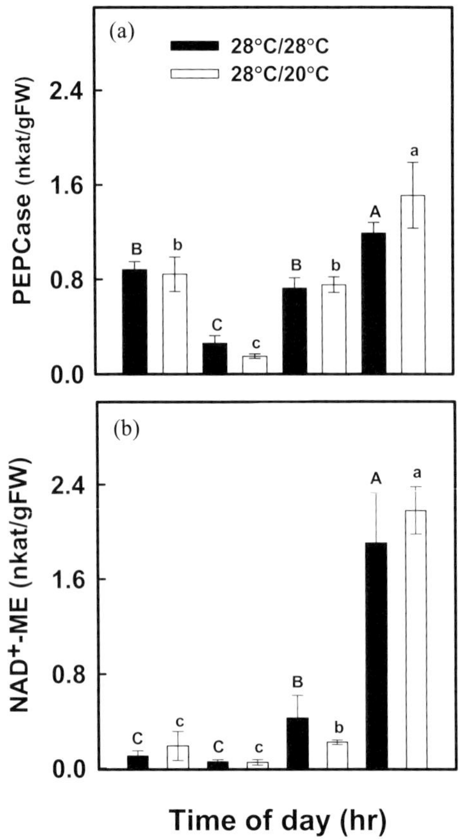

Fig. 10.4. The diurnal oscillation of PEPCase (a) and NAD^+-ME (b) activities of *P*. aphrodite under warm- (shaded bars) and cool-night (unshaded bars) conditions. All values are mean ± SE of four individual plants. The different capital and lowercase letters show statistically significant ($P < 0.05$) differences in the diurnal changes under warm- and cool-night conditions, respectively.

was significantly enhanced once the plants were exposed to a diurnal cycle of warm-day and cool-night temperatures over three weeks (Chen *et al.*, 2008). These data were consistent with the results published by many authors (Sakanishi *et al.*, 1980; Yoneda *et al.*, 1991; Lee and Lin, 1984) as well as the strategy of orchid growers to control the flowering date in commercial production (Blanchard and Runkle, 2006).

Sugars provide metabolic carbon and energy sources and adjust the cellular osmotic potentials (Gibson, 2005). Moreover, sugars also act as a signal to control reproductive development in plants (Gibson, 2005; Bernier *et al.*, 1993; Rolland *et al.*, 2006). According to the results of Lee and Lee (1996), when mature white *Phalaenopsis* plants were exposed to a fluctuating day/night temperature of 25°C/20°C for three weeks, it was found that the starch content decreased and sucrose concentration increased at 10 am (phase III). Kataoka *et al.* (2004) also quantified the sucrose content of *Phalaenopsis* leaves at midday (11 am) (phase III) and also found that a two-week warm treatment (27°C on average), when compared to the control (23°C), significantly reduced the level of sucrose, but not that of glucose or fructose, and delayed the spiking day. Therefore, they arrived at the conclusion that the sucrose content correlated markedly with the days to spiking. Indeed, during phase III (12 noon), samples from the constant warm treatment had a lower sucrose level (Fig. 10.3c) but there was no significant difference in glucose (Fig. 10.3a) and fructose (Fig. 10.3b) levels between the two treatments. Instead, in other phases, such as phase I (12 midnight) and phase IV (6 pm), glucose (Fig. 10.3a) and fructose (Fig. 10.3b) levels were higher under cool-night conditions. Sucrose maintained the same level in phase I (12 midnight) and phase II (6 am) (Fig. 10.4c) irrespective of the treatment. Since the levels of free sugars, especially sucrose, did not appear to have a pronounced difference between the two treatments throughout the daily course analysis (Fig. 10.3c), it seems pertinent to highlight that the cause-and-effect relationship between spike induction and sucrose level in *Phalaenopsis* may not always be so straightforward as suggested by Lee and Lee (1996) and Kataoka *et al.* (2004) So, the mechanism of spike emergence of *Phalaenopsis* still needs closer scrutiny.

10.5. Conclusion

Mature *Phalaenopsis aphrodite* subsp. *formosana* plants clearly exhibited a diurnal oscillation in stomatal conductance, net CO_2 uptake rate, malate and starch levels, and phosphoenolpyruvate carboxylase (EC 4.1.1.31) and NAD^+-malic enzyme (EC 1.1.1.39) activities.

Hence, *P. aphrodite* is an obligate crassulacean acid metabolism plant. In general, CAM plants, having a very flexible metabolic system, quickly respond to changes in environmental conditions. With a cool-night temperature, over 75 percent of accumulated malate of *P. aphrodite* was obtained from ambient CO_2. In contrast, at the higher night temperature, the higher respiration activity resulted in the majority of malate being synthesized by recycling respiratory CO_2, and only 34 percent was derived from ambient CO_2. Evidently, different night temperatures do indeed alter the metabolic responses of *P. aphrodite*.

Acknowledgments

The authors sincerely thank the Taiwan Sugar Research Institute for the use of its LI-6400 Portable Photosynthesis System (Li-Cor Inc., Lincoln, NE, USA) and Professor Yau-Lun Kuo (Graduate Institute of Bioresources, National Pingtung University of Science and Technology) for providing a light emitting diode source (6400-02, Li-Cor Inc., Lincoln, NE, USA) for gas exchange measurements. This work was supported by the grants 93AS–1.1.2–FD–Z3, 94AS–1.3.2–ED–Z3, and 95AS–1.3.2–FD–Z3 from the Council of Agriculture, Executive Yuan, Taiwan.

References

Arron GP, Spalting MH, Edwards GE. (1979) Isolation and oxidative properties of intact mitochondria from the leaves of *Sedum praealtum*, a crassulacean acid metabolism plant. *Plant Physiol* **64**: 182–186.

Baker NR. (2008) Chlorophyll fluorescence: A probe of photosynthesis *in vivo*. *Ann Rev Plant Biol* **59**: 89–113.

Bernier G, Havelange A, Houssa C *et al.* (1993) Physiological signals that induce flowering. *Plant Cell* **5**: 1147–1155.

Blanchard MG, Runkle ES. (2006) Temperature during the day, but not during the night, controls flowering of *Phalaenopsis* orchids. *J Exp Bot* **57**: 4043–4049.

Borland AM, Griffiths H. (1997) A comparative study on the regulation of C3 and C4 carboxylation processes in the constitutive crassulacean acid metabolism (CAM) plant *Kalanchoë daigremontiana* and C3-CAM intermediate *Clusia minor*. *Planta* **201**: 363–378.

Calatayud A, Roca D, Martinez PF. (2006) Spatial-temporal variations in rose leaves under water stress conditions studied by chlorophyll fluorescence imaging. *Plant Physiol Biochem* **44**: 564–573.

Carter P, Wilkins MB, Nimmo HG, Fewson CA. (1995) Effects of temperature on the activity of phosphoenolpyruvate carboxylsae and on the control of CO_2 fixation in *Bryophyllum fedtschenkoi*. *Planta* **196**: 375–380.

Chen LS, Lin Q, Nose A. (2002) A comparative study on diurnal changes in metabolite levels in the leaves of three crassulacean acid metabolism (CAM) species, *Ananas comosus, Kalanchoë daigremontiana* and *K. pinnata*. *J Exp Bot* **53**: 341–350.

Chen WH, Tseng YC, Liu YC *et al.* (2008) Influence of cool-night on the photosynthetic efficiency and the nonstructural carbohydrate and organic acid pools in *Phalaenopsis aphrodite*. *Plant Cell Rep* **27**: 1667–1675.

Conti S, Smirnoff N. (1994) Rapid triggering of malate accumulation in the C3/CAM intermediate plant *Sedum telephium*: Relationship with water status and phosphoenolpyruvate carboxylase. *J Exp Bot* **45**: 1613–1621.

Day DA. (1980) Malate decarboxylation by *Kalanchoë daigremontiana* mitochondria and its role in crassulacean acid metabolism. *Plant Physiol* **65**: 675–679.

Dittrich P. (1976) Nicotinamide adenine dinucleotide-specific "malic" enzyme in *Kalanchoë daigremontiana* and other plants exhibiting crassulacean acid metabolism. *Plant Physiol* **57**: 310–314.

Endo M, Ikusima I. (1989) Diurnal rhythm and characteristics of photosynthesis and respiration in the leaf and root of a *Phalaenopsis* plant. *Plant Cell Physiol* **30**: 43–47.

Gibson SI. (2005) Control of plant development and gene expression by sugar signaling. *Curr Opin Plant Biol* **8**: 93–102.

Guo WJ, Lee N. (2006) Effect of leaf and plant age and day/night temperature on net CO_2 uptake in *Phalaenopsis amabilis* var. *formosa*. *J Amer Soc Hort Sci* **131**: 320–326.

Haag-Kerwer A, Franco AC, Luttge U. (1992) The effect of temperature and light on gas exchange and acid accumulation in the C3-CAM plant *Clusia minor* L. *J Exp Bot* **43**: 345–352.

Herppich M, Willert DJV, Herppich WB. (1995) Diurnal rhythm in citric acid content-preceded onset of nighttime malic acid accumulation during metabolic changes from C3 to CAM in salt-stressed plants of *Mesembryanthemum crystallinum*. *J Plant Physiol* **147**: 38–42.

Kataoka K, Sumitomo K, Fudano T, Kawase K. (2004) Changes in sugar content of *Phalaenopsis* leaves before floral transition. *Sci Hort* **102**: 121–132.

Lee N, Lin GM. (1984) Effect of temperature on growth and flowering of *Phalaenopsis* white hybrid. *J Chin Soc Hortic Sci* **30**: 223–231.

Lee N, Lee CH. (1996) Changes in carbohydrates in *Phalaenopsis* during flower induction and inflorescence development. *J Chin Soc Hort Sci* **42**: 262–275.

Lootens P, Heursel J. (1998) Irradiance, temperature, and carbon dioxide enrichment affect photosynthesis in *Phalaenopsis* hybrids. *Hort Sci* **33**: 1183–1185.

Lüttge U. (1988) Day–night changes of citric-acid levels in crassulaceaen acid metabolism: Phenomenon and ecophysiological significance. *Plant Cell Environ* **11**: 445–451.

Lüttge U. (1990) Nocturnal citrate accumulation and its response to environmental stress in the CAM plant *Kalanchoë pinnata* (*Lam.*) Pers. *Plant Cell Environ* **13**: 977–982.
Lüttge U. (2004) Ecophysiology of crassulaceaen acid metabolism. *Ann Bot* **93**: 629–652.
Lüttge U, Beck F. (1992) Endogenous rhythms and chaos in crassulacean acid metabolism. *Planta* **188**: 28–38.
Maxwell K, Johnson GN. (2000) Chlorophyll fluorescence — A practical guide. *J Exp Bot* **51**: 659–668.
McWilliams EL. (1970) Comparative rates of dark CO_2 uptake and acidification in *Bromeliaceae, Orchidaceae,* and *Euphorbiaceae. Bot gaz* **131**: 285–290.
Neales TF. (1973) Effect of night temperature on the assimilation of carbon dioxide by mature pineapple plants, *Ananas comosus(l) merr. Aust J Biol Sci* **26**: 539–546.
Nimmo HG. (2000) The regulation of phosphenolpyruvate carboxylase in CAM plants. *Trends Plant Sci* **5**: 75–80.
Osmond CB. (1978) Crassulacean acid metabolism: A curiosity in context. *Annu Rev Plant Physiol* **29**: 379–414.
Ota K, Morioka K, Yamamoto Y. (1991) Effects of leaf age, inflorescence, temperature, light intensity and moisture conditions on CAM photosynthesis in *Phalaenopsis. J Japan Soc Hort Sci* **60**: 125–132.
Rolland F, Baena-Gonzalez E, Sheen J. (2006) Sugar sensing and signaling in plants: Conserved and novel mechanisms. *Annu Rev Plant Biol* **57**: 675–709.
Röhacek K, Bartak M. (1999) Technique of the modulated chlorophyll fluorescence: Basic concepts, useful parameters, and some applications. *Photosynthetica* **37**: 339–363.
Sakanishi Y, Imanishi H, Ishida G. (1980) Effect of temperature on growth and flowering of *Phalaenopsis amabilis. Bull Univ Osaka Prefect Ser B* **32**: 1–9.
Winter K, Schröppel-Meier G, Caldwell MM. (1985) Respiratory CO_2 as carbon source for nocturnal acid synthesis at high temperatures in three species exhibiting crassulaceaen acid metabolism. *Plant Physiol* **81**: 390–394.
Yoneda K, Momose H, Kubota S. (1991) Effects of day length and temperature on flowering in juvenile and adult *Phalaenopsis* plants. *J Jpn Soc Hort Sci* **60**: 651–657.

CHAPTER 11

Flower Development in the *Oncidium* Orchid

Yu-Yun Chang[*,#], Hsing-Fun Hsu[*,#]
and Chang-Hsien Yang[*,†]

An ABCDE model has been proposed to explain the interactions among five classes of MADS box gene during floral organ formation in angiosperms. *Ondicium* Gower Ramsey is a popular orchid with important economic value in cut-flower markets around the world. Compared with other orchids, the large and well-expanded lips and five small identical sepals/petals make *O.* Gower Ramsey a special case for the study of the diverse functions of MADS box genes during flower development. To date, we have already reported on the roles of A, B, C, D and E class MADS box gene homologs in regulating flower organ formation and initiation in *O.* Gower Ramsey. In this chapter, a comprehensive characterization and functional analysis of these five classes of MADS box genes from *O.* Gower Ramsey is provided. A potential strategy to modify the flower perianth formation and flower transition in *O.* Gower Ramsey by manipulating the expression of these MADS box genes is also discussed.

11.1. Introduction

11.1.1. *The ABCDE model for flower development*

Flowers are undoubtedly the most attractive parts of plants, with extraordinary morphological diversity in angiosperms. To study flower development in angiosperms, the now well-known ABCDE model was

[*]Graduate Institute of Biotechnology, National Chung Hsing University, Taichung 402, Taiwan.
[†]Corresponding author. E-mail: chyang@dragon.nchu.edu.tw
[#]These authors contributed equally.

proposed to explain gene interactions during floral organ formation (Coen and Meyerowitz, 1991; Litt and Kramer, 2010; Purugganan *et al.*, 1995; Rounsley *et al.*, 1995; Theissen, 2001; Theissen *et al.*, 2000; Theissen and Saedler, 1995, 2001). In the ABCDE model, the A function genes *APETALA1* (*AP1*) and *APETALA2* (*AP2*) control sepal formation; A-, B-[*APETALA3* (*AP3*), *PISTILLATA* (*PI*)] and E-(*SEPALLATA1/2/3*) function genes together regulate petal formation; B-, C-[*AGAMOUS* (*AG*)] and E-function genes control stamen formation; C- and E-function genes regulate carpel formation; and the D-function genes *FLORAL BINDING PROTEIN7 (FBP7)* and *FBP11* are involved in ovule development (Theissen, 2001; Theissen and Saedler, 2001). Most ABCDE genes encode MADS box proteins, playing a central role in transcriptional regulation during flower development (Coen and Meyerowitz, 1991; Litt and Kramer, 2009; Purugganan *et al.*, 1995; Rounsley *et al.*, 1995; Theissen, 2001; Theissen *et al.*, 2000; Theissen and Saedler, 1995, 2001; Weigel and Meyerowitz, 1994).

A- and E-function genes are both phylogenetically grouped in the *AP1/AGL9* family of MADS box genes (Theissen, 2001; Theissen and Saedler, 2001). A-function genes have been placed into the *SQUA*-like subfamily, which is further divided into *FUL/AGL8*-like and eu*AP1*-like genes in dicots, but remains in the paleo*AP1* subgroup in monocots (Litt and Irish, 2003; Vandenbussche *et al.*, 2003). E-function genes have been placed into the *SEP*-like subfamily, which contains two major clades, the *AGL9 (SEP3)* group and the *AGL2/3/4* (*SEP1/2/4*) group (Malcomber and Kellogg, 2005; Theissen, 2001; Theissen *et al.*, 2000; Zahn *et al.*, 2005a). In addition, other genes, such as *AGL6* in the *AGL6* subfamily and its putative orthologs identified from different plant species, have also been placed between *SQUA*-like and *SEP*-like subfamilies in the *AP1/AGL9* group (Hsu *et al.*, 2003; Ma *et al.*, 1991; Mena *et al.*, 1995; Mouradov *et al.*, 1998; Shindo *et al.*, 1999; Tandre *et al.*, 1995). Most genes in the *SQUA* subfamily are expressed in the early floral meristem and flower primordia, as has been shown for *AP1* and *SQUA* (Huijser *et al.*, 1992; Mandel *et al.*, 1992a).

The expression patterns of *SEP1* and *SEP2* are expressed in all four organ whorls, whereas *SEP3* and its orthologs are only expressed in the inner three whorls of the flower (Pelaz *et al.*, 2000; Rounsley *et al.*, 1995; Savidge *et al.*, 1995; Shitsukawa *et al.*, 2007; Zahn *et al.*, 2005a). The entire *SEP*-like subfamily consists of functionally redundant genes with the potentially conserved function of controlling the identities of all floral organs (Pelaz *et al.*, 2000; Theissen and Saedler, 2001; Zahn *et al.*, 2005a). Mutations of any one of these genes produced only subtle phenotypic differences (Pelaz *et al.*, 2000; Pelaz *et al.*, 2001; Savidge *et al.*, 1995; Zahn *et al.*, 2005a). However, all of the flower organs were converted into sepals in *sep1 sep2 sep3* triple mutants (Pelaz *et al.*, 2000), whereas *sep1 sep2 sep3 sep4* quadruple mutants had a complete transformation of all floral organs to vegetative leaves (Ditta *et al.*, 2004). The function of *SEP*-like genes in controlling the identities of all floral organs was further supported by the interaction between SEP and A-, B-, and C-function proteins in Arabidopsis and other species (de Folter *et al.*, 2005; Honma and Goto, 2001; Kotilainen *et al.*, 2000; Pelaz *et al.*, 2000; Pelaz *et al.*, 2001; Shitsukawa *et al.*, 2007; Theissen, 2001; Theissen *et al.*, 2000).

B-function MADS box genes, such as *APETALA3 (AP3)* and *PISTILLATA (PI)*, have been shown to have a major role in specifying petal and stamen development (Goto and Meyerowitz, 1994; Hernandez-Hernandez *et al.*, 2007; Irish, 2009; Jack, 2004; Kanno *et al.*, 2007; Kramer *et al.*, 1998; Krizek and Meyerowitz, 1996; Whipple *et al.*, 2007). In *Arabidopsis*, mutations in *AP3* or *PI* caused identical phenotypes of petal conversions into sepal structures and stamens into carpel structures (Bowman *et al.*, 1989; Goto and Meyerowitz, 1994; Jack *et al.*, 1992). Similar homeotic conversions for petals and stamens were observed in the mutants of the *AP3* and *PI* orthologs from a number of core eudicots (Angenent *et al.*, 1993; de Martino *et al.*, 2006; Liu *et al.*, 2004; Sommer *et al.*, 1990; Trobner *et al.*, 1992; van der Krol *et al.*, 1993; Vandenbussche *et al.*, 2004; Yu *et al.*, 1999), basal eudicot species (Drea *et al.*, 2007; Kramer *et al.*, 2007; Prasad and Vijayraghavan, 2003), and monocot species (Ambrose *et al.*, 2000; Nagasawa *et al.*, 2003; Prasad and Vijayraghavan, 2003; Yadav *et al.*, 2007; Yao *et al.*, 2008). It was proposed that B-function genes may have arisen from an ancestral gene through multiple gene duplication events (Doyle, 1994; Hernandez-Hernandez *et al.*, 2007; Kim *et al.*,

2004; Kramer *et al.*, 1998; Kramer and Irish, 1999; Lamb and Irish, 2003; Purugganan, 1997; Stellari *et al.*, 2004; Theissen *et al.*, 2000; Theissen *et al.*, 1996; Zahn *et al.*, 2005b). In the gymnosperms, there was a single putative B-function lineage that duplicated to generate the paleo*AP3* and *PI* lineages in angiosperms (Irish, 2009; Kramer *et al.*, 1998; Theissen *et al.*, 2000). The paleo*AP3* lineage is composed of *AP3* orthologs identified in lower eudicots, magnolid dicots, and monocots (Kramer *et al.*, 1998). Genes in this lineage contain the conserved paleo*AP3*- and *PI*-derived motifs in the C-terminal end of the proteins, which have been thought to be characteristics of the B-function ancestral gene (Hsu and Yang, 2002; Kramer *et al.*, 1998; Tzeng and Yang, 2001). The *PI* lineage is composed of *PI* orthologs that contain a highly conserved *PI* motif identified in most plant species (Kramer *et al.*, 1998). Subsequently, a second duplication at the base of the core eudicots produced the eu*AP3* and *TM6* lineages (Kramer *et al.*, 1998; Kramer and Irish, 1999). The paleo*AP3* motif in the C-terminal end of the proteins was retained in the *TM6* lineage and replaced by a conserved eu*AP3* motif in the eu*AP3* lineage of most eudicot species (Kramer *et al.*, 1998). In addition, many lineage-specific duplications for the paleo*AP3* lineage have occurred in plants such as orchids (Hsu and Yang, 2002; Kim *et al.*, 2007; Mondragon-Palomino *et al.*, 2009; Mondragon-Palomino and Theissen, 2008, 2009; Tsai *et al.*, 2004), Ranunculaceae, and Ranunculales (Di Stilio *et al.*, 2005; Kramer, 2009; Kramer *et al.*, 2003; Shan *et al.*, 2006). Unlike the A or C class MADS box proteins, which form homodimers that regulate flower development, the ability of B-function proteins to form homodimers has only been reported in gymnosperms and in the paleo*AP3* and *PI* lineages of some monocots (Hsu and Yang, 2002; Kanno *et al.*, 2003; Tsai *et al.*, 2004; Tsai *et al.*, 2005; Tzeng and Yang, 2001; Winter *et al.*, 2002). Proteins in the eu*AP3* lineage and most paleo*AP3* lineages were not able to form homodimers and had to interact with *PI* to form heterodimers in order to regulate petal and stamen development in various plant species (Kanno *et al.*, 2003; Moon *et al.*, 1999a; Riechmann *et al.*, 1996; Schwarz-Sommer *et al.*, 1992; Trobner *et al.*, 1992; Vandenbussche *et al.*, 2004; Winter *et al.*, 2002; Yao *et al.*, 2008).

C- and D-function genes play conservative roles in the control of reproductive tissue formation. They are thought to have arisen from gene duplications during the evolution of angiosperms (Kramer *et al.*, 2004; Theissen *et al.*, 2000). This interpretation is supported by the presence of two highly conserved *AG* motifs (I and II) in the C-terminal region of C/D function proteins (Kramer *et al.*, 2004; Song *et al.*, 2006; Yun *et al.*, 2004b). The C-function gene *AG* of *Arabidopsis* is the first reported MADS box gene and is responsible for stamen and carpel development (Yanofsky *et al.*, 1990). In *Arabidopsis, ag* mutants appear to have normal sepals and petals, but the stamens are transformed into petals and the carpels are transformed into additional *ag* mutant flowers. These phenotypes produced by mutation reveal that *AG* plays important roles in controlling the differentiation of stamens and carpels (Coen and Meyerowitz, 1991; Mizukami *et al.*, 1996). The best-known D class MADS box genes are *FBP7* and *FBP11* of petunia, which are expressed specifically in the ovules (Angenent and Colombo, 1996; Angenent *et al.*, 1995; Colombo *et al.*, 1995; Colombo *et al.*, 1997a; Colombo *et al.*, 1997b). Altered ovule formation is observed in co-suppression mutants of *FBP7* or *FBP11* (Angenent *et al.*, 1995; Colombo *et al.*, 1995; Colombo *et al.*, 1997a). Based on similarity in protein sequence and expression pattern, several putative orthologs of *FBP7/11* have been identified in other plant species (Lopez-Dee *et al.*, 1999; Pinyopich *et al.*, 2003; Rounsley *et al.*, 1995; Schmidt *et al.*, 1993; Song *et al.*, 2006; Theissen and Saedler, 1995; Tzeng *et al.*, 2002; Yun *et al.*, 2004b; Hsu *et al.*, 2010). However, the functions of these genes remain uncertain due to the lack of analysis of corresponding mutants or functional analysis. On the basis of the sequence similarity, C- and D-function genes are considered to be possibly produced by duplicated events from an ancestral gene (Theissen *et al.*, 2000). The characterization of C/D class genes has been reported in representative monocot species (Kang *et al.*, 1995; Schmidt *et al.*, 1993; Song *et al.*, 2006; Tzeng *et al.*, 2002; Yun *et al.*, 2004a; Yun *et al.*, 2004b; Hsu *et al.*, 2010). However, it remains uncertain whether C- and D-function or other related MADS box genes share similar functions in the specification of reproductive organ formation.

11.1.2. *Flower development in nongrass monocots*

A general flower structure consists of a number of perianth organs surrounding the male and female reproductive organs. Differentiated perianths are often composed of outer sepals and inner petals, but some species, such as grasses, may possess a novel perianth composed of paleas, lemmas and lodicules. The flowers of nongrass monocots such as lilies, tulips and orchids consist of two whorls of petaloid organs, termed tepals, surrounding the reproductive organs. Unlike lilies or tulips, orchids have flowers of spectacular morphological diversity. The perianths of orchids can be distinguished into three types: outer tepals (T1–T3; corresponding to "sepals") in floral whorl 1, and lateral inner tepals (t1 and t2; corresponding to "petals") and a median inner tepal (t3) named the labellum or lip in whorl 2 (Mondragon-Palomino and Theissen, 2008, 2009; Rudall and Bateman, 2002). The lip is highly evolved in its structure and morphogenesis, including the formation of the callus, spur and gland, and the phenomenon of resupination (Rudall and Bateman, 2002). In orchids, the reproductive organ (a gynostemium or column) is a fusion of male and female organs in the center of the flower, with species-specific variations resulting from the adaptation to specific pollinators (Cozzolino and Widmer, 2005).

The classic ABC model is hard to apply in the floral organ identity of nongrass monocots. Therefore a modified ABC model has been proposed to explain this floral morphology in nongrass monocots: the expression of class B genes was extended to whorl 1, with the result that the floral organs of the first and second whorls have almost identical petaloid structure (van Tunen *et al.*, 1993). This modified ABC model has been supported by the identification and characterization of putative class B genes from Liliaceae plants such as tulip (Kanno *et al.*, 2003), *Lilium regale* (Theissen *et al.*, 2000), *L. longiflorum* (Tzeng and Yang, 2001) and Asparagales plants (Kanno *et al.*, 2007; Nakada *et al.*, 2006; Nakamura *et al.*, 2005; Tsaftaris *et al.*, 2006). The detection of their expression in whorl 1 outer tepals fits the modified ABC model nicely. However, the expression of B-function genes in asparagus is restricted to whorls 2 and 3, despite the presence of almost identical petaloid perianths in whorls 1 and 2 (Park *et al.*, 2004; Park *et al.*, 2003).

In addition, despite its expression being detected in all four whorls, the LMADS1 protein of *L. longiflorum* accumulates only in whorls 2 and 3 (Tzeng and Yang, 2001). These results cannot be explained by the modified ABC model for the development of two whorls of petaloid tepals. Interestingly, four clades of paleo*AP3* genes and one *PI* gene have been identified in several orchids (Tsai *et al.*, 2004; Kim *et al.*, 2007; Mondragon-Palomino and Theissen, 2008; 2009; Mondragon-Palomino *et al.*, 2009). Although studies have been described the possible interactions among these orchid B class genes, the exact mechanism that involves these genes in regulating the differentiation and formation of the sepal/petal/lip of orchids remains unclear and needs to be clarified by further experimental investigation.

11.1.3. *The Oncidium flower as a model system*

Oncidium Gower Ramsey is a popular orchid with delicate flower shapes, and with a high economic value in the international cut-flower markets. The climate in Taiwan is ideal for the growth of *O.* Gower Ramsey all over this island. Most *O.* Gower Ramsey flowers are cultured in greenhouses in a monitored environment to ensure a high quality production. In an effort to develop the international floral market, many investments have been made into biotechnology development for *Oncidium* research in Taiwan.

In an inflorescence of *O.* Gower Ramsey, short branches containing 10 to 15 flowers grow from the nodes toward the apex. The flower of *O.* Gower Ramsey consists of three outer tepals (T1–T3) in the first floral whorl, and two lateral inner tepals (t1 and t2) and a median inner tepal (t3), termed the labellum or lip, in the second whorl with red-brown around the callus, and a column (reproductive organs) that has the basal part of the lip attached. Sepals and petals are yellow with red-brown stripes and blotches toward the base of the segment. The anther cap covers the pollinarium (male reproductive organ), which consists of two pollinia tapering into a stalk connected to a viscidium (Hsu and Yang, 2002).

The flowers of *O.* Gower Ramsey commonly have normal orchid perianths. However, there are floral terata with petals transformed into lip-like structures occurring sporadically in plant cultures (Mondragon-Palomino and Theissen, 2009; Rudall and Bateman, 2002, 2003). These floral terata may involve homeotic transformations of one or more tepals. The homeotic mutations probably result in the evolutionary diversity of orchid flowers. Therefore, changes in the expression of *DEF*-like genes or genes controlling their expression may affect floral morphology. The homeotic gene expression vibrations in intensities or territories may change organ identity, such as transformation of lateral inner tepals into lip-like organs or *vice versa* (Mondragon-Palomino and Theissen, 2009). Compared with other orchids, the large and well-expanded lips and five small identical sepals/petals make *O.* Gower Ramsey a special case for the study of the diverse functions of MADS box genes during flower development. Various studies conducted by own laboratory have reported on the roles of A, B, C, D and E class MADS box gene homologs in regulating flower organ formation and initiation in *O.* Gower Ramsey (Table 11.1) (Chang *et al.*, 2009; Chang *et al.*, 2010; Hsu *et al.*, 2010; Hsu *et al.*, 2003; Hsu and Yang, 2002).

Table 11.1 Characterization of MADS Box Genes in *Oncidium*

Genes	Amino Acid	Ortholog (Arabidopsis MADS Box Genes)
A class		
OMADS1	240	*AGL6*
OMADS7	243	*AGL6*
OMADS10	242	*AP1*
B class		
OMADS3	204	*AP3*
OMADS5	227	*AP3*
OMADS8	210	*PI*
OMADS9		*AP3*
C class		
OMADS4	233	*AG*
D class		
OMADS2	234	*STK*
E class		
OMADS6	243	*SEP3*
OMADS11	248	*SEP1*

11.2. Characterization of *Oncidium* MADS Box Genes

11.2.1. *A/E class genes and flower transition*

To investigate the role of the *AP1/AGL9*-like MADS box genes in the regulation of floral transition and flower development in the orchid (*O*. Gower Ramsey), five *AP1/AGL9*-like genes, *OMADS1, 6, 11, 7* and 10, were identified and characterized (Chang *et al.*, 2009; Hsu *et al.*, 2003). The sequence alignment and phylogenetic tree analysis indicated that *OMADS6* is a *SEP3* ortholog; *OMADS11* is closely related to *SEP1/2* orthologs within the E-function genes; *OMADS1* and 7 are closely related to *AGL6*-like genes and *OMADS10* is a putative paleo*AP1* ortholog (Table 11.1).

The expression patterns for these five orchid *AP1/AGL9*-like genes are interesting. First, the patterns for *OMADS6, 11* and *7* are almost identical (Chang *et al.*, 2009) (Table 11.2). In contrast, *OMADS1* and *10* showed a completely different expression pattern from *OMADS6/11/7*; for example, their expression was only strongly detected in the lip and carpel of mature flowers. In addition, *OMADS10* was also expressed in vegetative leaves. This indicated the possibility of functional diversification of *OMADS1/10* to *OMADS6/11/7*. *OMADS10* (*AP1*-like) might be separated from *OMADS1/6/11/7* (*AGL9*-like) through an early duplication event in the orchid. This assumption was supported by phylogenetic analysis.

It has been suggested that a duplication event near the base of the angiosperms gave rise to the *SEP3* and *SEP1/2* lineages about 300 million years ago (Zahn *et al.*, 2005a). Our results indicated that there were at least three duplication events that further generated *OMADS1/6/11/7* in the orchid. *OMADS1/7* (*AGL6*-like) might be separated from *OMADS6/11* through an early duplication event, while the second duplication event separated *OMADS6* (*SEP3*-like) and *OMADS11* (*SEP1/2*-like), and the third duplication event further separated *OMADS1* and *OMADS7*. The expression pattern of *OMADS7* was largely different from *OMADS1* (Hsu *et al.*, 2003), indicating that *OMADS7* should have retained a similar function to *OMADS6/11* in the orchid after it separated from *OMADS1* through gene duplication.

Table 11.2 Comparison of Gene Expression and the Effects Caused by Ectopic Expression in Arabidopsis for OMADSs Genes

		Gene Expression								Ectopic Expression in Arabidopsis				
Group	Gene	lf[a]	fb	mf	s	p	l	an	o	Early Flowering	Carpelloid Sepal	Staminoid Petal	Petaoid Speal	Curly Leaf
A	OMADS1	–[b]	+	+	–	–	+	–	+	+	+	+	–	+
	OMADS7	–	+	+	+	+	+	(+)	+	+	+	–	–	+
	OMADS10	+	–	(+)	–	–	+	–	+	+	–	–	–	–
B	OMADS3	+	+	+	+	+	+	+	+	(+)	–	–	–	–
	OMADS5	–	+	+	+	+	–	–	–	–	–	–	–	–
	OMADS8	+	+	+	+	+	+	+	(+)	(+)	–	–	+	(+)
	OMADS9	–	+	+	–	+	+	–	–	–	–	–	–	–
C	OMADS4	–	+	+	–	–	–	+	+	(+)	–	–	–	(+)
D	OMADS2	–	+	+	–	–	–	–	+	+	–	–	–	+
E	OMADS6	–	+	+	+	+	+	+	+	+	+	+	–	+
	OMADS11	–	+	+	+	+	+	–	+	+	–	–	–	+

[a]lf: leaf, fb: floral bud, mf: mature flower, s: sepal, p: petal, l: lip, an: anther, o: ovary.
[b]-: no, (+): weak, +: strong expression or phenotype observed.

The second interesting characteristics of *OMADS6, 11, 7* and *10* are their spatial and temporal expression patterns. The mRNA expression pattern for *OMADS6* was slightly different from *SEP3*. *SEP3* and its orthologs were only expressed in the inner three whorls of the flower (Angenent *et al.*, 1992; Angenent *et al.*, 1994; Pelaz *et al.*, 2000; Pnueli *et al.*, 1994; Rounsley *et al.*, 1995; Savidge *et al.*, 1995; Shitsukawa *et al.*, 2007; Zahn *et al.*, 2005a). In contrast, mRNA for *OMADS6* was detected in all four floral organ whorls (Chang *et al.*, 2009) and was similar to that observed for *SEP1/2* (Mandel and Yanofsky, 1998; Pelaz *et al.*, 2000; Rounsley *et al.*, 1995). This may be due to the great similarity between the sepals and petals within the flowers. Therefore, the genes that control petal formation in the orchid are very likely expressed in the sepal. This assumption was supported by a similar result observed for *LMADS3*, a lily *SEP3* ortholog, which is also expressed in all four flower organs (Tzeng *et al.*, 2003). Furthermore, although *OMADS6* expression was detected in all flower organs, relatively low expression in the stamen compared to that of other *SEP*-like genes was observed.

The mRNA expression pattern for *OMADS11* was similar to *OMADS6* and was different from *SEP1/2* since its expression was almost undetectable in the stamen (Chang *et al.*, 2009). Since *SEP1/2/3* are functionally redundant genes (Pelaz *et al.*, 2000; Theissen and Saedler, 2001), the slight difference in expression between *OMADS6/SEP3* and *OMADS11/SEP1/2* may reflect diversity of the E group genes in various plant species during evolution.

Expression only in vegetative leaves and in the lip and carpel of mature flowers distinguished *OMADS10* from most of the other genes in the *SQUA* subfamily (Chang *et al.*, 2009). Most genes in the *SQUA* subfamily were expressed in the early floral meristem and were absent in vegetative leaves while there were diverse gene expression patterns in flower organs (Huijser *et al.*, 1992; Kempin *et al.*, 1995; Kyozuka *et al.*, 2000; Mandel *et al.*, 1992b; Moon *et al.*, 1999b; Yu and Goh, 2000). In leaves, the expression of several *SQUA*-like genes of monocots, such as *OsMADS18* in rice and *LMADS5/6* in the lily, have been reported (Chen *et al.*, 2008; Fornara *et al.*, 2004; Masiero *et al.*, 2002). Interestingly, in flowers, *LMADS5/6* mRNA was also strongly expressed in carpels and almost undetectable in the other three flower organs, similarly to *OMADS10*. This revealed functional conservation for *SQUA*-like genes in the monocots lily and orchid.

The possible roles for *OMADS1, 6, 11, 7* and *10* in flower initiation and flower formation were further interpreted through functional transgenic analysis (Chang *et al.*, 2009). The early flowering phenotype in 35S::*OMADS1, 6, 11, 7* and *10* transgenic *Arabidopsis* plants were similar to those observed in transgenic plants that ectopically expressed A-, E- or *AGL6*-like-function MADS box genes (Chen *et al.*, 2008; Ferrario *et al.*, 2003; Fornara *et al.*, 2004; Honma and Goto, 2001; Hsu *et al.*, 2003; Jeon *et al.*, 2000; Kyozuka *et al.*, 1997; Mandel and Yanofsky, 1995; Pelaz *et al.*, 2001; Shitsukawa *et al.*, 2007; Sung *et al.*, 1999; Tzeng *et al.*, 2003; Xu *et al.*, 2008). Although 35S::*OMADS10* caused a relatively weak early flowering phenotype, our results suggested that *OMADS10* should also be involved in floral initiation similar to *OMADS6, 11* and 7.

Interestingly, the alterations of flower organs observed in the 35S::*OMADS1, 6, 11, 7* and *10* transgenic plants are different (Chang *et al.*, 2009). Flower organ conversions, such as carpelloid sepals and staminoid petals, were observed in 35S::*OMADS1* and 35S::*OMADS6* transgenic plants. Only carpelloid sepals were produced in the first whorl of 35S::*OMADS7* flowers. In contrast, no flower organ conversion was observed in 35S::*OMADS11* and 35S::*OMADS10* transgenic flowers. This result may reveal the possible functional diversification of *OMADS1, 6, 11, 7* and *10* in regulating flower formation. When the two putative orchid *AGL6*-like genes *OMADS7* and *OMADS1* were compared, their expression patterns were significantly different. However, ectopic expression of these two genes caused very similar effects on flower transition and formation in transgenic *Arabidopsis* (Chang *et al.*, 2009; Hsu *et al.*, 2003). These results indicated that a functional diversification occurred among *AP1/AGL9* genes in monocot orchids during evolution, although many characteristics were still similar in these genes.

In summary, five *AP1/AGL9* MADS box genes, *OMADSI, 6, 11, 7* and *10,* that specify flower development have been characterized from the orchid (*O.* Gower Ramsey). The similarities and differences of the expression patterns and effects caused by the ectopic expression of the orchid *AP1/AGL9*-like genes are summarized in Table 11.2. The characteristics of these five genes provide useful additional information in the understanding of the relationships among the *AP1/AGL9* MADS box genes in regulating flower transition and formation.

11.2.2. *B class genes and perianth formation*

In angiosperms, the flowers generally consist of perianth organs surrounding the reproductive organs that have evolved to facilitate pollinator attraction. Orchids, like monocot lilies or tulips, have two trimerous whorls of petaloid organs; the reproductive organs are centrally arranged, constituting column (a compound structure formed by adnation of male and female organs). The first whorl is formed by what are often termed sepals in the outer petaloid organs, and the inner whorl consists of two lateral petals and one labellum, or lip (Mondragon-Palomino and Theissen, 2008; Rudall and Bateman, 2002). B class genes are considered to regulate the petal formation in tulips or lilies and result in outer and inner whorls having the same petaloid structure (Kanno *et al.*, 2003). Recently, it has been proposed that the perianth identity of orchids is specified by the combination of one *PI* gene and four paleo*AP3* (clade 1–4) genes; these genes show highly conserved and clade-specific expression patterns (Tables 11.3 and 11.4).

To investigate the role of B class MADS box genes in the regulation of flower development in the orchid (*O.* Gower Ramsey), four genes, *OMADS3, 5, 8,* and *9,* were identified and characterized (Chang *et al.*, 2010; Hsu and Yang, 2002). Based on sequence alignment, the conserved motifs in the C-terminal regions of the proteins, and phylogenetic tree analysis, *OMADS5* belongs to clade 1, *OMADS3* belongs to clade 2, and *OMADS9* belongs to clade 3 of the paleo*AP3* genes in orchids. *OMADS8* is a *PI*-like gene and is closely related to monocot *PI* orthologs (Tables 11.1 and 11.3, and Fig. 11.1) (Chang *et al.*, 2010; Hsu and Yang, 2002).

The mRNA for *OMADS5* was only strongly detected in sepals and petals (Chang *et al.*, 2010), similar to other clade-1 paleo*AP3* genes,

Table 11.3 Annotation of B Class Genes were Classified by the BLAST in NCBI and Phylogenetic Analysis

	AP3 Ortholog				
Species	**Clade 1**	**Clade 2**	**Clade 3**	**Clade 4**	**PI Ortholog**
Cymbidium hybrid	ChMADS2				
Dendrobium crumenatum	DcOAP		DcOAP3B		
Dendrobium devonianum	DdAP3				
Dendrobium moniliforme				DMMADS4	
Gongora galeata	GogalDEF1	GogalDEF2	GogalDEF3		
Habenaria radiata			HrDEF		HrGLO1 HrGLO2
***Oncidium* Gower Ramsey**	**OMADS5**	**OMADS3**	**OMADS9**		**OMADS8**
Phalaenopsis equestris	PeMADS2	PeMADS5	PeMADS3	PeMADS4	PeMADS6
Phalaenopsis hybrid					PhaPI10
Phragmipedium longifolium		PhlonDEF1	PhlonDEF3	PholnDEF4	PhlonGLO1
Vanilla planifolia	VaplaDEF1		VaplaDEF3	VaplaDEF2	VaplaGOL1
Spiranthes odorata	SpodoDEF4	SpodoDEF2	SpodoDEF1	SpodoDEF3	SpodoGLO1

such as *PeMADS2* of *Phalaenopsis* (Tsai *et al.*, 2004) (Tables 11.2 and 11.4, and Fig. 11.1). The *OMADS9* mRNA was only strongly detected in petals and lips, similar to clade-3 paleo*AP3* genes (Tables 11.2 and 11.4, and Fig. 11.1) (Kim *et al.*, 2007; Tsai *et al.*, 2004; Xu *et al.*, 2006). The expression of both *OMADS5* and *9* was absent in stamens and leaves. Interestingly, the expression patterns of *OMADS5* and *9* were largely different from *OMADS3*, which was expressed in all flower organs and leaves (Tables 11.2 and 11.4, and Fig. 11.1) (Hsu and Yang, 2002). This indicated the possibility of functional and transcriptional diversification of *OMADS5*, *9*, and *3*.

Table 11.4 The Expression Pattern of B Class Genes in Orchids

B Class Genes	Leaves	Sepal	Petal	Lip	Column	Anther	Ovary	Ref.
Clade 1								
DcOAP	+	+	+	+	+	+		Xu *et al.*, 2006
OMADS5	–	+	+	–	–	–	–	Chang *et al.*, 2010
PeMADS2	–	+	+	(+)	(+)	–	–	Tsai *et al.*, 2004
Clade 2								
OMADS3	+	+	+	+	+	+	+	Hsu *et al.*, 2003
PeMADS5	–	+	++	+	(+)	–	–	Tsai *et al.*, 2004
Clade 3								
DcOAP3B	–	–	+	+	+	+		Xu *et al.*, 2006
HrDEF		–	+	(+)	(+)			Kanno *et al.*, 2007
OMADS9	–	–	+	+	–	–	–	Chang *et al.*, 2010
PeMADS3	–	–	+	+	(+)	–	–	Tsai *et al.*, 2004
Clade 4								
DMMADS4		–	+	+	+			Sirisawat *et al.*, 2009
PeMADS4	–	–	–	+	+	–	–	Tsai *et al.*, 2004

The possible role of *OMADS5* and *OMADS9* in regulating perianth formation is interesting. The difference in petal and lip formation may be due to the expression of *OMADS5* in the petal and the absence of *OMADS5* expression in the lip. This suggests a possible negative role for *OMADS5* in regulating lip formation. In the absence of *OMADS5* expression, the organs will be converted into lips in the presence of *OMADS9* expression. This assumption was supported by the expression analysis of *OMADS5* and 9 in peloric mutants of orchids. *OMADS5* mRNA was significantly reduced in both lip-like petals and

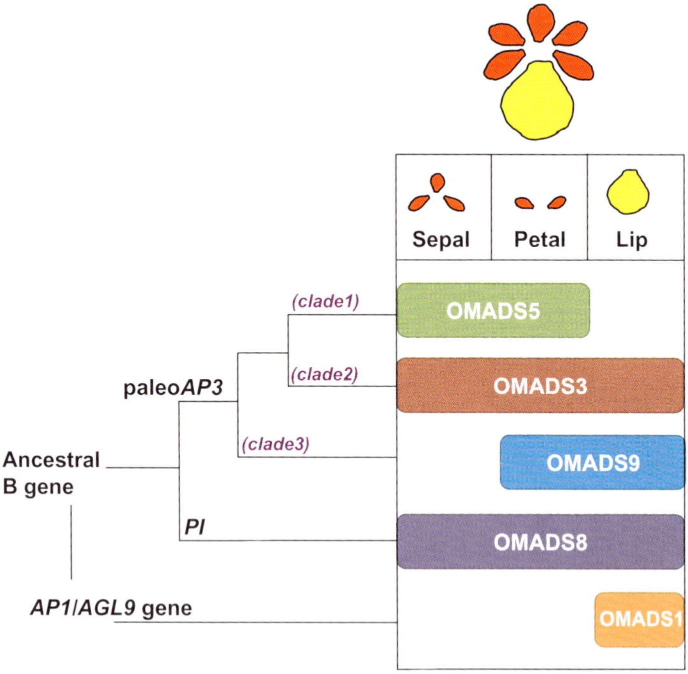

Fig. 11.1. Gene duplications and expression patterns of the B class genes in *O*. Gower Ramsey. A major duplication event from an ancestral B gene generated the paleo*AP3* and *PI* lineages. In orchid *O*. Gower Ramsey, there is only one *PI* gene, *OMADS8,* and two more duplications occurred in the paleo*AP3* gene that generated at least three paleo*AP3*-like genes, *OMADS5* (clade 1), *OMADS3* (clade 2), and *OMADS9* (clade 3). *OMADS8* and *OMADS3* were expressed in the sepal/petal/lip; *OMADS5* was expressed in the sepal/petal while *OMADS9* was expressed in the petal/lip. The *AGL6*-like gene *OMADS1* was expressed in the lip and carpel.

lip-like sepals of peloric mutant flowers (Chang *et al.*, 2010). These results indicated that the loss of *OMADS5* (clade 1) expression is likely the main reason for the lip specification and for the conversion of the sepals/petals into lips in *O*. Gower Ramsey.

Similar to *OMADS3*, the mRNA for *OMADS8* was detected in all four floral organ whorls as well as in vegetative leaves and roots (Tables 11.2 and 11.4, and Fig. 11.1) (Chang *et al.*, 2010). The expression in organs other than flowers has also been reported in some B class genes that showed a similar expression pattern to *OMADS8/3* by being expressed in both flower organs and leaves (Adam *et al.*, 2007; Kanno *et al.*, 2003; Munster *et al.*, 2001; Yu *et al.*, 1999). The expression of *OMADS3* and *OMADS8* in sepals, petals, and lips suggested that *OMADS8* and *OMADS3* may play similarly general and necessary roles in regulating the formation of these three flower organs in *O.* Gower Ramsey. Further yeast two-hybrid analyses indicated that OMADS3 formed homodimers and heterodimers with OMADS5, 8, and 9. OMADS8 only formed heterodimers with OMADS3, whereas OMADS5 and 9 formed homodimers individually and heterodimers with each other and with OMADS3 (Chang *et al.*, 2010). The ability to form homodimers supports the possibility that *OMADS3, 5,* and *9* also likely retained the characteristic of ancestral forms of the B genes for the paleo*AP3* lineage in monocots.

Based on the expression patterns and the protein interactions among these four orchid B class genes, a model was proposed to describe the interaction of these four B class proteins in the regulation of the formation of sepals, petals, and lips in *O.* Gower Ramsey (Fig. 11.2) (Chang *et al.*, 2010). A complex composed of at least OMADS3/8 and/or 9 is needed for formation of sepals, petals, and lips. Adding OMADS5 to the complex may convert organs into sepals/petals; any complex without OMADS5 controls lip formation. Since an increase in cell proliferation rather than cell expansion was the main reason for the large size of the lips in *Oncidium* (Chang *et al.*, 2010), it is reasonable to propose that *OMADS5* may play a role in suppressing lip formation by inhibiting cell proliferation in orchids like *Oncidium*. In addition to these B class proteins, other A/E class proteins are expected to participate in these complexes. For example, *OMADS1*, an *AGL6*-like gene of *O.* Gower Ramsey, is predominantly expressed in lips and could form heterodimers with *OMADS3* (Hsu *et al.*, 2003). Since most of the A/E genes of *O.* Gower Ramsey have been isolated and characterized in our laboratory (Chang *et al.*, 2009), further investigation of the role of these A/E genes in the interaction with B genes should lead to deeper understanding of sepal/petal/lip formation in *O.* Gower Ramsey.

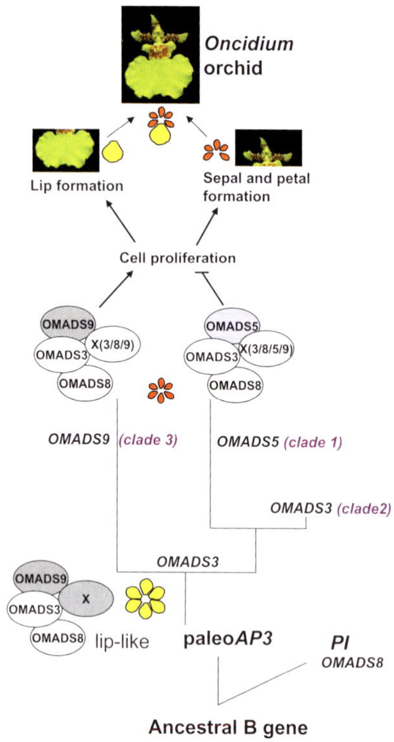

Fig. 11.2. Possible evolutionary relationships between B class genes in the regulation of sepal/petal/lip formation for *O*. Gower Ramsey. B group genes in monocots are thought to have been produced by a major duplication event from an ancestral gene that generated the paleo*AP3* and *PI* lineages. In *O*. Gower Ramsey, there is only one *PI* gene, *OMADS8* and at least two more duplications occurred in the paleo*AP3* gene *OMADS3* that generated the paleo*AP3*-like genes *OMADS5* and *OMADS9*. The ancestral complex containing OMADS3/8 and other A/E proteins (X) is sufficient to convert the first and second whorl organs into well-expanded lip-like sepals/petals (in yellow). After gene duplication, *OMADS9* may retain the ancestral B gene function of possibly promoting cell proliferation, whereas *OMADS5* might have been altered in relation to this function. This caused the possible suppression of cell proliferation for any complex that includes OMADS5 (for example, OMADS3/8/5/9/X) and might result in the formation of six short sepals/petals (in red) in the orchid flowers. During evolution, one of the short sepals/petals was converted into a well-expanded lip structure (in yellow) due to the loss of *OMADS5* expression, resulting in the formation of one expanded lip (in yellow) (OMADS3/8/9/X) and five short sepals/petals (in red) (OMADS3/8/5/9/X) as seen in the modern *O*. Gower Ramsey.

Functional analysis produced useful results for the interpretation of the possible role of *OMADS5, 8,* and 9 in flower formation. The production of elongated petal-like sepals in *35S::OMADS8* transgenic *Arabidopsis* plants was similar to that observed in transgenic *Arabidopsis* plants ectopically expressing *PI* (Lamb and Irish, 2003) or its orthologs, such as *DcOPI* in *Dendrobium* (Xu *et al.*, 2006) and *LMADS8, 9* in lily (unpublished data). This result provides further evidence to support the assumption that *OMADS8* simply functions as a B gene in the *PI* lineage. By contrast, no homeotic conversion in the floral organs was observed in the flowers ectopically expressing either *OMADS9* or *OMADS5*. This may be because the orchid OMADS9 and OMADS5 proteins are unable to interact, or are insufficient for interaction, with the *Arabidopsis* B/A/E class protein complex that regulates sepal and petal formation.

11.2.3. *C, D class genes and sexual organ formation*

The unique sexual organ of *Oncidium*, the column or gynostemium, is a compound structure formed by adnation of the male and female parts. The anther cap covers the pollinarium (male reproductive organ), which consists of two pollinia tapering into a stalk connected to a viscidium on the basal part of the lip. The orchid ovary is located behind the flower, and ovule development is delayed until anthesis of the flower and is triggered by pollination (Nadeau *et al.*, 1996; O'Neill *et al.*, 1993; Zhang and O'Neill, 1993). The unusual structures and development process of the orchid's sexual organ make it worthy of study.

To investigate the roles of floral homeotic genes in the regulation of sexual organ development in *O.* Gower Ramsey, we have identified and characterized one C-function MADS box gene (*OMADS4*) and one D-function MADS box gene (*OMADS2*) through sequence alignment and phylogenetic analysis (Hsu *et al.*, 2010). The assignment of these two MADS box genes to the C and D classes is further supported by analysis of the conserved motifs in the C-terminal regions of the proteins. Two highly conserved motifs specific to C/D proteins, AG motif I and AG motif II (Kramer *et al.*, 2004; Song *et al.*, 2006; Yun *et al.*, 2004b), are found in the C-terminal regions of the OMADS2 and OMADS4 proteins (Hsu *et al.*, 2010). In addition, a conserved

Chang, 2000). The colorful flower and the lip are the major contributing organs. New *Oncidium* varieties are difficult to produce, and hence molecular breeding is an attractive approach to create new flower forms. The B class MADS box genes that determine perianth formation have been identified in *Oncidium* (Chang *et al.*, 2010). We have shown that *OMADS3* and *OMADS8* are the essential genes in flower formation, whereas *OMADS5* and *OMADS9* regulate the perianth determination (Chang *et al.*, 2010). Since *OMADS5* is believed to play a negative role in regulating lip formation, it is possible to manipulate perianth formation by simple alteration of *OMADS5* expression. It is expected that knockdown of *OMADS5* expression using RNAi (*pOMADS5::OMADS5-RNAi*) specifically in the sepals and petals will potentially convert these two organs into lips, and then produce flowers with multiple (six) lips (Fig. 11.3b). Alternatively, introducing the *pOMADS9::OMADS5-RNAi* construct into the *Oncidium* orchid will specifically convert petals into lips and generate flowers with three lips (Fig. 11.3c). By contrast, ectopic expression of *OMADS5* (*pOMADS9::OMADS5*) in lips will convert lips into sepals/petals and result in a flower with six sepals/petals (Fig. 11.3d). In future, it will be possible to genetically after *Oncidium* to exhibit not only various colors and sizes but also different flower styles.

11.3.2. *Other applications of Oncidum MADS box genes*

In higher plants, controlling flowering time is essential to ensure efficient plant reproduction and to vary the harvest peak. There are two major harvest seasons for *Oncidium* in Taiwan: May to June and September to November. The initiation of the reproductive phase involving the emergence and growth of the flower bud occurs two months prior to blooming. Variation in the environmental conditions including light and temperature regulates both the flowering time and flower quality. Some relevant genes with roles in the polysaccharide mobilization pathway and genes concerned with flowering time have been shown to be involved in the *Oncidium* flowering process (Hou and Yang, 2009; Wang *et al.*, 2008). Ectopic expression of the

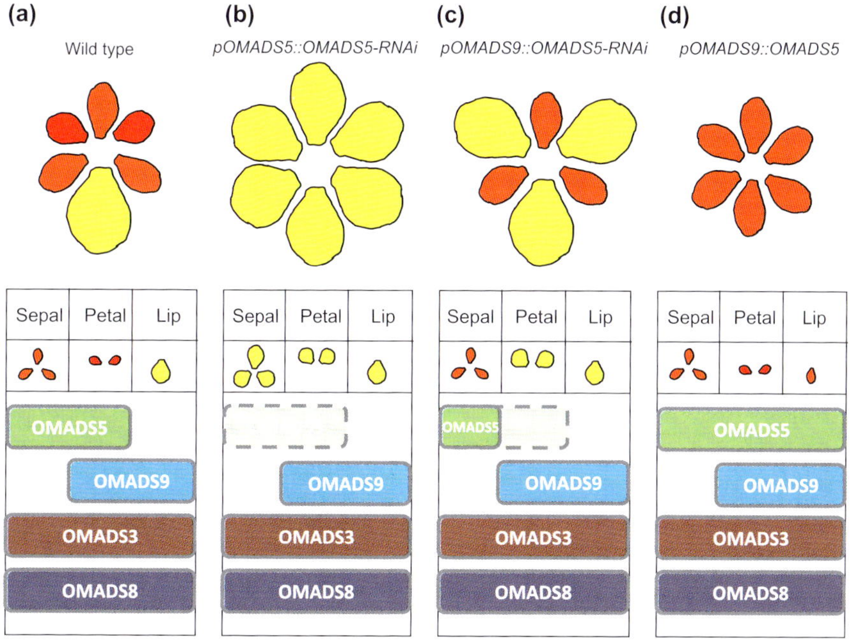

Fig. 11.3. The manipulation of *Oncidium* perianth by altering the expression of the B genes. (a) Normal expression patterns of the B class genes in *O.* Gower Ramsey. (b) The transformation of *pOMADS5::OMADS5-RNAi* construct into *Oncidium* will potentially convert the petals and sepals into lips, and then produce flowers with six lips. (c) The transformation of *pOMADS9::OMADS5-RNAi* construct into *Oncidium* will specifically convert petals into lips and generate flowers with three lips. (d) Ectopic expression of *OMADS5* (*pOMADS9::OMADS5*) in lips will convert lips into sepals/petals and result in a flower with six sepals/petals.

AP1/AGL9 MADS box genes from *Oncidium* caused earlier flowering in transgenic *Arabidopsis* (Chang *et al.*, 2009; Hsu *et al.*, 2003). Furthermore, ectopic expression of the *AP1/AGL9* MADS box gene *OMADS1* from *Oncidium* also promoted flowering in transgenic tobacco and *Eustoma grandiflorum* (Hsu *et al.*, 2003, Thiruvengadam and Yang, 2009). Thus, it will be possible in the future to manipulate the *Oncidium* flowering time by controlling the expression of the *AP1/AGL9* MADS box gene fused with a suitable includable promoter.

11.4. Conclusions and Future Prospects

In general, the mutational robustness of organisms comes after gene duplications and consequently promotes evolutionary innovations (Wagner, 2008). According to the "orchid code" hypothesis, orchid systems represent a close relationship between gene duplications and the origin of evolutionary innovations that are directly involved in the process of speciation (Mondragon-Palomino and Theissen, 2009; Wagner, 2008). Evolutionary developmental biology ("evo-devo") offers the important idea that the developmental genetic systems of organisms can influence the tempo and direction of evolutionary change (Brakefield, 2003). Based on recently proposed molecular mechanisms, the origin and evolution of MADS box floral organ identity genes may reflect that of the flowers of angiosperms. In the future, the orchid flower could be a valuable system in evo-devo studies to improve the understanding of the association between developmental gene evolution and changes in morphology, and thereby the origin of speciation. Other than the model plant *Arabidopsis*, many plant species exist with two or more homologs in one class of MADS box genes (de Martino *et al.*, 2006; Kanno *et al.*, 2007; Kanno *et al.*, 2003; Nakada *et al.*, 2006; Song *et al.*, 2006; Theissen *et al.*, 2000; Tsaftaris *et al.*, 2006; Tsai *et al.*, 2004; Xu *et al.*, 2006; Yun *et al.*, 2004a; Yun *et al.*, 2004b), as does *O.* Gower Ramsey. For example, we have identified three orthologs of *DEF*-like B-function MADS box genes, and five *AP1/AGL9* function MADS box genes in *O.* Gower Ramsey (Chang *et al.*, 2009; Chang *et al.*, 2010; Hsu *et al.*, 2003; Hsu and Yang, 2002). Once the study systems for validating *O.* Gower Ramsey gene function, such as transformation and virus-induced gene silencing, are well-established, the full scope of floral homeotic gene interactions will be revealed with applications to altering flowering time and modifying floral morphology.

To gain insight into the molecular mechanism of B-function MADS box protein complex formation in *Oncidium*, we will investigate the effects of C-terminal sequence on the dimerization behavior of paleoAP3 homologs (OMADS3 and OMADS5) and aim to isolate the protein complexes involved in the specific petaloid organ development. Unlike OMADS5, OMADS3 is a unique paleo*AP3* homolog lacking any conserved motifs in the C-terminal region (Chang *et al.*, 2010; Hsu and Yang, 2002). The importance of the C-terminal domain in the dimerization ability of OMADS3 will be examined by yeast two-hybrid analysis with serial deletions in the C-terminal region. Our preliminary data suggest that a critical subdomain in the C-terminal, instead of the sequence length, would determine the dimerization ability of OMADS3 (unpublished data). Using COILS prediction software, we obtained a clue to the relationship between C-terminal sequence and the coiled-coil structures in the K-box domain. The loss of the consensus subdomain in the C-terminal may destroy the coiled-coil structure, and thereby abolish the dimerization ability of OMADS3/OMADS5. Whether C-terminal sequences affect OMADS3/OMADS5 physiological function in floral organ development will be analyzed by complementary experiments. We are also intrigued by the mRNA expression patterns of *OMADS3/5/8/9* in the sepal, petal and lip (Chang *et al.*, 2010; Hsu and Yang, 2002). By *in vitro* co-immunoprecipitation assay or proximity ligation *in situ* assay (Soderberg *et al.*, 2006), we hope to identify the sepal/petal/lip-specific interacting complexes in *Oncidium*. To know how these *Oncidium* B-function MADS box proteins form cooperative complexes during sepal/petal/lip development will lead to a better understanding of the "orchid code" for *Oncidium* floral organ identity.

Acknowledgments

The authors thank Mr. T-F Chou for helping to grow the orchid plants used in this research. This work was supported by grants to C-H Yang from the National Science Council, Taiwan (NSC95-2317-B-005-006 and NSC96-2317-B-005-019). This work was also supported by a 5Y/50B grant from the Ministry of Education, Taiwan.

References

Adam H, Jouannic S, Orieux Y *et al.* (2007) Functional characterization of MADS box genes involved in the determination of oil palm flower structure. *J Exp Bot* **58**: 1245–1259.

Ambrose BA, Lerner DR, Ciceri P *et al.* (2000) Molecular and genetic analyses of the silky1 gene reveal conservation in floral organ specification between eudicots and monocots. *Mol Cell* **5**: 569–579.

Angenent GC, Busscher M, Franken J *et al.* (1992) Differential expression of two MADS box genes in wild-type and mutant petunia flowers. *Plant Cell* **4**: 983–993.

Angenent GC, Colombo L. (1996) Molecular control of ovule development. *Trends Plant Sci* **1**: 228–232.

Angenent GC, Franken J, Busscher M *et al.* (1993) Petal and stamen formation in petunia is regulated by the homeotic gene fbp1. *Plant J* **4**: 101–112.

Angenent GC, Franken J, Busscher M *et al.* (1995) A novel class of MADS box genes is involved in ovule development in petunia. *Plant Cell* **7**: 1569–1582.

Angenent GC, Franken J, Busscher M *et al.* (1994) Co-suppression of the petunia homeotic gene fbp2 affects the identity of the generative meristem. *Plant J* **5**: 33–44.

Benedito VA, Visser PB, van Tuyl JM *et al.* (2004) Ectopic expression of LLAG1, an AGAMOUS homologue from lily (*Lilium longiflorum* Thunb.), causes floral homeotic modifications in *Arabidopsis*. *J Exp Bot* **55**: 1391–1399.

Bowman JL, Smyth DR, Meyerowitz EM. (1989) Genes directing flower development in *Arabidopsis*. *Plant Cell* **1**: 37–52.

Brakefield PM. (2003) The power of evo-devo to explore evolutionary constraints: Experiments with butterfly eyespots. *Zoology (Jena)* **106**: 283–290.

Chang YY, Chiu YF, Wu JW, Yang CH. (2009) Four orchid (*Oncidium* Gower Ramsey) AP1/AGL9-like MADS box genes show novel expression patterns and cause different effects on floral transition and formation in *Arabidopsis thaliana*. *Plant Cell Physiol* **50**: 1425–1438.

Chang YY, Kao NH, Li JY *et al.* (2010) Characterization of the possible roles for B class MADS box genes in regulation of perianth formation in orchid. *Plant Physiol* **152**: 837–853.

Chen JT, Chang C, Chang WC. (1999) Direct somatic embryogenesis on leaf explants of *Oncidium* Gower Ramsey and subsequent plant regeneration. *Plant Cell Rep* **19**: 143–149.

Chen JT, Chang WC. (2000) Plant regeneration via embryo and shoot bud formation from flower-stalk explants of *Oncidium* Sweet Sugar. *Plant Cell Tiss Org Cult* **62**: 95–100.

Chen MK, Lin IC, Yang CH. (2008) Functional analysis of three lily (*Lilium longiflorum*) APETALA1-like MADS box genes in regulating floral transition and formation. *Plant Cell Physiol* **49**: 704–717.

Coen ES, Meyerowitz EM. (1991) The war of the whorls: Genetic interactions controlling flower development. *Nature* **353**: 31–37.

Colombo L, Franken J, Koetje E *et al.* (1995) The petunia MADS box gene FBP11 determines ovule identity. *Plant Cell* **7**: 1859–1868.

Colombo L, Franken J, Van der Krol AR *et al.* (1997a) Downregulation of ovule-specific MADS box genes from petunia results in maternally controlled defects in seed development. *Plant Cell* **9**: 703–715.

Colombo L, van Tunen AJ, Dons HJM, Angenent GC. (1997b) Molecular control of flower development in *Petunia hybrida*. *Adv Bot Res* **26**: 229–250.

Cozzolino S, Widmer A. (2005) Orchid diversity: An evolutionary consequence of deception? *Trends Ecol Evol* **20**: 487–494.

de Folter S, Immink RG, Kieffer M *et al.* (2005) Comprehensive interaction map of the *Arabidopsis* MADS box transcription factors. *Plant Cell* **17:** 1424–1433.

de Martino G, Pan I, Emmanuel E *et al.* (2006) Functional analyses of two tomato APETALA3 genes demonstrate diversification in their roles in regulating floral development. *Plant Cell* **18**: 1833–1845.

Di Stilio VS, Kramer EM, Baum DA. (2005) Floral MADS box genes and homeotic gender dimorphism in *Thalictrum dioicum* (Ranunculaceae) — A new model for the study of dioecy. *Plant J* **41**: 755–766.

Ditta G, Pinyopich A, Robles P *et al.* (2004) The SEP4 gene of *Arabidopsis thaliana* functions in floral organ and meristem identity. *Curr Biol* **14**: 1935–1940.

Doyle JJ. (1994) Evolution of a plant homeotic multigene family: Toward connecting molecular systematic and molecular developmental genetics. *Syst Biol* **43**: 307–328.

Drea S, Hileman LC, de Martino G, Irish VF. (2007) Functional analyses of genetic pathways controlling petal specification in poppy. *Development* **134**: 4157–4166.

Ferrario S, Immink RG, Shchennikova A *et al.* (2003) The MADS box gene FBP2 is required for SEPALLATA function in petunia. *Plant Cell* **15**: 914–925.

Fornara F, Parenicova L, Falasca G *et al.* (2004) Functional characterization of OsMADS18, a member of the AP1/SQUA subfamily of MADS box genes. *Plant Physiol* **135**: 2207–2219.

Goto K, Meyerowitz EM. (1994) Function and regulation of the *Arabidopsis* floral homeotic gene PISTILLATA. *Genes Dev* **8**: 1548–1560.

Hernandez-Hernandez T, Martinez-Castilla LP, Alvarez-Buylla ER. (2007) Functional diversification of B MADS-box homeotic regulators of flower development: Adaptive evolution in protein–protein interaction domains after major gene duplication events. *Mol Biol Evol* **24**: 465–481.

Honma T, Goto K. (2001) Complexes of MADS-box proteins are sufficient to convert leaves into floral organs. *Nature* **409**: 525–529.

Hou CJ, Yang CH. (2009) Functional analysis of *FT* and *TFL1* orthologs from orchid (*Oncidium* Gower Ramsey) that regulate the vegetative to reproductive transition. *Plant Cell Physiol* **50**: 1544–1557.

Hsu HF, Hsieh WP, Chen MK *et al.* (2010) C/D Class MADS box genes from two monocots orchid (*Oncidium* Gower Ramsey) and lily (*Lilium longiflorum*) exhibit different effects on floral transition and formation in *Arabidopsis thaliana*. *Plant Cell Physiol* **51**: 1029–1045.

Hsu HF, Huang CH, Chou LT, Yang CH. (2003) Ectopic expression of an orchid (*Oncidium* Gower Ramsey) *AGL6*-like gene promotes flowering by activating flowering time genes in *Arabidopsis thaliana*. *Plant Cell Physiol* **44**: 783–794.

Hsu HF, Yang CH. (2002) An orchid (*Oncidium* Gower Ramsey) *AP3*-like MADS gene regulates floral formation and initiation. *Plant Cell Physiol* **43**: 1198–1209.

Huijser P, Klein J, Lonnig WE *et al.* (1992) Bracteomania, an inflorescence anomaly, is caused by the loss of function of the MADS-box gene squamosa in *Antirrhinum majus*. *EMBO J* **11**: 1239–1249.

Irish VF. (2009) Evolution of petal identity. *J Exp Bot* **60**: 2517–2527.

Jack T. (2004) Molecular and genetic mechanisms of floral control. *Plant Cell* **16** Suppl: S1–17.

Jack T, Brockman LL, Meyerowitz EM. (1992) The homeotic gene APETALA3 of *Arabidopsis thaliana* encodes a MADS box and is expressed in petals and stamens. *Cell* **68**: 683–697.

Jeon JS, Lee S, Jung KH *et al.* (2000) Production of transgenic rice plants showing reduced heading date and plant height by ectopic expression of rice MADS-box genes. *Mol Breeding* **6**: 581–592.

Kang HG, Noh YS, Chung YY *et al.* (1995) Phenotypic alterations of petal and sepal by ectopic expression of a rice MADS box gene in tobacco. *Plant Mol Biol* **29**: 1–10.

Kanno A, Nakada M, Akita Y, Hirai M. (2007) Class B gene expression and the modified ABC model in nongrass monocots. *Sci World J* **7**: 268–279.

Kanno A, Saeki H, Kameya T *et al.* (2003) Heterotopic expression of class B floral homeotic genes supports a modified ABC model for tulip (*Tulipa gesneriana*). *Plant Mol Biol* **52**: 831–841.

Kater MM, Colombo L, Franken J *et al.* (1998) Multiple AGAMOUS homologs from cucumber and petunia differ in their ability to induce reproductive organ fate. *Plant Cell* **10**: 171–182.

Kempin SA, Mandel MA, Yanofsky MF. (1993) Conversion of perianth into reproductive organs by ectopic expression of the tobacco floral homeotic gene NAG1. *Plant Physiol* **103**: 1041–1046.

Kempin SA, Savidge B, Yanofsky MF. (1995) Molecular basis of the cauliflower phenotype in *Arabidopsis*. *Science* **267**: 522–525.

Kim S, Yoo MJ, Albert VA *et al.* (2004) Phylogeny and diversification of B-function MADS-box genes in angiosperms: Evolutionary and functional implications of a 260-million-year-old duplication. *Am J Bot* **91**: 2102–2118.

Kim SY, Yun PY, Fukuda T *et al.* (2007) Expression of a *DEFICIENS*-like gene correlates with the differentiation between sepal and petal in the orchid, *Habenaria radiata* (Orchidaceae). *Plant Sci* **172**: 319–326.

Kotilainen M, Elomaa P, Uimari A *et al.* (2000) GRCD1, an AGL2-like MADS box gene, participates in the C function during stamen development in *Gerbera hybrida*. *Plant Cell* **12**: 1893–1902.

Kramer EM. (2009) Aquilegia: A new model for plant development, ecology, and evolution. *Annu Rev Plant Biol* **60**: 261–277.

Kramer EM, Di Stilio VS, Schlüter PM. (2003) Complex patterns of gene duplication in the *APETALA3* and *PISTILLATA* lineages of the Ranunculaceae. *Int J Plant Sci* **164**: 1–11.

Kramer EM, Dorit RL, Irish VF. (1998) Molecular evolution of genes controlling petal and stamen development: Duplication and divergence within the APETALA3 and PISTILLATA MADS-box gene lineages. *Genetics* **149**: 765–783.

Kramer EM, Holappa L, Gould B *et al.* (2007) Elaboration of B gene function to include the identity of novel floral organs in the lower eudicot *Aquilegia*. *Plant Cell* **19**: 750–766.

Kramer EM, Irish VF. (1999) Evolution of genetic mechanisms controlling petal development. *Nature* **399**: 144–148.

Kramer EM, Jaramillo MA, Di Stilio VS. (2004) Patterns of gene duplication and functional evolution during the diversification of the AGAMOUS subfamily of MADS box genes in angiosperms. *Genetics* **166**: 1011–1023.

Krizek BA, Meyerowitz EM. (1996) The *Arabidopsis* homeotic genes APETALA3 and PISTILLATA are sufficient to provide the B class organ identity function. *Development* **122**: 11–22.

Kyozuka J, Harcourt R, Peacock WJ, Dennis ES. (1997) Eucalyptus has functional equivalents of the *Arabidopsis* AP1 gene. *Plant Mol Biol* **35**: 573–584.

Kyozuka J, Kobayashi T, Morita M, Shimamoto K. (2000) Spatially and temporally regulated expression of rice MADS box genes with similarity to *Arabidopsis* class A, B and C genes. *Plant Cell Physiol* **41**: 710–718.

Lamb RS, Irish VF. (2003) Functional divergence within the APETALA3/PISTILLATA floral homeotic gene lineages. *Proc Natl Acad Sci USA* **100**: 6558–6563.

Litt A, Irish VF. (2003) Duplication and diversification in the APETALA1/FRUITFULL floral homeotic gene lineage: Implications for the evolution of floral development. *Genetics* **165**: 821–833.

Litt A, Kramer EM. (2010) The ABC model and the diversification of floral organ identity. *Semin Cell Dev Biol* **21**: 129–137.

Liu Y, Nakayama N, Schiff M *et al.* (2004) Virus induced gene silencing of a DEFICIENS ortholog in *Nicotiana benthamiana*. *Plant Mol Biol* **54**: 701–711.

Lopez-Dee ZP, Wittich P, Enrico Pe M *et al.* (1999) OsMADS13, a novel rice MADS-box gene expressed during ovule development. *Dev Genet* **25**: 237–244.

Ma H, Yanofsky MF, Meyerowitz EM. (1991) AGL1-AGL6, an *Arabidopsis* gene family with similarity to floral homeotic and transcription factor genes. *Genes Dev* **5**: 484–495.

Malcomber ST, Kellogg EA. (2005) SEPALLATA gene diversification: Brave new whorls. *Trends Plant Sci* **10**: 427–435.

Mandel MA, Bowman JL, Kempin SA *et al.* (1992a) Manipulation of flower structure in transgenic tobacco. *Cell* **71**: 133–143.

Mandel MA, Gustafsonbrown C, Savidge B, Yanofsky MF. (1992b) Molecular characterization of the *Arabidopsis* floral homeotic gene APETALA1. *Nature* **360**: 273–277.

Mandel MA, Yanofsky MF. (1995) A gene triggering flower formation in *Arabidopsis*. *Nature* **377**: 522–524.

Mandel MA, Yanofsky MF. (1998) The *Arabidopsis AGL9* MADS box gene is expressed in young flower primordia. *Sex Plant Reprod* **11**: 22–28.

Masiero S, Imbriano C, Ravasio F *et al.* (2002) Ternary complex formation between MADS-box transcription factors and the histone fold protein NF-YB. *J Biol Chem* **277**: 26429–26435.

Mena M, Mandel MA, Lerner DR *et al.* (1995) A characterization of the MADS-box gene family in maize. *Plant J* **8**: 845–854.

Mizukami Y, Huang H, Tudor M *et al.* (1996) Functional domains of the floral regulator AGAMOUS: Characterization of the DNA binding domain and analysis of dominant negative mutations. *Plant Cell* **8**: 831–845.

Mizukami Y, Ma H. (1992) Ectopic expression of the floral homeotic gene AGAMOUS in transgenic *Arabidopsis* plants alters floral organ identity. *Cell* **71**: 119–131.

Mondragon-Palomino M, Hiese L, Harter A *et al.* (2009) Positive selection and ancient duplications in the evolution of class B floral homeotic genes of orchids and grasses. *BMC Evol Biol* **9**: 81.

Mondragon-Palomino M, Theissen G. (2008) MADS about the evolution of orchid flowers. *Trends Plant Sci* **13**: 51–59.

Mondragon-Palomino M, Theissen G. (2009) Why are orchid flowers so diverse? Reduction of evolutionary constraints by paralogues of class B floral homeotic genes. *Ann Bot* **104**: 583–594.

Moon YH, Jung JY, Kang HG, An G. (1999a) Identification of a rice APETALA3 homologue by yeast two-hybrid screening. *Plant Mol Biol* **40**: 167–177.

Moon YH, Kang HG, Jung JY *et al.* (1999b) Determination of the motif responsible for interaction between the rice APETALA1/AGAMOUS-LIKE9 family proteins using a yeast two-hybrid system. *Plant Physiol* **120**: 1193–1204.

Mouradov A, Glassick TV, Hamdorf BA *et al.* (1998) Family of MADS-box genes expressed early in male and female reproductive structures of monterey pine. *Plant Physiol* **117**: 55–62.

Munster T, Wingen LU, Faigl W *et al.* (2001) Characterization of three GLOBOSA-like MADS-box genes from maize: Evidence for ancient paralogy in one class of floral homeotic B-function genes of grasses. *Gene* **262**: 1–13.

Nadeau JA, Zhang XS, Li J, O'Neill SD. (1996) Ovule development: Identification of stage-specific and tissue-specific cDNAs. *Plant Cell* **8**: 213–239.

Nagasawa N, Miyoshi M, Sano Y *et al.* (2003) SUPERWOMAN1 and DROOPING LEAF genes control floral organ identity in rice. *Development* **130**: 705–718.

Nakada M, Komatsu M, Ochiai T *et al.* (2006) Isolation of *MaDEF* from *Muscari armeniacum* and analysis of its expression using laser microdissection. *Plant Sci* **170**: 143–150.

Nakamura T, Song IJ, Fukuda T *et al.* (2005) Characterization of TrcMADS1 gene of *Trillium camtschatcense* (Trilliaceae) reveals functional evolution of the SOC1/TM3-like gene family. *J Plant Res* **118**: 229–234.

O'Neill SD, Nadeau JA, Zhang XS *et al.* (1993) Interorgan regulation of ethylene biosynthetic genes by pollination. *Plant Cell* **5**: 419–432.

Park JH, Ishikawa Y, Ochiai T *et al.* (2004) Two GLOBOSA-like genes are expressed in second and third whorls of homochlamydeous flowers in *Asparagus officinalis* L. *Plant Cell Physiol* **45**: 325–332.

Park JH, Ishikawa Y, Yoshida R *et al.* (2003) Expression of AODEF, a B-functional MADS-box gene, in stamens and inner tepals of the dioecious species *Asparagus officinalis* L. *Plant Mol Biol* **51**: 867–875.

Pelaz S, Ditta GS, Baumann E *et al.* (2000) B and C floral organ identity functions require SEPALLATA MADS-box genes. *Nature* **405**: 200–203.

Pelaz S, Gustafson-Brown C, Kohalmi SE *et al.* (2001) APETALA1 and SEPALLATA3 interact to promote flower development. *Plant J* **26**: 385–394.

Pinyopich A, Ditta GS, Savidge B *et al.* (2003) Assessing the redundancy of MADS-box genes during carpel and ovule development. *Nature* **424**: 85–88.

Pnueli L, Hareven D, Rounsley SD *et al.* (1994) Isolation of the tomato AGAMOUS gene TAG1 and analysis of its homeotic role in transgenic plants. *Plant Cell* **6**: 163–173.

Prasad K, Vijayraghavan, U. (2003) Double-stranded RNA interference of a rice PI/GLO paralog, OsMADS2, uncovers its second-whorl-specific function in floral organ patterning. *Genetics* **165**: 2301–2305.

Purugganan MD. (1997) The MADS-box floral homeotic gene lineages predate the origin of seed plants: Phylogenetic and molecular clock estimates. *J Mol Evol* **45**: 392–396.

Purugganan MD, Rounsley SD, Schmidt RJ, Yanofsky MF. (1995) Molecular evolution of flower development: Diversification of the plant MADS-box regulatory gene family. *Genetics* **140**: 345–356.

Riechmann JL, Wang M, Meyerowitz EM. (1996) DNA-binding properties of *Arabidopsis* MADS domain homeotic proteins APETALA1, APETALA3, PISTILLATA and AGAMOUS. *Nucleic Acids Res* **24**: 3134–3141.

Rounsley SD, Ditta GS, Yanofsky MF. (1995) Diverse roles for MADS box genes in *Arabidopsis* development. *Plant Cell* **7**: 1259–1269.

Rudall PJ, Bateman RM. (2002) Roles of synorganisation, zygomorphy and heterotopy in floral evolution: The gynostemium and labellum of orchids and other lilioid monocots. *Biol Rev Camb Philos Soc* **77**: 403–441.

Rudall PJ, Bateman RM. (2003) Evolutionary change in flowers and inflorescences: Evidence from naturally occurring terata. *Trends Plant Sci* **8**: 76–82.

Rutledge R, Regan S, Nicolas O *et al.* (1998) Characterization of an AGAMOUS homologue from the conifer black spruce (*Picea mariana*) that produces floral homeotic conversions when expressed in *Arabidopsis*. *Plant J* **15**: 625–634.

Savidge B, Rounsley SD, Yanofsky MF. (1995) Temporal relationship between the transcription of two *Arabidopsis* MADS box genes and the floral organ identity genes. *Plant Cell* **7**: 721–733.

Schmidt RJ, Veit B, Mandel MA *et al.* (1993) Identification and molecular characterization of ZAG1, the maize homolog of the *Arabidopsis* floral homeotic gene AGAMOUS. *Plant Cell* **5**: 729–737.

Schwarz-Sommer Z, Hue I, Huijser P *et al.* (1992) Characterization of the *Antirrhinum* floral homeotic MADS-box gene deficiens: Evidence for DNA binding and autoregulation of its persistent expression throughout flower development. *EMBO J* **11**: 251–263.

Shan H, Su K, Lu W *et al.* (2006) Conservation and divergence of candidate class B genes in *Akebia trifoliata* (Lardizabalaceae). *Dev Genes Evol* **216**: 785–795.

Shindo S, Ito M, Ueda K *et al.* (1999) Characterization of MADS genes in the gymnosperm *Gnetum parvifolium* and its implication on the evolution of reproductive organs in seed plants. *Evol Dev* **1**: 180–190.

Shitsukawa N, Tahira C, Kassai K *et al.* (2007) Genetic and epigenetic alteration among three homoeologous genes of a class E MADS box gene in hexaploid wheat. *Plant Cell* **19**: 1723–1737.

Soderberg O, Gullberg M, Jarvius M *et al.* (2006) Direct observation of individual endogenous protein complexes *in situ* by proximity ligation. *Nat Methods* **3:** 995–1000.

Sommer H, Beltran JP, Huijser P *et al.* (1990) Deficiens, a homeotic gene involved in the control of flower morphogenesis in *Antirrhinum majus*: The protein shows homology to transcription factors. *EMBO J* **9**: 605–613.

Song IJ, Nakamura T, Fukuda T *et al.* (2006) Spatiotemporal expression of duplicate AGAMOUS orthologues during floral development in *Phalaenopsis. Dev Genes Evol* **216**: 301–313.

Stellari GM, Jaramillo MA, Kramer EM. (2004) Evolution of the APETALA3 and PISTILLATA lineages of MADS-box-containing genes in the basal angiosperms. *Mol Biol Evol* **21**: 506–519.

Sung SK, Yu GH, An G. (1999) Characterization of MdMADS2, a member of the SQUAMOSA subfamily of genes, in apple. *Plant Physiol* **120**: 969–978.

Tandre K, Albert VA, Sundas A, Engstrom P. (1995) Conifer homologues to genes that control floral development in angiosperms. *Plant Mol Biol* **27**: 69–78.

Theissen G. (2001) Development of floral organ identity: Stories from the MADS house. *Curr Opin Plant Biol* **4**: 75–85.

Theissen G, Becker A, Di Rosa A *et al.* (2000) A short history of MADS-box genes in plants. *Plant Mol Biol* **42**: 115–149.

Theissen G, Kim JT, Saedler H. (1996) Classification and phylogeny of the MADS-box multigene family suggest defined roles of MADS-box gene subfamilies in the morphological evolution of eukaryotes. *J Mol Evol* **43**: 484–516.

Theissen G, Saedler H. (1995) MADS-box genes in plant ontogeny and phylogeny: Haeckel's 'biogenetic law' revisited. *Curr Opin Genet Dev* **5**: 628–639.

Theissen G, Saedler H. (2001) Plant biology. Floral quartets. *Nature* **409**: 469–471.

Thiruvengadam M, Yang CH. (2009) Ectopic expression of two MADS box genes from orchid (*Oncidium* Gower Ramsey) and lily (*Lilium longiflorum*) alters flower transition and formation in *Eustoma grandiflorum. Plant Cell Rep* **28**: 1463–1473.

Trobner W, Ramirez L, Motte P *et al.* (1992) GLOBOSA: A homeotic gene which interacts with DEFICIENS in the control of *Antirrhinum* floral organogenesis. *EMBO J* **11**: 4693–4704.

Tsaftaris AS, Polidoros AN, Pasentsis K, Kalivas A. (2006) Tepal formation and expression pattern of B-class paleo*AP3*-like MADS-box genes in crocus (*Crocus sativus* L.). *Plant Sci* **170**: 238–246.

Tsai WC, Kuoh CS, Chuang MH *et al.* (2004) Four DEF-like MADS box genes displayed distinct floral morphogenetic roles in *Phalaenopsis* orchid. *Plant Cell Physiol* **45**: 831–844.

Tsai WC, Lee PF, Chen HI *et al.* (2005) PeMADS6, a GLOBOSA/PISTILLATA-like gene in *Phalaenopsis equestris* involved in petaloid formation, and correlated with flower longevity and ovary development. *Plant Cell Physiol* **46:** 1125–1139.

Tzeng TY, Chen HY, Yang CH. (2002) Ectopic expression of carpel-specific MADS box genes from lily and lisianthus causes similar homeotic conversion of sepal and petal in *Arabidopsis*. *Plant Physiol* **130**: 1827–1836.

Tzeng TY, Hsiao CC, Chi PJ, Yang CH. (2003) Two lily SEPALLATA-like genes cause different effects on floral formation and floral transition in *Arabidopsis*. *Plant Physiol* **133**: 1091–1101.

Tzeng TY, Yang CH. (2001) A MADS box gene from lily (*Lilium longiflorum*) is sufficient to generate dominant negative mutation by interacting with PISTILLATA (PI) in *Arabidopsis thaliana*. *Plant Cell Physiol* **42**: 1156–1168.

van der Krol AR, Brunelle A, Tsuchimoto S, Chua NH. (1993) Functional analysis of petunia floral homeotic MADS box gene pMADS1. *Genes Dev* **7**: 1214–1228.

van Tunen AJ, Eikeboom W, Angenent GC. (1993) Floral organogenesis in *Tulipa*. *Flow Newsl* **16**: 33–38.

Vandenbussche M, Theissen G, Van de Peer Y, Gerats T. (2003) Structural diversification and neo-functionalization during floral MADS-box gene evolution by C-terminal frameshift mutations. *Nucleic Acids Res* **31**: 4401–4409.

Vandenbussche M, Zethof J, Royaert S *et al.* (2004) The duplicated B-class heterodimer model: Whorl-specific effects and complex genetic interactions in *Petunia hybrida* flower development. *Plant cell* **16**: 741–754.

Wagner A. (2008) Gene duplications, robustness and evolutionary innovations. *Bioessays* **30**: 367–373.

Wang CY, Chiou CY, Wang HL *et al.* (2008) Carbohydrate mobilization and gene regulatory profile in the pseudobulb of *Oncidium* orchid during the flowering process. *Planta* **227**: 1063–1077.

Weigel D, Meyerowitz EM. (1994) The ABCs of floral homeotic genes. *Cell* **78:** 203–209.

Whipple CJ, Zanis MJ, Kellogg EA, Schmidt RJ. (2007) Conservation of B class gene expression in the second whorl of a basal grass and outgroups links the origin of lodicules and petals. *Proc Natl Acad Sci USA* **104**: 1081–1086.

Winter KU, Becker A, Munster T *et al.* (1999) MADS-box genes reveal that gnetophytes are more closely related to conifers than to flowering plants. *Proc Natl Acad Sci USA* **96**: 7342–7347.

Winter KU, Saedler H, Theissen G. (2002) On the origin of class B floral homeotic genes: Functional substitution and dominant inhibition in *Arabidopsis* by expression of an orthologue from the gymnosperm *Gnetum*. *Plant J* **31**: 457–475.

Xu Y, Teo LL, Zhou J *et al.* (2006) Floral organ identity genes in the orchid *Dendrobium crumenatum*. *Plant J* **46**: 54–68.

Xu Y, Zhang L, Xie H, *et al.* (2008) Expression analysis and genetic mapping of three *SEPALLATA*-like genes from peach (*Prunus persica* (L.) Batsch). *Tree Genet Genomes* **4**: 693–703.

Yadav SR, Prasad K, Vijayraghavan U. (2007) Divergent regulatory OsMADS2 functions control size, shape and differentiation of the highly derived rice floret second-whorl organ. *Genetics* **176**: 283–294.

Yanofsky MF, Ma H, Bowman JL *et al.* (1990) The protein encoded by the *Arabidopsis* homeotic gene agamous resembles transcription factors. *Nature* **346**: 35–39.

Yao SG, Ohmori S, Kimizu M, Yoshida H. (2008) Unequal genetic redundancy of rice PISTILLATA orthologs, OsMADS2 and OsMADS4, in lodicule and stamen development. *Plant Cell Physiol* **49**: 853–857.

Yu D, Kotilainen M, Pollanen E *et al.* (1999) Organ identity genes and modified patterns of flower development in *Gerbera hybrida* (Asteraceae). *Plant J* **17**: 51–62.

Yu H, Goh CJ. (2000) Identification and characterization of three orchid MADS-box genes of the AP1/AGL9 subfamily during floral transition. *Plant Physiol* **123**: 1325–1336.

Yun PY, Ito T, Kim SY *et al.* (2004a) The *AVAG1* gene is involved in the development of the reproductive organs in the ornamental asparagus, *Asparagus virgatus. Sex Plant Reprod* **17**: 1–8.

Yun PY, Kim SY, Ochiai T *et al.* (2004b) *AVAG2* is a putative D-class gene from an ornamental asparagus. *Sex Plant Reprod* **17**: 107–116.

Zahn LM, Kong H, Leebens-Mack JH *et al.* (2005a) The evolution of the SEPALLATA subfamily of MADS-box genes: A preangiosperm origin with multiple duplications throughout angiosperm history. *Genetics* **169**: 2209–2223.

Zahn LM, Leebens-Mack J, DePamphilis CW *et al.* (2005b) To B or not to B a flower: The role of DEFICIENS and GLOBOSA orthologs in the evolution of the angiosperms. *J Hered* **96**: 225–240.

Zhang XS, O'Neill SD. (1993) Ovary and gametophyte development are coordinately regulated by auxin and ethylene following pollination. *Plant Cell* **5**: 403–418.

CHAPTER 12

The Relationship Between Ethylene and Orchid Ovule Development

Wen-Chieh Tsai[*,‡], Yu-Yun Hsiao[†], Zhao-Jun Pan[†], Yun-Wen Chen[*] and Hong-Hwa Chen[*,†]

Orchids are known for their unique reproductive biology and ecological strategies. One of the remarkable reproductive characters of most orchid species is that ovule development is precisely initiated following pollination. Thus, orchids are attractive systems for investigating both the initiation and subsequent development of the ovule. Ethylene plays a critical role in ovary maturation and ovule differentiation in orchids. Genes corresponding to ethylene biosynthesis in orchid reproductive organs have been identified and characterized. The phosphorylation states of proteins involved in ethylene biosynthesis or ethylene signaling may be correlated with ovule development. Recently, an orchid *GLOBOSA*-like MADS-box gene was isolated from *Phalaenopsis* and shown to have a role in ovule development via auxin/ethylene signaling. Adopting functional genomics strategies will help us to get insights into the regulation network between the ethylene signaling pathway and the ovule developmental program in orchids.

12.1. Introduction

12.1.1. *Ethylene biosynthesis and ethylene signal transduction*

The phytohormone ethylene is a gaseous plant hormone that plays a role in physiological processes throughout the life cycle of plants. It affects myriad developmental processes and fitness responses including germination, formation of the apical hook in dark-grown seedlings,

[*]Institute of Tropical Plant Sciences, National Cheng Kung University, Tainan 701, Taiwan.
[†]Department of Life Sciences, National Cheng Kung University, Tainan 701, Taiwan.
[‡]Corresponding author. E-mail: tsaiwe@mail.ncku.edu.tw

the regulation of cell expansion, leaf and flower senescence, fruit ripening and programmed cell death. Ethylene also regulates plant response to biotic stresses such as those induced by pathogen attractions, and to abiotic stresses such as those induced by flooding or drought (Abeles *et al.*, 1992; Roman *et al.*, 1995; Penninckx *et al.*, 1998; O'Donnell *et al.*, 2003).

Ethylene can be produced by almost all parts of higher plants, although the rate of production depends on the type of tissue and the stage of development. The amino acid methionine is the precursor of ethylene, and 1-aminocyclopropane-1-carboxylic (ACC) serves as an intermediate in the conversion to ethylene. The CH3-S group of methionine is recycled via the Yang cycle (Fig. 12.1). The rate-limiting step in the pathway is the conversion of S-Adenosyl-methionine (AdoMet) to ACC, which is catalyzed by the enzyme ACC synthase. The last step in the pathway, the conversion of ACC to ethylene, requires oxygen and is catalyzed by the enzyme ACC oxidase.

Many key components of the ethylene signal transduction pathway were identified from a simple genetic screen that made use of ethylene's effect on dark-grown seedlings known as the "triple response." This response is characterized by the inhibition of hypocotyl and root cell elongation, radial swelling of the hypocotyl, and an exaggerated apical hook. Over the past decade, genetic screens that are based on the triple-response phenotype have been extensively conducted on *Arabidopsis*. More than a dozen unique mutants have been identified and these can be divided into two distinct categories: constitutive triple-response mutants, including *ctr1* (*constitutive triple response1*) (Kieber *et al.*, 1993), *ran1* (*responsive to antagonist1*) (Hirayama *et al.*, 1999; Woeste and Kieber, 2000), and *eto1* (*ethylene overproduction1*) (Stepanova, 2000); and ethylene-insensitive mutants, including *etr1* (*ethylene receptor1*) (Bleecker *et al.*, 1988; Chang *et al.*, 1993), *etr2* (Sakai, 1998), *ein2* (*ethylene insensitive2*) (Guzmán and Ecker, 1990; Alonso *et al.*, 1999), *ein3* (Roman *et al.*, 1995; Chao *et al.*, 1997), *ein4, ein5, ein6* (Roman *et al.*, 1995), *hls1* (*hookless1*) (Guzmán and Ecker, 1990), and *eir1* (*ethylene insensitive root1*) (Roman, 1995).

A combination of genetic and molecular analyses of these mutants has defined a pathway for ethylene signaling transduction leading

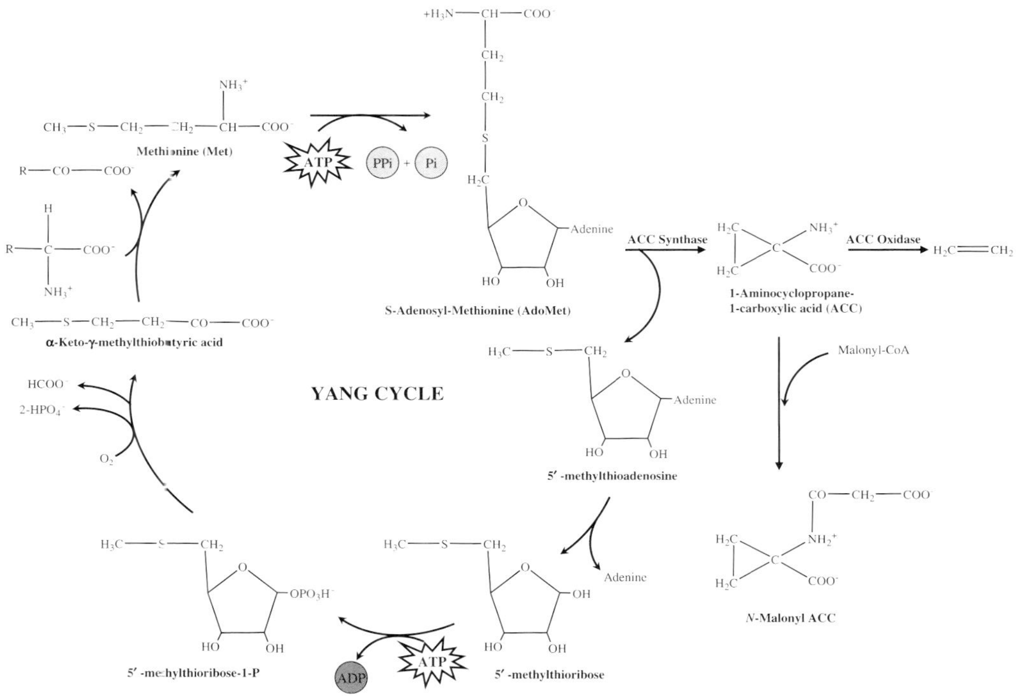

Fig. 12.1. Ethylene biosynthetic pathway and the Yang cycle. Methionine is the precursor of ethylene. The rate-limiting step in the pathway is the conversion of AdoMet to ACC, which is catalyzed by the enzyme ACC synthase. Conversion of ACC to ethylene is the last step in the pathway, which is catalyzed by the enzyme ACC oxidase. The CH_3-S group of methionine is recycled via the Yang cycle.

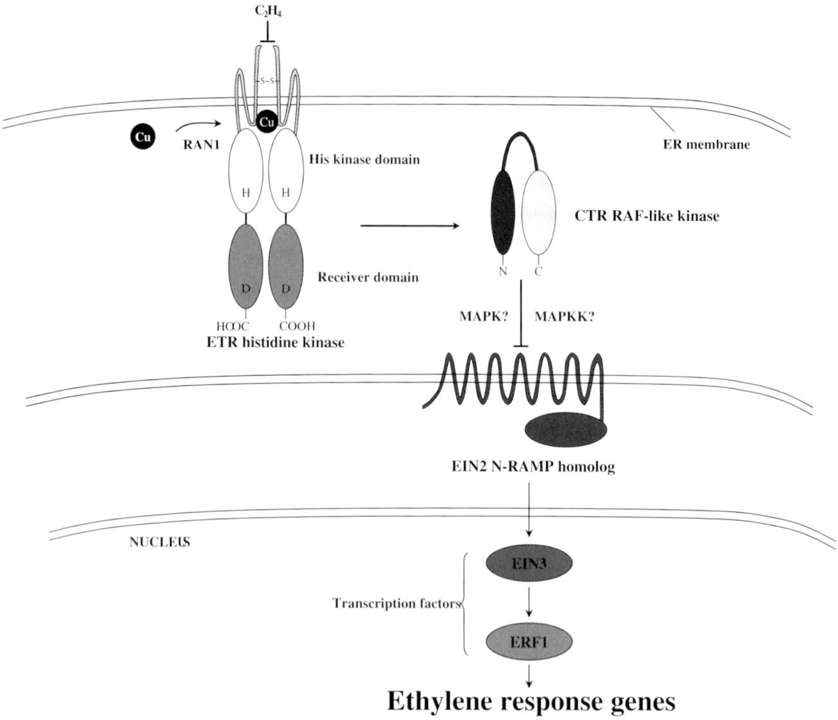

Fig. 12.2. Model of ethylene signaling in *Arabidopsis*. Ethylene binds to the ETR1 receptor, which is an integral membrane protein of the endoplasmic reticulum membrane. The receptor is a dimer, held together by disulfide bonds. Ethylene binds within the transmembrane domain, through a copper cofactor. Downstream signal transduction components include CTR1, a member of the Raf family of protein kinase; and EIN2, a channel-like transmembrane protein. The pathway activates a cascade of transcription factors, including the EIN3 and EREBP families, which then modulate gene expression.

from initial hormone perception to transcription regulation (Fig. 12.2). An important feature of the ethylene signaling pathway is that it contains both positive and negative regulators. According to the basic working model, ethylene is perceived by a family of membrane-associated receptors, including ETR1/ETR2, ETHYLENE RESPONSE SENSOR1 (ERS1)/ERS2 and EIN4 in *Arabidopsis* (Chen *et al.*, 2005). The ethylene receptors activate CTR1, a Raf-like serine/threonine (Ser/Thr) kinase, in the absence of ethylene. CTR1 then actively suppresses the downstream signaling process, such that EIN2 and the EIN3/EIL transcription factors remain inactive. EIN2 is an integral membrane protein whose function is not understood (Alonso *et al.*, 1999). The nuclear protein EIN3 is a transcription factor that regulates the expression of its immediate target genes such as *ETHYLENE RESPONSE FACTOR1* (*ERF1*) (Chao *et al.*, 1997; Solano *et al.*, 1998). ERF1 belongs to a large family of APETALA2-domain-containing transcription factors that bind to promoters of ethylene-inducible genes (Hao *et al.*, 1998). Upon binding ethylene, the receptors no longer suppresses the pathway. This relief of suppression allows for the activation of EIN2, induction of the transcriptional cascade, and the establishment of ethylene responses.

12.1.2. *Orchid reproductive biology*

Like all other living organisms, present-day orchids have evolved from ancestral forms as a result of selection pressure and adaptation. Containing more than 20,000 species, the Orchidaceae, classified in class Liliopsida, order Asparagales, is one of the largest angiosperm families. Associated with this enormous size is an extraordinary floral diversity. Orchids are extremely rich in species, and speciation rates are presumed to be exceptionally high (Gill, 1989), suggesting that orchids are still actively evolving. Although this spectacular diversification has been linked to the intimate and sometimes bizarre interaction of many species with their pollinators (Darwin, 1885), we still face the challenge of explaining how these mechanisms work and what the driving forces are behind their evolution.

Orchidaceae represents an unusually coherent group among monocots, possessing several reliable floral morphological synapomorphies, including a gynostemium, or column, fused by the style and at least part of the androecium; a highly evolved petal, the labellum; and resupination caused by 180° torsion of the pedicel (Rudall and Bateman, 2002). In addition to floral morphological synapomorphies, several unique reproductive strategies contribute to the success of the orchid family.

These include mature pollen grains packaged as pollinia, pollination-regulated ovary/ovule development, synchronized timing of micro- and megagametogenesis for effective fertilization, and the release of thousands or millions of immature embryos (seeds without endosperm) in a mature capsule (Yu and Goh, 2001). In most flowering plants, such as the model plants *Arabidopsis* and rice, the ovary matures during flower development and contains ovules, fully formed prior to pollination, that are ready for fertilization, which generally occurs soon after pollination (Bouman, 1984). The fact that pollination triggers ovule development in orchids indicates that these plants provide a unique system for studying of the mechanisms of initiation and continued development of ovules.

12.2. The Unique Control System for Orchid Ovule Development

12.2.1. *Categories of ovule development in Orchidaceae*

The most diagnostic synapomorphies of angiosperms are male reproductive organs with two pairs of pollen sacs and a carpel enclosing one or more ovules (Crane *et al.*, 1995). Ovules are the sites of processes essential for sexual plant reproduction, including the formation of the megagametophyte, fertilization, embryogenesis, and, finally, the formation of the persistent propagule — the seed. Different stages of ovule development at the time of anthesis have been reported for orchid species. In certain orchids such as *Cattleya, Sophronitis, Epidendron, Laelia, Phalaenopsis, Dendrobium,* and *Doritis,* ovules are completely absent in unpollinated ovaries, and their development is triggered following pollination (Duncan and Curtis, 1942, 1943; Israel and Sagawa, 1965; Yasugi, 1983; Yeung and Law, 1989; Zhang and O'Neill, 1993). In other orchid genera, the ovules are weakly developed or they contain well-differentiated archesporial or megaspore mother cells. This group includes *Cypripedium, Paphiopedium, Phragmipedium, Herminium, Epipactis,* and *Platanthera* (Duncan and Curtis, 1942; Sood and Mohana, 1986; Fredrikson, 1991, 1992). Other orchid species possess ovules that are ready for fertilization at the time of anthesis. Among these are *Epipogium aphyllum* (Afzelius, 1954), and *Hetaeria shikokiana* (Tohda, 1967). In *H. shikokiana,* fertilization can take place within four days of anthesis (Tohda, 1967).

12.2.2. *Orchid ovule morphogenesis*

Orchids are unusual among flowering plants in that in many species the ovule is not mature at the time of pollination. Pollination-regulated ovule initiation and development in *Phalaenopsis* has been characterized in detail (Zhang and O'Neill, 1993; Nadeau *et al.,* 1996). Within two days of pollination and before pollen tube germination, cell proliferation is initiated along the placental ridges of the ovary (Fig. 12.3a, and the star in Fig. 12.3e). The placental ridges continue to elongate and branch to form thousands of finger-like ovule primordia by about 40 days post-pollination (Fig. 12.3b, and the star in Fig. 12.3f). At this stage, the inner integument appears as a collar-like growth near the tip of the primordia (Fig. 12.3c, and the arrow in Fig. 12.3g), and the outer integument is initiated shortly thereafter (Fig. 12.3c, and the arrowhead in Fig. 12.3g). The archesporial cell enlarges further to form the megasporocyte which, following meiosis, gives rise to the expanding megaspore (the star in Fig. 12.3g). Subsequent mitotic divisions result in the development of a *Polygonum*-type embryo sac with synergids and egg cells (Fig. 12.3d, and labels S and E in Fig. 12.3h). The elapsed time between pollination and maturation of the ovule and final fertilization in *Phalaenopsis* is about 80 days (Zhang and O'Neill, 1993; Nadeau *et al.,* 1996).

12.2.3. *Benefit of a unique control system for orchid ovule development*

In a number of orchid species, the unique regulation of megasporogenesis and megagametogenesis by pollination, as well as long-lived flowers, is related to their reproductive ecology. A corollary to this absolute requirement for pollination to induce post-pollination development is the low probability of pollination by highly specific pollinators.

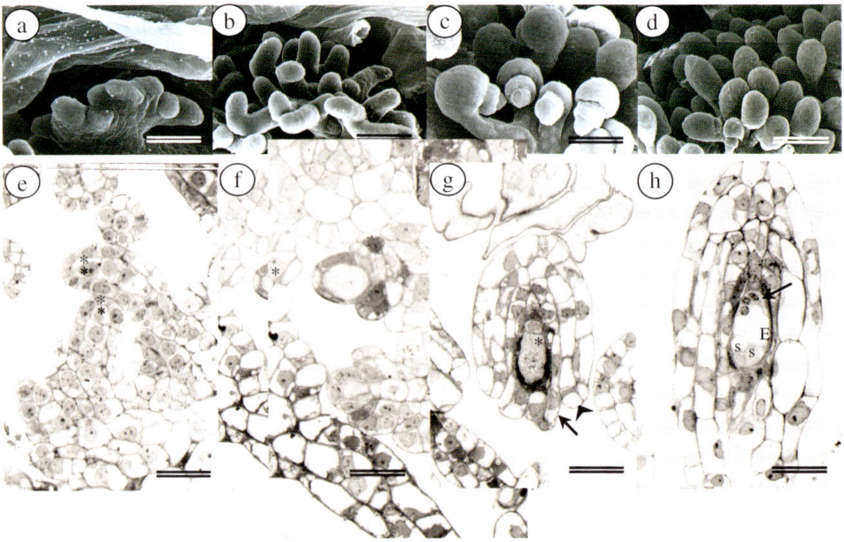

Fig. 12.3. (a)–(d) Scanning electron micrographs and (e)–(h) transverse sections of ovule development of the *Phalaenopsis* orchid. Placental protuberances after pollination are shown in (a) and (e). The uppermost nucellar cell ultimately differentiates into the archesporial cell [indicated by stars in (e)]. Elongation and branching forms thousands of finger-like ovule primordia as shown in (b) and (f). Archesporial cells enlarge and transform directly into megasporocytes [indicated by the star in (f)]. (c) and (g) show the development of the inner integument (arrow) and the outer integument (arrowhead). (g) shows the expanding megaspore (star). The mature embryo sac is shown in (d) and (h). The embyro sac in (h) shows synergids (S), the egg cell (E) and antipodals (arrow). The bar corresponds to 60 μm in (a)–(c), 200 μm in (d), 40 μm in (e)–(g), and 20 μm in (h). Adapted from Tsai *et al.* (2008).

Darwin proposed that the extraordinary longevity of orchid flowers was an adaptation to pollination by highly specific insect vectors, with a long life span increasing the likelihood that a specific pollinator would eventually find a particular flower with which it coevolved (Darwin, 1862). The orchid reproductive strategy invests in female reproductive development only after pollination, when fertilization is being assured.

12.3. Ethylene is Involved in Orchid Ovule Development

12.3.1. *Ethylene function in plant physiology and development*

Although many physiological and developmental processes in higher plants are controlled by the gaseous plant hormone ethylene, such as seed germination, root hair development, root nodulation, flower senescence, abscission, fruit ripening, and response to wounding, (Abeles *et al.*, 1992; Theologis, 1992; Johnson and Ecker, 1998), the role of ethylene in ovule development has, so far, received very little attention. Actually, in many flowers the gynoecium is an active site of ethylene production (Burg and Burg, 1966; Hall and Forsyth, 1967). Pollination causes an initial, substantial increase in ethylene production in the stigma and style within hours and a subsequent increase in ethylene production by other floral organs (Lipe and Morgan, 1973; Mor *et al.*, 1985; Lovell *et al.*, 1987; De Martinis *et al.*, 2002). Because ovule initiation triggers maturation of the ovary and differentiation of ovules in the orchid, these flowers provide a unique opportunity to dissect the signals that evoke ovule development and to elucidate the molecular mechanisms underlying ovule development.

12.3.2. *Evidence of ethylene involved in orchid ovule development*

Zhang and O'Neill (1993) applied the inhibitors of ethylene biosynthesis on the stigma to examine the role of ethylene on ovary/ovule development in orchids. Their results supported the idea that pollination-regulated ethylene production is related to the coordination of the senescence process and plays an important role in post-pollination ovule development in orchids (Zhang and O'Neill, 1993). A clear demonstration that ethylene controls ovule development came from the silencing of the ACC oxidase (ACO) gene expression, specifically, a pistil-specific *ACO*, in transgenic tobacco (De Martinis and Mariani, 1999). Flowers from transgenic plants show female sterility due to arrested ovule development (De Martinis and Mariani, 1999).

However, ethylene alone is not sufficient to trigger initiation of ovary development and ovule differentiation in orchid flowers. The development of the ovary/ovule has an absolute requirement for the participation of pollen-borne auxin (Zhang and O'Neill, 1993). Even before auxin was identified as a component of orchid pollen, exogenous application of auxin was found to mimic pollination to induce ethylene production and initiate most pollination-regulated developmental events (Curtis, 1943; Burg and Dijkman, 1967; Arditti *et al.*, 1971; Tsai, 2005). Temporal and spatial regulation of the processes that regulate ethylene and auxin levels in the flower may be important for the post-pollination developmental response of the ovary/ovule in orchids. Recently, the role of ethylene and auxin in stigma-to-ovule signaling was investigated in maize (Mól *et al.*, 2004). This study revealed that both ethylene synthesis and polar auxin transport in the silks were necessary to evoke accelerated differentiation of the egg cells in maize ovules after pollination with fresh pollen (Mól *et al.*, 2004). The orchestration of ethylene and auxin played in ovule development may be common in plants.

12.3.3. *Expression of Phal-ACS is regulated by pollination*

Pollination not only promotes ethylene biosynthesis in flower organs but also increases the accumulation of mRNA coding for ethylene biosynthetic enzymes (O'Neill *et al.*, 1993; Larsen *et al.*, 1995; Clark *et al.*, 1997; Bui and O'Neill, 1998; Jones and Woodson, 1999; Ketsa and Rugkong, 2000). In many species, ACC synthase is encoded by multiple genes that exhibit differential tissue specificity and/or differential regulation by environmental or hormonal stimuli (Kende, 1993). To investigate whether ACC synthase (ACS) genes were expressed in the stigma of the pollinated *Phalaenopsis* flower and whether these genes were regulated by the pollination signal, Bui and O'Neill identified and characterized ACS genes expressed in the orchid flower (Bui and O'Neill, 1998). Three *ACS* genes, *Phal-ACS1*, *Phal-ACS2*, and *Phal-ACS3*, were identified and found to be differentially expressed in the pollinated orchid flower (Bui and O'Neill, 1998). Pollination signaling elicits high levels of both *Phal-ACS2* and *Phal-ACS3* transcripts in stigma and ovary. With both genes up-regulated primarily by auxin, this hormone was then considered the primary pollination signal (Bui and O'Neill, 1998). In contrast, the expression of *Phal-ACS1* is regulated by ethylene, the secondary pollination signal, and may serve to amplify or sustain the primary pollination signal by regulating autocatalytic ethylene production (O'Neill, 1997; Bui and O'Neill, 1998). Although auxin and ethylene are the signals for post-pollination development of the ovary/ovule in orchids, we still do not know which *Phal-ACS* gene is directly involved in ovary/ovule development or whether all three have redundant functions.

12.3.4. *Phosphorylation status of Phal-ACS correlated with ethylene production in the ovary*

Pollination induces significant increases in the level of protein phosphorylation and ethylene production in orchid flowers (Porat *et al.*, 1995). To study the possible role of protein phosphorylation in the regulation of ethylene production in the orchid flower, Wang *et al.*, (2001) used okadaic acid (OA), a specific inhibitor of type 1 or type 2A serine/threonone protein phosphatases, to relate its effect on the expression of *Phal-ACS* genes. Differential expression levels of the *Phal-ACS* genes were observed in the OA-treated stigmas (Wang *et al.*, 2001). The OA-induced level of *Phal-ACS1* transcripts in the stigma, labellum, and ovary was higher than that of *Phal-ACS2* and *Phal-ACS3*. These results suggest that the hyperphosphorylation status of (an) unidentified protein(s) is involved in up-regulating the expression of *Phal-ACS1* resulting in increased ethylene production (Wang *et al.*, 2001). So far, evidence is lacking for the direct involvement of protein phosphatases in the ethylene signal transduction pathway. However, results of studies of general kinase and phosphatase inhibitors have suggested that protein phosphorylation/dephosphorylation is implicated in the regulation of ethylene induction in plants under stress (Spanu *et al.*, 1994; Tuomainen *et al.*, 1997; Felix *et al.*, 2000). In addition, phosphorylation of ACS2 and ACS6 by MPK6 leads to the accumulation of the ACS protein and thus elevated levels of cellular ACS activity and ethylene production in *Arabidopsis* (Liu and Zhang, 2004). Possible mechanisms underlying the phosphorylation and dephosphorylation regulation of

ACS include changes in ACS activity or stability, or both (Spanu *et al.*, 1994). It is possible that pollination could induce increases in the level of *Phal-ACS* phosphorylation to produce a significant amount of ethylene to initiate ovary/ovule development in orchids.

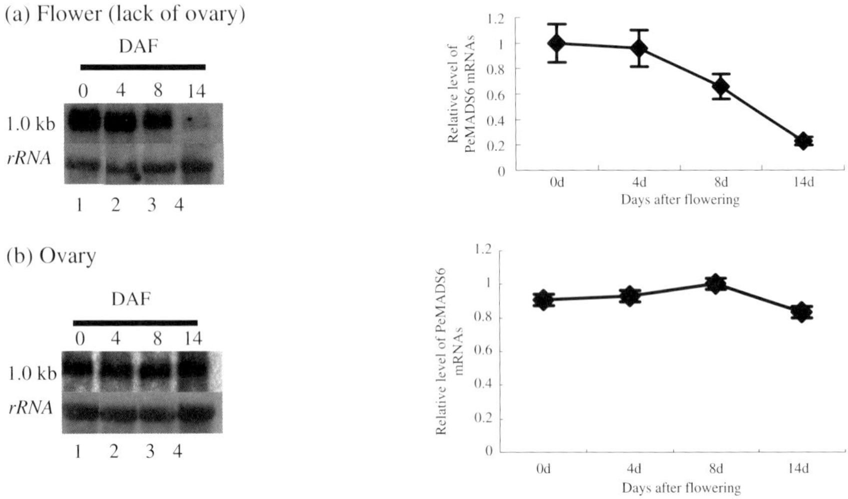

Fig. 12.4. Accumulation of *PeMADS6* mRNA in the flowers and immature ovaries during the natural senescence process and after pollination. (a), (b) Accumulation of *PeMADS6* mRNA in (a) the flowers and (b) the immature ovaries during the natural senescence process. Each lane contained 10 μg of total RNA from tissues at day 0 (Fig. 12.4a and 12.4b, lane 1), the day of flowering; 4 days after flowering (DAF) (Fig. 12.4a and 12.4b, lane 2), 8 DAF (Fig. 12.4a and 12.4b, lane 3); and 14 DAF (senescence day) (Fig. 12.4a and 12.4b, lane 4). The accumulation of *PeMADS6* mRNA in (c) the flowers and (d) the immature ovaries after pollination. Each lane contained 10 μg of total RNA from tissues at 1 h (Fig. 12.4c and 12.4d, lane 1), 6 h (Fig. 12.4c and 12.4d, lane 2), 12 h (Fig. 12.4c and 12.4d, lane 3), 1 d (Fig. 12.4c and 12.4d, lane 4), 2 d (Fig. 12.4c and 12.4d, lane 5), and 4 d (Fig. 12.4c and 12.4d, lane 6) after pollination (DAP). RNA gel blots were scanned with a densitometer, and the obtained values were used to generate a graph of relative *PeMADS6* mRNA levels. Each point is the average of the signals detected from three independent experiments. Standard errors are indicated as vertical bars. Adapted from Tsai *et al.* (2005).

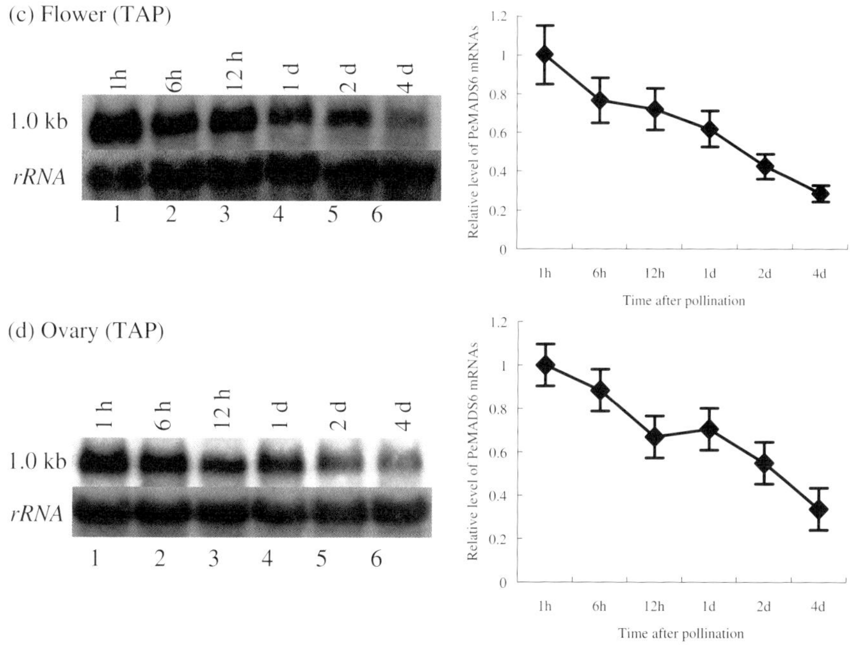

Fig. 12.4. *(Continued)*

12.4. Orchid *GLOBOSA*-like Gene, *PeMADS6*, Involved in Ovary/ovule Development via a Mode of Action by Auxin/ethylene

12.4.1. *Expression of PeMADS6 is correlated to pollination*

Recently, a *GLOBOSA*-like MADS-box B-function gene, *PeMADS6*, was identified from *Phalaenopsis* orchids (Tsai *et al.*, 2005). *PeMADS6* was expressed not only in the four whorls of the orchid flower but also in immature ovaries, from flower opening to senescence (Tsai *et al.*, 2005). In mature flowers, the level of *PeMADS6* transcripts decreased during the progressing senescence and was almost eliminated when flowers were senescent on day 14 post-anthesis (Fig. 12.4a). In contrast, the expression of *PeMADS6* in immature ovaries remained unchanged during the natural senescence process (Fig. 12.4b). Interestingly, the expression of *PeMADS6* decreased both in flowers and immature ovaries during the progressing pollination-induced senescence (Fig. 12.4c and 12.4d). These results suggest that the accumulation of *PeMADS6* mRNA was reduced in flowers during natural or pollination-induced senescence. In addition, pollination represses *PeMADS6* expression in the ovary (Tsai *et al.*, 2005).

12.4.2. *Expression of PeMADS6 is regulated by auxin and ethylene*

Interestingly, applying auxin as naphthaleneacetic acid (NAA) to the orchid stigma not only induced stigma closure, ovary swelling and perianth senescence, similar to a post-pollination syndrome (Fig. 12.5), but also reduced *PeMADS6* expression in the ovary after two-day treatment (Fig. 12.6a). Several studies have demonstrated that pretreatment of the orchid flower with the ethylene biosynthesis inhibitor aminoethoxyvinylglycine (AVG) inhibits ethylene production induced not only by pollination, but also by auxin (Fredrikson, 1991; O'Neill *et al.*, 1993; Tsai *et al.*, 2005). To examine the separate regulations of the *PeMADS6* expression by both auxin and ethylene, auxin-induced ethylene production was inhibited by pretreating the stigma with AVG, and then treating with auxin (Tsai *et al.*, 2005). Interestingly, the inhibition of the accumulated *PeMADS6* transcript was still observed in the ovary in the presence of AVG, suggesting a key role for auxin in the regulation of *PeMADS6* expression in the orchid ovary (Fig. 12.6b). Thus, pollination inhibits *PeMADS6* expression in the ovary via the auxin signaling pathway to promote ovary/ovule development and this is not mediated by auxin-induced ethylene. However, it is not exclusive that ethylene may play a role in inhibiting *PeMADS6* expression in the ovary to affect its development (Fig. 12.7).

To gain more insight into the regulation of *PeMADS6* expression, transcription factor binding sites on its *cis*-element were identified. A 1.5-kb fragment upstream from the translational start site of *PeMADS6* was isolated by using genome walking. Interstingly, both a putative auxin response element (ARE) and an ethylene response

Fig. 12.5. Structure of *Phalaenopsis equestris* flower. (a) The front view of *Phalaenopsis* showing the flower organs. Inset: The column showing the stigmatic cavity. (b) Back view of the same flower showing the inferior ovary. (c) The post-pollination syndrome of flower at 4 DAP: the swelling ovary and the senescent perianth. Inset: The swelling of the stigma enclosing the pollinia. (d) Morphological changes four days after treatment using auxin (NAA): the swelling ovary and senescent perianth. Inset: The enclosing stigma. (e) Senescence-related degradative changes in all organs of the flower four days after treatment using ACC. The whole flower and ovary senescence was observed. Adapted from Tsai *et al.* (2005).

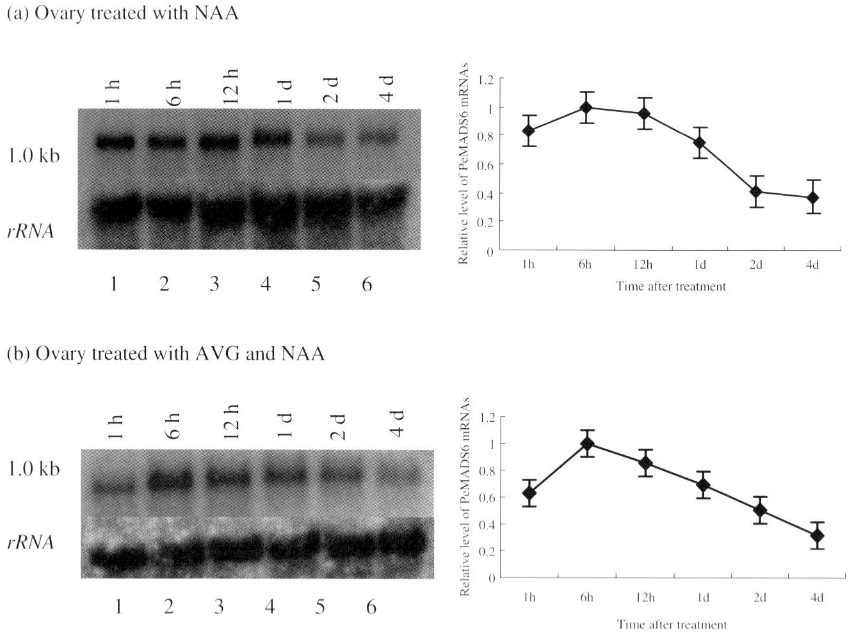

Fig. 12.6. Time courses of the accumulation of *PeMADS6* mRNAs in immature ovaries after stigma were treated with (a) auxin, or (b) AVG in combination with NAA. Each lane contained 10 μg of total RNA from the ovaries at 1 h (Fig. 12.6a and 12.6b, lane 1), 6 h (Fig. 12.6a and 12.6b, lane 2), 12 h (Fig. 12.6a and 12.6b, lane 3), 1 d (Fig. 12.6a and 12.6b, lane 4), 2 d (Fig. 12.6a and 12.6b, lane 5), and 4 d (Fig. 12.6a and 12.6b, lane 6) after treatment. RNA gel blots were scanned with a densitometer, and the obtained values were used to generate a graph of relative *PeMADS6* mRNA levels. Each point is the average of the signals detected from three independent experiments. Standard errors are indicated as vertical bars. Adapted from Tsai *et al.* (2005).

element (ERE) in addition to the MADS-box protein consensus binding motif, the CArG box, were identified. The putative ARE and ERE were located at nucleotide −1223 and nucleotide −778 upstream from the translational start site of *PeMADS6*, respectively. It is plausible that the coordination of both auxin and ethylene plays an important role in the regulation of *PeMADS6* expression. Moreover, two auxin response factor (ARF)-related expressed sequence tags (ESTs) and five

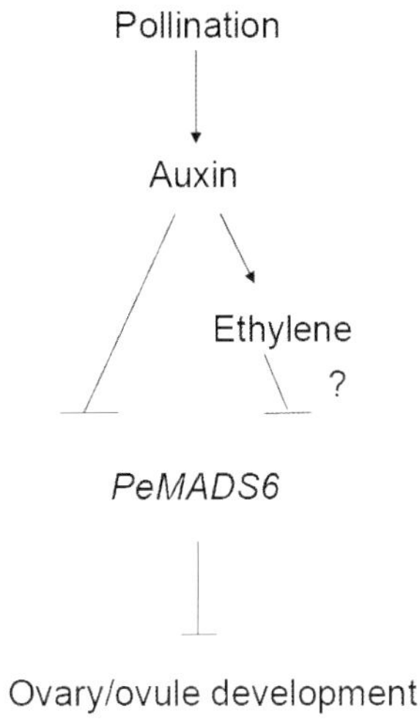

Fig. 12.7. The relationship between pollination and *PeMADS6* expression regulates *Phalaenopsis* ovary/ovule development.

ethylene response element binding protein (EREBP)-related ESTs exist in an in-house *Phalaenopsis equestris* flower bud EST database (Tsai and Chen, 2006). Understanding the response of these genes followed by the stimulation of pollination, the protein-DNA interaction between ARF and ARE, and between EREBP and ERE of *PeMADS6* will guide us to explore further the molecular mechanism of orchid ovule development.

12.5. Future Prospects

Due to their unique floral morphology as well as reproductive biology, orchids are attractive models for botanists to study floral morphogenesis, ovary/ovule morphogenesis and embryogenesis. The ethylene perception and transduction pathway components involved in plant ovary/ovule development still require full elucidation. Although in *Arabidopsis* the gene encoding ethylene receptor ETR2 shows an enhanced expression pattern in developing petals and in the ovule (Sakai *et al.*, 1998), after the early stages of megasporogenesis, the *etr2* mutant does not show any phenotype after the early stages

(Hua and Meyerowitz, 1998). It is possible that the redundancy of five members of the ethylene receptor genes masks the effect of ethylene on ovule development. Targets of the ethylene signaling cascade include EREBPs. It has been shown that genes necessary for ovule and female gametophyte development, such as *AINTEGUMENTA* (*ANT*), encode putative transcription factor that belong to EREBPs. In *Arabidopsis*, severe *ant* alleles, such as *ant-4* and *ant-5*, lead to the complete absence of integuments, with the resulting ovules having only a rudimentary bulge in the chalazal region (Baker *et al.*, 1997). APETALA2 (AP2), another EREBPs member, is also required for ovule development (Modrusan *et al.*, 1994). However, whether ethylene plays a role in ovule morphogenesis was not concluded in these studies.

The precise pollination-regulated ovary/ovule development of orchids offers an attractive model system to study how the ovary/ovule perceives the ethylene signal and then initiates the developmental program. However, owing to the large genome size, long life cycle, and inefficient transformation system of orchids, few studies on orchid biology exist. Recently, the genetic architecture of the sophisticated floral organization of orchids has begun to be investigated (Tsai *et al.*, 2004, 2005, 2006). Given that orchids represent the most successful and diverse plant families worldwide, the development of genomic resources is imperative (Cozzolino and Widmer, 2005). Thanks to the advanced progress of genomics and bioinformatics, plentiful gene information and integrated bioinformatic tools are ready for use in studying orchid biology (Wang *et al.*, 2005; Hsiao *et al.*, 2006; Tsai *et al.*, 2006). These resources will accelerate the cloning of ethylene perception, signal transduction, and ethylene-inducible transcription factor genes involved in ovule development.

Another intriguing question is how ethylene simultaneously regulates development in orchids, i.e., how ethylene ensures the coordination of life and death for various organs. Perhaps transcriptomics and proteomics approaches coupled with a more extensive use of mutants may help to elucidate the regulatory networks of ethylene and better define the role of ethylene. More recently, virus-induced gene-silencing vectors were developed for studying orchid gene function and revealed their powerful ability for dissecting orchid floral organ identity genes (Lu *et al.*, 2007). Adopting these tools to dissect the details of the signaling components controlling the ovary/ovule developmental program using the orchid reproductive system will advance the understanding in this field.

Acknowledgments

The authors would like to thank Dr. S. F. Yang for his kind suggestions and input while we carried out our experiments on the auxin/ethylene effect on *PeMADS6* gene expression in orchid ovule development.

References

Abeles FB, Morgan PW, Saltveit Jr. ME. (1992) *Ethylene in Plant Biology*. Academic Press, San Diego.

Afzelius K. (1954) Embryo-sac development in *Epipogium aphyllum. Svensk Bot Tidskr* **48**: 513–520.

Alonso JM, Hirayama T, Roman G *et al.* (1999) EIN2, a bifunctional transducer of ethylene and stress responses in *Arabidopsis*. *Science* **284**: 2148–2152.

Arditti J, Jeffrey DC, Flick BH. (1971) Post-pollination phenomena in orchid flowers. III. Effects and interactions of auxin, kinetin or gibberellin. *New Phytol* **70**: 1125–1141.

Baker SC, Robinson-Beers K, Villanueva JM *et al.* (1997) Interactions among genes regulating ovule development in *Arabidopsis thaliana*. *Genetics* **145**: 1109–1124.

Bleecker AB, Estelle MA, Somerville C, Kende H. (1988) Insensitivity to ethylene conferred by a dominant mutation in *Arabidopsis thaliana*. *Science* **241**: 1086–1089.

Bouman F. (1984) The ovule. In: *Embryology of Angiosperms*. Springer-Verlag, Berlin.

Bui AQ, O'Neill SD. (1998) Three 1-aminocyclopropane-1-carboxylate synthase genes regulated by primary and secondary polliantion signals in orchid flowers. *Plant Physiol* **116**: 419–428.

Burg SP, Burg EA. (1966) The interaction between auxin and ethylene and its role in plant growth. *Proc Natl Acad Sci USA* **55**: 262–269.

Burg SP, Dijkman MJ. (1967) Ethylene and auxin participation in pollen induced fading of *Vanda* orchid blossoms. *Plant Physiol* **42**: 1648–1650.

Chang C, Kwok SF, Bleecker AB, Meyerowitz EM. (1993) *Arabidopsis* ethylene response gene *ETR1*: Similarity of product to two-component regulators. *Science* **262**: 539–544.

Chao Q, Rothenberg M, Solano R *et al.* (1997) Activation of the ethylene gas response pathway in *Arabidopsis* by the nuclear protein ETHYLENE-INSENSITIVE3 and related proteins. *Cell* **89**: 1133–1144.

Chen Y-F, Etheridge N, Schaller GE. (2005) Ethylene signal transduction. *Ann Bot* **95**: 901–905.

Clark DG, Richards C, Hilioti Z *et al.* (1997) Effect of pollination on accumulation of ACC synthase and ACC oxidase transcripts, ethylene production and flower petal abscission in geranium (*Pelargonium x hortorum* L.H. Bailey). *Plant Mol Biol* **34**: 855–865.

Cozzolino S, Widmer A. (2005) Orchid diversity: An evolutionary consequence of deception? *Trends Ecol Evol* **20**: 487–494.

Crane PR, Friis EM, Pedersen KR. (1995) The origin and early diversification of angiosperms. *Nature* **374**: 27–33.

Curtis JT. (1943) An unusual pollen reaction in *Phalaenopsis*. *Am Orchid Soc Bull* **21**: 98–100.

Darwin C. (1862) *Fertilization of Orchids by Insects*. John Murray, London.

Darwin C. (1885) *On the Various Contrivances by Which Orchids are Fertilised by Insects*, second ed., John Murray, London.

De Martinis D, Cotti G, Te Lintel Hekker S *et al.* (2002) Ethylene response to pollen tube growth in *Nicotiana tabacum* flower. *Planta* **214**: 806–812.

De Martinis D, Mariani C. (1999) Silencing gene expression of the ethylene-forming enzyme results in a reversible inhibition of ovule development in transgenic tobacco plant. *Plant Cell* **11**: 1061–1071.

Duncan RE, Curtis JT. (1942) Intermittent growth of fruits of *Cypripedium* and *Paphiopedium*. A correlation of the growth of orchid fruits with their internal development. *Bull Torrey Club* **69**: 353–359.

Duncan RE, Curtis JT. (1943) Growth of fruits in *Cattleya* and allied genera in the Orchidaceae. *Bull Torrey Club* **70**: 104–119.

Felix G, Regenass M, Boller T. (2000) Sensing of osmotic pressure changes in tomato cells. *Plant Physiol* **124**: 1169–1179.

Fredrikson M. (1991) An embryological study of *Platanthera bifolia* (Orchidaceae). *Plant Syst Evol* **714**: 213–220.

Fredrikson M. (1992) The development of female gametophyte of *Epipactis* (Orchidaceae) and its inference for reproductive ecology. *Am J Bot* **79**: 61–68.

Gill DE. (1989) Fruiting failure, pollinator inefficiency, and speciation in orchids. In: *Speciation and Its Consequences*. Sinauer Associates, Sunderland, MA.

Guzmán P, Ecker JR. (1990) Exploiting the triple response of *Arabidopsis* to identify ethylene-related mutants. *Plant Cell* **2**: 513–523.

Hall IV, Forsyth FR. (1967) Production of ethylene by flowers following pollination and treatments with water and auxin. *Can J Bot* **45**: 1163–1166.

Hao D, Ohme-Takagi M, Sarai A. (1998) Unique mode of GCC box recognition by the DNA-binding domain of ethylene-response-element-binding factor (ERF domain) in plants. *J Biol Chem* **273**: 26857–26861.

Hirayama T, Kieber JJ, Hirayama N *et al.* (1999) RESPONSIVE-TO-ANTAGONIST1, a Menkes/Wilson disease-related copper transporter, is required for ethylene signaling in *Arabidopsis*. *Cell* **97**: 383–393.

Hsiao YY, Tsai WC, Kuoh CS *et al.* (2006) Comparison of transcripts in *Phalaenopsis bellina* and *Phalaenopsis equestris* (Orchidaceae) flowers to deduce monoterpene biosynthesis pathway. *BMC Plant Biol* **6**: 14.

Hua J, Meyerowitz EM. (1998) Ethylene responses are negatively regulated by a receptor gene family in *Arabidopsis thaliana*. *Cell* **94**: 261–271.

Israel HW, Sagawa Y. (1965) Post-pollination ovule development in *Dendrobium* orchids III. Fine structure of meiotic prophase I. *Caryologia* **18**: 15–23.

Johnson PR, Ecker JR. (1998) The ethylene gas signal transduction pathway: A molecular perspective. *Annu Rev Genet* **32**: 227–254.

Jones ML, Woodson RW. (1999) Differential expression of three members of the 1-aminocyclopropane-1-carboxylate synthase gene family in carnation. *Plant Physiol* **119**: 755–764.

Kende H. (1993) Ethylene biosynthesis. *Annu Rev Plant Physiol Plant Mol Biol* **44**: 283–307.

Ketsa S, Rugkong A. (2000) Ethylene production, senescence and ethylene sensitivity of *Dendrobium* 'Pompadour' flowers following pollination. *J Hort Sci Biotechnol* **75**: 149–153.

Kieber JJ, Rothenberg M, Roman G *et al.* (1993) CTR1, a negative regulator of the ethylene response pathway in *Arabidopsis*, encodes a member of the Raf family of protein kinases. *Cell* **72**: 427–441.

Larsen PB, Ashworth EN, Jones ML, Woodson WR. (1995) Pollination-induced ethylene in carnation: Role of pollen tube growth and sexual compatibility. *Plant Physiol* **108**: 1405–1412.

Lipe JA, Morgan PW. (1973) Location of ethylene production in cotton flowers and dehiscing fruits. *Planta* **115**: 93–96.

Liu Y, Zhang S. (2004) Phosphorylation of 1-aminocyclopropane-1-carboxylic acid synthase by MPK6, a stress-responsive mitogen-activated protein kinase, induces ethylene biosynthesis in *Arabidopsis*. *Plant Cell* **16**: 3386–3399.

Lovell PJ, Lovell PH, Nichols R. (1987) The control of senescence in petunia (*Petunia hybrida*). *Ann Bot* **60**: 49–59.

Lu HC, Chen HH, Tsai WC *et al.* (2007) Strategies for functional validation of genes involved in reproductive stages of orchids. *Plant Physiol* **143**: 558–569.

Mól R, Filek M, Machackova I, Matthys-Rochon E. (2004) Ethylene synthesis and auxin augmentation in pistil tissues are important for egg cell differentiation after pollination in maize. *Plant Cell Physiol* **45**: 1396–1405.

Modrusan Z, Reiser L, Feldmann KA *et al.* (1994) Homeotic transformation of ovules into carpel-like structures in *Arabidopsis*. *Plant Cell* **6**: 333–349.

Mor Y, Halevy AH, Spiegelstein H, Mayak S. (1985) The site of 1-aminocyclopropane 1-carboxylic acid synthesis in senescing carnation petals. *Physiol Plant* **65**: 196–202.

Nadeau JA, Zhang XS, Li J, O'Neill SD. (1996) Ovule development: Identification of stage-specific and tissue-specific cDNAs. *Plant Cell* **8**: 213–239.

O'Donnell PJ, Schmelz E, Block A *et al.* (2003) Multiple hormones act sequentially to mediate a susceptible tomato pathogen defense response. *Plant Physiol* **133**: 1181–1189.

O'Neill SD. (1997) Pollination regulation of flower development. *Annu Rev Plant Physiol Plant Mol Biol* **48**: 547–574.

O'Neill SD, Nadeau JA, Zhang XS *et al.* (1993) Interorgan regulation of ethylene biosynthetic genes by pollination, plantcell **5**: 419–432.
Penninckx IA, Thomma BP, Buchala A *et al.* (1998) Concomitant activation of jasmonate and ethylene response pathways is required for induction of a plant defensin gene in *Arabidopsis. Plant Cell* **10**: 2103–2113.
Porat R, Halevy AH, Serek MB, Borochov A. (1995) An increase in ethylene sensitivity following pollination is the initial event triggering an increase in ethylene production and enhanced senescence of *Phalaenopsis* orchid flowers. *Physiol Plant* **93**: 778–784.
Roman G, Lubarsky B, Kieber JJ *et al.* (1995) Genetic analysis of ethylene signal transduction in *Arabidopsis thaliana*: Five novel mutant loci integrated into a stress response pathway. *Genetics* **139**: 1393–1409.
Rudall PJ, Bateman RM. (2002) Roles of synorganisation, zygomorphy and heterotopy in floral evolution: The gynostemium and labellum of orchids and other lilioid monocots. *Biol Rev* **77**: 403–441.
Sakai H, Hua J, Chen QG *et al.* (1998) *ETR2* is an *ETR1*-like gene involved in ethylene signalling in *Arabidopsis. Proc Natl Acad Sci USA* **95**: 5812–5817.
Solano R, Stepanova A, Chao Q, Ecker JR. (1998) Nuclear events in ethylene signaling: A transcriptional cascade mediated by ETHYLENE-INSENSITIVE3 and ETHYLENE-RESPONSE-FACTOR1. *Genes Dev* **12**: 3703–3714.
Sood SK, Mohana Rao PR. (1986) Development of male and female gametophytes in *Herminium angustifolium* (Orchidaceae). *Phytomorphology* **36**: 11–15.
Spanu P, Grosskopf DG, Felix G, Boller T. (1994) The apparent turnover of 1-aminocyclopropane-1-carboxylate synthase in tomato cells is regulated by protein phosphorylation and dephosphorylation. *Plant Physiol* **106**: 529–535.
Stepanova AN, Ecker JR. (2000) Ethylene signaling: From mutants to molecules. *Curr Opin Plant Biol* **3**: 353–360.
Theologis A. (1992) One rotten apple spoils the whole bushel: The role of ethylene in fruit ripening. *Cell* **70**: 181–184.
Tohda, H. (1967) An embryological study of *Hetaeria shikokiana*, a saprophytic orchid in Japan. *Sci Rep Tohoku Univ Ser IV (Biol)* **33**: 83–95.
Tsai WC, Chen HH. (2006) The orchid MADS-box genes controlling floral morphogenesis. *Sci World J* **6**: 1933–1944.
Tsai WC, Hsiao YY, Lee SH *et al.* (2006) Expression analysis of the ESTs derived from the flower buds of *Phalaenopsis equestris. Plant Sci* **170**: 426–432.
Tsai WC, Hsiao YY, Pan ZJ *et al.* (2008) The role of ethylene in orchid ovule development. *Plant Sci* **175**: 98–105.
Tsai WC, Kuoh CS, Chuang MH *et al.* (2004) Four *DEF*-Like MADS box genes displayed distinct floral morphogenetic roles in *Phalaenopsis* orchid. *Plant Cell Physiol* **45**: 831–844.
Tsai WC, Lee PF, Chen HI *et al.* (2005) *PeMADS6*, a GLOBOSA/PISTILLATA-like gene in *Phalaenopsis equestris* involved in petaloid formation, and correlated with flower longevity and ovary development. *Plant Cell Physiol* **46**: 1125–1139.
Tuomainen J, Betz C, Kangasjarvi J *et al.* (1997) Ozone induction of ethylene emission in tomato plants: Regulation by differential accumulation of transcripts for the biosynthetic enzymes. *Plant J* **12**: 1151–1162.

Wang HC, Kuo HC, Chen HH *et al.* (2005) KSPF: Using gene sequence patterns and data mining for biological knowledge management. *Expert Syst Appl* **28**: 537–545.

Wang NN, Yang SF, Charng YY. (2001) Differential expression of 1-aminocyclopropane-1-carboxylate synthase genes during orchid flower senescence induced by the protein phosphatase inhibitor okadaic acid. *Plant Physiol* **126**: 253–260.

Woeste KE, Kieber JJ. (2000) A strong loss-of-function mutation in *RAN1* results in constitutive activation of the ethylene response pathway as well as a rosette-lethal phenotype. *Plant Cell* **12**: 443–455.

Yasugi S. (1983) Ovule and embryo development in *Doritis pulcherrima* (Orchidaceae). *Am J Bot* **70**: 555–560.

Yeung EC, Law SK. (1989) Embryology of *Epidendron ibaguense*. I. Ovule development. *Can J Bot* **67**: 2219–2226.

Yu H, Goh CJ. (2001) Molecular genetics of reproductive biology in orchids. *Plant Physiol* **127**: 1390–1393.

Zhang XS, O'Neill SD. (1993) Ovary and gametophyte development are coordinately regulated following pollination by auxin and ethylene. *Plant Cell* **5**: 403–418.

CHAPTER 13

The Application of the Chloroplast Genome of *Oncidium* in Plant Identification and Breeding in Oncidiinae

Fu-Hui Wu[*], Ming-Tsair Chan[*], De-Chih Liao[*], Chen-Tran Hsu[*], Yi-Wei Lee[*], Henry Daniell[†], Melvin R. Duvall[‡] and Choun-Sea Lin[*,§]

Oncidium spp. is an important orchid species among cut flowers. However, determining their phylogenetic relationships and identifying varieties are very difficult. Molecular markers can provide useful tools for phylogenetic resolution, the chloroplast being one resource for those markers. The complete chloroplast genome of *Onc.* Gower Ramsey (Accession no. GQ324949) was determined. The size of the *Oncidium* chloroplast genome is 146,484 bp. Interesting, *Onc.* Gower Ramsey chloroplast-encoded *NADH dehydrogenase* (*ndh*) genes, except *ndhE*, have no functions. Based on this information, eight regions of the *Onc.* Gower Ramsey chloroplast genome were amplified by polymerase chain reaction for phylogenetic analysis to identify the relationships between the 15 Oncidiinae hybrids. The results confirm that the chloroplast genome provides a useful molecular marker for species identification.

13.1. Introduction

13.1.1. *Plant identification and breeding in Oncidium*

In the Oncidiinae subtribe of the Orchidaceae family, consisting of about 70 closely related genera with over 1,000 species, *Oncidium*

[*]Agricultural Biotechnology Research Center, Academia Sinica, Taipei, Taiwan.
[†]Department of Molecular Biology and Microbiology, College of Medicine, Biomolecular Science Building, University of Central Florida, Orlando, USA.
[‡]Department of Biology, Northern Illinois University, DeKalb, USA.
[§]Corresponding author. E-mail: cslin99@gate.sinica.edu.tw

is its largest genus (Webster, 1992). In cellular biology, ecology and morphology, *Oncidium* is also the most diverse genus in Orchidaceae. Previous taxonomy of the Oncidiinae was based on flower morphology (Chase and Palmer, 1989). However, flower morphology is affected by environmental and other factors. Therefore, for some Oncidiinae species taxonomic positions are debated. Furthermore, as Oncidiinae can be crossed intergenerically, there have been many studies on intergeneric hybrids (Webster, 1992). As a result, 20 percent of the *Oncidium* group has been redistributed into other genera.

The DNA sequence of an organism provides taxonomic and evolutionary information. The DNA sequence reflects the genotype and records the genomic events that led to evolution. As polymerase chain reaction (PCR) has become a routine technique, taxonomists have used DNA isolated from different organelles for evolutionary research. There are different molecular marker techniques available for conducting useful genetic analysis by PCR, including restriction fragment length polymorphism (RFLP), random amplified polymorphic DNA (RAPD), amplification fragment length polymorphism (AFLP), and simple sequence repeat (SSR). All of these methods have been employed in orchid taxonomy and evolutionary studies. Chou *et al.* (2002) used RAPD markers to investigate the relationships between 24 species of Oncidiinae. However, they could not identify the relationships among these species.

13.1.2. *PCR is a simple method for chloroplast genome sequencing*

The chloroplast DNA (cpDNA) is in many respects the genome of choice for orchid taxonomic studies (Chase and Palmer, 1989). There are many advantages to using cpDNA for taxonomy and evolutionary research (Wu *et al.*, 2010). Chloroplast DNA is a circular DNA of 120∼220 kb that consists of two inverted repeats (IRa and IRb), a large single-copy region (LSC), and a small single-copy region (SSC). Therefore, we can design primers for other cpDNA sequencing by PCR (Dhingra and Folta, 2005). This approach has been demonstrated for the sequencing of two bamboo cpDNA genomes (Wu *et al.*, 2009a).

The chloroplast genome of one member of the Orchidaceae family, *Phalaenopsis aphrodite*, has already been published (Chang *et al.*, 2006). Hence, it would be interesting to ascertain the complete cpDNA of other orchids.

13.2. The Chloroplast Genome of *Oncidium* Gower Ramsey

13.2.1. *The structure and characterization of the chloroplast genome of Oncidium Grower Ramsey*

The size of the *Onc.* Gower Ramsey chloroplast genome is 146,484 bp (Wu *et al.*, 2010). The genome includes a pair of IRs of 25,755 bp each, an SSC region of 12,650 bp, and a LSC region of 82,324 bp. The *Onc.* Gower Ramsey chloroplast genome contains 101 different genes, of which 16 are duplicated in the IR, giving a total of 133 genes. There are 29 distinct tRNAs, six of which are duplicated in the IR. Sixteen genes contain one or two introns, with six of the introns in tRNAs. Coding regions make up 49.94 percent of the chloroplast genome (41.86 percent protein-coding genes, 8.08 percent RNA genes) and non-coding regions, which contain intergenic spacer (IGS) regions and introns, comprise 50.06 percent. The overall guanine-cytosine (GC) and adenine-thymine (AT) content of the chloroplast genome is 37.32 percent and 62.68 percent, respectively (Wu *et al.*, 2010).

13.2.2. *Broad phylogenetic analysis*

Our broad phylogenetic analysis resulted in two trees, an ML tree with $-\ln L = 4.12281.26$ (Wu *et al.*, 2010) and an MP tree of 75,521 steps with a consistency index (excluding uninformative characters) of 0.3649 and a retention index of 0.5997 (MP tree not shown). The topologies of the monocot subtrees were identical for the two analyses in which *Oncidium* was maximally supported as the sister to *Phalaenopsis* and the two orchids were united with *Yucca*, the other representative of Asparagales, with maximum support. These results are consistent with most single- and multi-gene studies of monocots [summarized in Chase *et al.*, (2006)], but with higher levels of support since the 61 conserved protein-coding sequences contained over 14,500 parsimony-informative characters (Wu *et al.*, 2010).

13.3. Analysis of Variability in *ndh* Genes of 15 *Oncidium* Species

13.3.1. *The ndh gene deletion in Oncidiinae cpDNA*

In higher plant chloroplasts, the NAD(P)H dehydrogenase (NDH) complex participates in PSI cyclic electron flow and chlororespiration (Sazanov *et al.*, 1998). Eleven subunits of the chloroplast *ndh* genes (*ndhA–ndhK*) are encoded in the chloroplast genome. However, in *Onc.* Gower Ramsey, of the 11 chloroplast-encoded *ndh* genes, only *ndhE* theoretically translates into a functional protein (Wu *et al.*, 2010). This *ndh* gene deletion was also observed in *P. aphrodite* (Chang *et al.*, 2006). Unlike *P. aphrodite*, in all 15 of the Oncidiinae varieties studied here, the *ndhJ* gene was truncated and the *ndhK* gene was absent. In *ndhC*, like in *P. aphrodite*, a frame shift occurred, creating a stop codon in the middle of the gene in all Oncidiinae, including *Onc.* Gower Ramsey, which results from the deletion of 17 bp (Wu *et al.*, 2010).

In *P. aphrodite*, the *ndhB* of Oncidiinae does not function due to a deletion in the first exon. Six orchid *ndhB* genes published in the NCBI database could translate into the putative functional NdhB protein. Based on these results, the *ndhB* gene is non-functional in some orchid species. *ndhF*, which is located in the LSC-IRa junction, was deleted in the Oncidiinae varieties. Notably, the nucleotide deletions in the *ndhF-rps15* region were different. Of the 12 Oncidiinae *ndhD* genes that have been cloned, all of them, including the one from *P. aphrodite*, are truncated. The overall pattern can be classified into two types: a deletion occurring in the 3′-end of *ndhD* (the *Beallara* species) and a deletion in the 5′-end in the rest of those tested clones (Wu *et al.*, 2010).

The sequences of *Onc.* Gower Ramsey, *Onc.* Sweet Sugar, *Odontoglossum* Margarete Holm, and *Odontocidium.* Golden Gate indicated that the translation capacity of *ndhE* in these species was retained. Of the species with modified *ndhE* genes, *Odm.* Violetta von Holm, and *Odcdm.* Wildcat contained frame shifts, *Degarmoara.* Flying High had 30-bp deletions, and there were bigger deletions in the four *Beallara* varieties. There was also a deletion in *ndhE* of *P. aphrodite* (Wu *et al.*, 2010).

Although there were no deletions in the *ndhG* genes of nine varieties, these varieties had three mutated stop codons within *ndhG* , rendering it inactive (Wu *et al.*, 2010). There was a deletion of about 250 bp in the *ndhG* gene of the *Beallara* species, which was shorter than in *P. aphrodite* (Wu *et al.*, 2010). The region encompassing *ndhI* is the most complicated of the chloroplast *ndh* genes. Generally, genes from the same genus had the same pattern (*Oncidium* and *Beallara,* an exception is *Odm.* Violetta von Holm; Wu *et al.*, 2010). In *Onc.* Gower Ramsey, the *ndhI* gene was partially deleted, as in *P. aphrodite.* In the *Beallara* species, *Zelenkocidium* Little Angel and *Odm.* Violetta von Holm, the *ndhI* gene was completely absent. Unlike in *P. aphrodite,* truncated *ndhA–ndhH* genes still existed in most of the Oncidiinae species in this study. Apart from frame shifts in *Oncdm.* Garfield *ndhA* and *Oncdm.* Golden Gate *ndhH,* the other *ndhA–ndhH* genes in five genera all show deletions of varying types (Wu *et al.*, 2010).

ndh gene deletion was also found to occur in an autotrophic species, like *Pinus thunbergii* (Wakasugi *et al.*, 1994). Another Coniferales species, *Keteleeria davidiana,* also contained non-functional *ndh* genes (Wu *et al.*, 2009b). In three gnetophytes, *Welwitschia mirabilis, Ephedra equisetina,* and *Gnetum parvifolium,* all 11 *ndh* genes are non-functional, with 10 being absent and one, *ndhB,* being a pseudogene (McCoy *et al.*, 2008; Wu *et al.*, 2009b). *ndh* gene deletion also occurs in other orchids (Neyland and Urbatsch, 1996). In *P. aphrodite,* the *ndhA, ndhF,* and *ndhH* genes were completely absent, and only remnants of the other eight subunit sequences were found; they are either truncated or frame-shifted, suggesting that they are non-functional (Chang *et al.*, 2006). In Wu *et al.* (2010), we demonstrated that this phenomenon was common in orchids. The deletion pattern differs not only between *Oncidium* and *Phalaenopsis*, but even within the 15 Oncidiinae species analyzed (Wu *et al.*, 2010).

13.3.2. *Chloroplast genome deletion in plants*

The deletion of *ndh* genes or other chloroplast-encoded genes occurs in many plants. In the parasitic plant *Epifagus virginiana*, the plastid genome has lost most of the 30 or more chloroplast genes needed for photosynthesis and most of a large family of plastid genes, the *ndh* genes (dePamphlis and Palmer, 1990). In the parasitic liverwort *Aneura mirabilis*, all *ndh* genes are either absent or pseudogenes. Five of 15 *psb* genes are pseudogenes, as are two of six *psa* genes and two of six *pet* genes. Pseudogenes of *cys*A, *cys*T, *ccs*A, and *ycf*3 have also been detected (Wickett *et al.*, 2008). *Cuscuta exaltata*, another parasitic plant, also lacks all *ndh* genes (Haberhausen and Zetsche, 1994). Interestingly, in *Cuscuta obtusiflora*, not only the *ndh* genes, but also the plastid-encoded RNA polymerase genes are deleted (McNeal *et al.*, 2007). According to these data, different functional gene deletions and chloroplast genome structures have occurred. As a result, the genome sizes of parasitic plants are diverse (Revill *et al.*, 2005; McNeal *et al.*, 2007).

Chloroplast gene deletions also occurs in achlorophyllous orchids. In the leafless mycoheterotrophic orchid *Cyrtosia septentrionalis*, *psaB*, *rbcL*, and *psbC* are pseudogenes (Cameron, 2004; Cameron and Molina, 2006), with the *rbcL* sequence at least 295 bp shorter than in photosynthetic orchids and with *psaB* 12 percent shorter than in other orchids (Cameron, 2004) due to an approximately 160-bp deletion located near the 3′-end of the gene (Cameron and Molina, 2006). Additionally, there were no *psaB*, *rbcL*, *psbB* and *psbC* PCR products obtained from other chlorophyll-lacking orchids (Cameron, 2004; Cameron and Molina, 2006).

From a physiological viewpoint, since parasitic plants obtain organic nutrients from the host plant, they do not need photosynthesis to produce carbohydrates. However, this cannot explain autotrophic plants such as *Oncidium* and *Pinus*, which have the capacity for photosynthesis. Note that this *ndh* deletion does not occur in all gymnosperm species. The *ndh* genes do exist in the chloroplast genomes of *Cryptomeria* and *Cycas* (Hirao *et al.*, 2008; Wu *et al.*, 2007). These autotrophic plants still require *ndh* proteins for photorespiration or photosynthesis. The ancestor of chloroplasts was once a free-living cyanobacterium. During evolution, the cyanobacteria integrated into the heterotrophic cell, retaining portions of their genome (Martin *et al.*, 1998), while other genes of theirs were transferred to the nucleus (Palmer, 1985, Martin *et al.*, 2002). Thus, one possibility is that ancestral plastid *ndh* genes were transferred to the nucleus, remaining functional to this day (Wakasugi *et al.*, 1994; Chang *et al.*, 2006). The presence of *ndh* homologs encoded within the nucleus was confirmed using primers designed to the chloroplast *ndhA*, *ndhF*, and *ndhH* in PCR assays of total DNA. The resulting sequences are in frame and imply that the ancestral functional *ndh* copies of the plastid may have been transferred to the nuclear genome (Chang *et al.*, 2006).

In studies of the gnetophytes, the cpDNA are seen to be the smallest and most compact (119,726 bp to 109,518 bp) of all photosynthetic land plants (McCoy *et al.*, 2008; Wu *et al.*, 2009b). This small size is also true of *Pinus thunbergii* (119,707 bp), another gymnosperm (Wakasugi *et al.*, 1994). However, when compared with other published cpDNAs, *Onc.* Gower Ramsey is of the regular, longer type. The reduction in size of gnetophyte and other plant cpDNA is due to shorter introns and intergenic spaces (Wu *et al.*, 2009b). The deletion of *ndh* genes in *Onc.* Gower Ramsey does not affect the overall cpDNA length.

13.4. Phylogenetic Analysis of 15 *Oncidium* Species

13.4.1. *Phylogenetic analysis of orchids*

Oncidium is very interesting in breeding terms. Because it is easy for inter-species or inter-genus crosses, there are many artificial hybrids in *Oncidium*. These hybrids are not distinct phenotypically. They are partially named for their parental background. First, regardless of whether *Oncidium* is the female or male parent, if crossed with *Odontoglossum* (*Odm.*), the offspring belongs to *Odontocidium* (*Odcdm.*). This means that species in the same genus may have different chloroplast DNA.

Second, differentially named genera may come from the same female parent. For example, *Odm. bictoniense* is the female parent of hybrids from four different genera: *Odcdm.* Golden Gate, *Odontioda* (*Oda.*) Stirbic, *Oda.* Bictaw, and *Odm.* Margarete Holm. This result suggests that species in different genera may have the same chloroplast genome. Third, species in the same genus may be derived from different genetic backgrounds. For example, *Beallara* could come from four kinds of crosses: *Bllra.* × *Miltonia* (*Milt.*), *Miltassia* (*Mtssa.*) × *Oda.*, or *Bllra.* × *Oda.* For these reasons, we cannot identify a female parent by its scientific name.

To further complicate matters, the names of genera and taxonomy changes are frequent in *Oncidiinae*. For example, *Milt. schroederiana* has been transferred to *Oncidium* as *Onc. schroederianum*, and *Colmanara* Wildcat has been changed to *Odcdm.* Wildcat, and *Zelenkocidium* Little Angel changed from *Oncidium*. The varieties are the hybrids of other hybrids. Some of the parental information has been lost. For example, one important parent of *Oncidiinae* breeding is *Bllra.* Tahoma Glacier, which is a hybrid of 56 species, 25 of which are original species, 31 of which are hybrids; some of them are unknown. Clearly, then, phylogenetic analysis of the orchidaceae is a very important issue.

13.4.2. *Multiple sequence combination is a good method for phylogenetic analysis*

At the beginning of our study, there were 11 cpDNA regions selected for phylogenetic analysis, based on the fact that universal primers could be used (Kress *et al.* 2005). In some species, we could not obtain the sequences of *psaB, LSC-IRa, trnF-trnL* and *ndhF* by PCR; these regions were eliminated. In the end, there were eight cpDNA regions that were usable for phylogenetic analysis through PCR [including non-coding regions (*trnH-psbA, trnF-ndhJ*, SSC-IRb) and coding regions (*accD, matK, rbcL, rpoB, rpoC1*)] for 15 Oncidiinae species (Wu *et al.* 2010).

As our results in this report show, it is difficult to conduct a phylogenetic analysis at the species level using a single cpDNA region.

For instance, *rbcL* is highly conserved in Oncidiinae and, as expected, these species cannot be distinguished, since bootstrap scores are lower then 50 percent and are not correlated with the parent relationship (data not shown). For *accD,* only the species belonging to *Beallara* and *Oncidium* can be separated out, as the pattern and relationship of other species is not correlated with the parent relationship. Accordingly, the *rbcL* and *accD* regions were not suitable for phylogenetic study of Oncidiinae at the genus and species level.

Among the eight sequences studied here, the phylogenetic analysis using *matK* was best correlated with the parent relationship. There are at least three advantages of the *matK* region for phylogenetic analysis: (i) this region is variable at the interspecies level (Lahaye *et al.*, 2008); (ii) this region is easy to amplify using published primer sequences (Cuenoud *et al.*, 2002); (iii) there is a rich amount of sequence information on Oncidiinae *matK,* including the number of sequences (695) and the length of the sequences (791 bp). There are 180 sequences related to the species in this study, which were eight genera and 125 species. The phylogenetic analysis of these sequences and 15 varieties were similar to what was expected. The female parent of *Beallara* is *Miltassia,* which makes the grand female parent *Brassia,* the sequences of *Beallara* were highly correlated with other *Brassia* species, and most closely with the female parent *Brassia verrucosa.* In *Odontoglossum,* there are fewer than five sequences in the database for *matK* and other cpDNA regions. The results indicated that the *Odontoglossum matK* were dispersed around the *Oncidium* group. To study the phylogenetic relationship between *Odontoglossum* and *Oncidium,* a greater number of *Odontoglossum matK* genes need to be sequenced.

The *trnH-psbA* region is another divergent cpDNA region useful for phylogenetic analysis. In DNA barcoding research, the 450 bp derived from *trnH-psbA* is the most variable plastid region in angiosperms (Kress *et al.,* 2005) and orchids (Lahaye *et al.,* 2008). This region could easily be amplified across a broad range of land plants. On the other hand, various expansions or contractions of inverted repeats (IRs) in chloroplast genomes lead to diverse IR-LSC junctions (Palmer and Stein DB, 1986, Plunkett and Downie, 2000). The structural changes in cpDNA provide useful phylogenetic inferences (Raubeson and Jansen, 2005) compared the organization of genes flanking two IR-LSC junctions in 123 angiosperms and classified them into three types. *Oncidium*, like other orchids (*Phalaenopsis* and *Dendrobium*), belongs to type III, which contains a *trnH-rps19* gene cluster near the IR-LSC junctions. According to these data, the LSC-IR junction regions are information-rich and could be used for phylogenetic analysis.

However, *Odcdm.* Wildcat and *Dgmra.* Flying High were more closely related to *Beallara* species when the *trnH-psaA* region was used (data not shown). Through comparison with the Oncidiinae *trnH-psaA* sequence database, *Odontoglossum* is still grouped with *Oncidium. Zelenkocidium* Little Angel is closer to *Brassia*, than to *Oncidium.* According to these results, a single region may not contain enough information for inter-species phylogenetic analysis, creating further debate about the current taxonomy of Oncidiinae.

To solve this problem, we used multiple cpDNA regions for phylogenetic analysis. First, eight sequences in each species were combined (the total length is 7,042 bp) as the correct reference (Wu *et al.*, 2010). To manage labor and supply costs, we wanted to identify the smallest regions that would result in the same performance as using all eight regions. As a consequence, we combined the cpDNA diverse regions, *trnH-psbA* (Kress *et al.*, 2005), *matK* and *trnL-ndhJ* (Lahaye *et al.*, 2008), in the phylogenetic analysis. The results indicated that the *trnH-psbA* + *ndhJ* combination was most similar to the results of all eight cpDNA regions (Wu *et al.*, 2010).

During phylogenetic analysis, the correlation is dependent on the length and property of DNA or amino acid information. In Poaceae phylogenetic analysis, there was an unexpected result when using protein-coding regions. Using full cpDNA genome sequences, the results indicated that the relationship between bamboo and rice was closer than that of the other Poaceae species. Because the information on orchid cpDNA is limited, combining of several sequences derived by PCR using universal primers could be a successful strategy, as has been done for *rbcL* + *matK* and *psaB* + *rbcL* in subfamilial clades of Orchidaceae (Freudenstein *et al.*, 2004, Cameron, 2004), *matK* + *trnL-F* + *nrITS* in *Aerides*, *psbB* + *psbC* in *Vanilla* (Cameron and Molina,

2006), and *atpF* intron + *rps16* introns, *trnL* – *trnF* and *trnS* – *trnG* in *Serapias* (Bellusci *et al.*, 2008). In DNA barcoding research, the performance by using combination sequences was also investigated (Lahaye *et al.*, 2008). According to our results, two variable cpDNAs, *trnH* – *psbA* and *trnL* – *ndhJ*, could provide sufficient information for genus-to-species level phylogenetic analysis. Although *matK* is diverse and suitable for DNA barcoding, our results indicated that the relationship of *Odm.* Violetta von Holm was different. More samples and sequences of *matK* will be needed to evaluate its suitability for phylogenetic analysis.

13.5. Conclusions

In this chapter, we used *P. aphrodite* as a prototype for primer design to complete the *Onc.* Gower Ramsey cpDNA. These primers and information will be useful for further orchid cpDNA sequencing. The results show that *ndh* genes were non-functional in Oncidiinae. This phenomenon was not only observed in non-photosynthetic plants, but also in other orchids, as in pine and *Phalaenopsis*. By using a PCR approach, the *ndh* genes in different Oncidiinae plants were identified. The data indicated that *ndh* genes were non-functional in all of the tested plants. The genes and cpDNA regions that would be the most useful for phylogenetic analysis were determined. We suggest using *trnH* – *psbA* + *trnF* – *ndhJ* non-protein coding regions for phylogenetic investigations. After checking these regions in popular commercial varieties, most are conformed to known inheritance information. However, there are some contentions that need further clarification. We can conclude that complete cpDNA information is useful for phylogenetic and evolutionary studies in *Oncidium* breeding and variety identification.

13.6. Future Prospects

There are some questions remaining. The first is the placement of *Odm.* Violetta von Holm, which has the female parent *Odm. bictoniense.* However, regardless of the cpDNA used, *Odm.* Violetta von Holm did not correlate with *Odcdm.* Golden Gate or *Odm.* Margarete Holm, which are both derived from the same female parent. Second is *Dgmra.* Flying High, which has the female parent *Mtssa.* Jetsetter. Theoretically, the cpDNA of *Dgmra.* Flying High will be closely related to the *Beallara* species, which are derived from a *Miltassia* female parent. The data indicated that *Dgmra.* Flying High is more similar to *Odontoglossum*. These data show that the recent taxonomy of Oncidiinae must be confirmed by cpDNA.

Acknowledgments

This work was supported by the Development Program of Industrialization for Agricultural Biotechnology, Taiwan. Support for M. R. Duvall was also obtained from the Plant Molecular Biology Center, Northern Illinois University, USA.

References

Bellusci F, Pellegrino G, Palermo AM, Musacchio A. (2008) Phylogenetic relationships in the orchid genus *Serapias* L. based on noncoding regions of the chloroplast genome. *Mol Phylogenet Evol* **47**: 986–991.

Cameron KM. (2004) Utility of plastid *psaB* gene sequences for investigating intrafamilial relationships within Orchidaceae. *Mol Phylog Evol* **31**: 1157–1180.

Cameron KM, Molina C. (2006) Photosystem II gene sequences of *psbB* and *psbC* clarify the phylogenetic position of *Vanilla* (Vanilloideae, Orchidaceae). *Caldistics* **22**: 239–248.

Chang CC, Lin HC, Lin IP *et al.* (2006) The chloroplast genome of *Phalaenopsis aphrodite* and comparative analysis of evolutionary rate with that of grasses. *Mol Biol Evol* **23**: 279–291.

Chase MW, Fay MF, Devey DS *et al.* (2006) Multigene analyses of monocot relationships: A summary. *Aliso* **22**: 63–75.

Chase W, Palmer JD. (1989) Chloroplast DNA systematics of lilioid monocots: Resources, feasibility, and an example from the Orchidaceae. *Am J Bot* **76**: 1720–1730.

Cuenoud P, Savolainen V, Chatrou LW *et al.* (2002) Molecular phylogenetics of Caryophyllales based on nuclear 18S rDNA and plastid *rbcL*, *atpB*, and *matK* DNA sequences. *Am J Bot* **89**: 132–144.

dePamphlis CW, Palmer JD. (1990) Loss of photosynthetic and chlororespiratory genes from the plastid genome of a parasitic flowering plant. *Nature* **348**: 337–339.

Dhingra A, Folta KM. (2005) ASAP: Amplification, sequencing, and annotation of plastomes. *BMC Genomics* **6**: 176.

Freudenstein JV, van den Berg C, Goldman DH *et al.* (2004) An expanded plastid DNA phylogeny of Orchidaceae and analysis of Jackknife branch support strategy. *Am J Bot* **91**: 149–157.

Haberhausen G, Zetsche K. (1994) Functional loss of *ndh* genes in an otherwise relatively unaltered plastid genome of the holoparasitic flowering plant *Cuscuta reflexa. Plant Mol Biol* **24**: 217–222.

Hirao T, Watanabe A, Kurita M *et al.* (2008) Complete nucleotide sequence of the *Cryptomeria japonica* D Don chloroplast genome and comparative chloroplast genomics: Diversified genomic structure of coniferous species. *BMC Plant Biol* **8**: 70.

Kresss WJ, Wurdack KJ, Zimmer EA *et al.* (2005) Use of DNA barcodes to identify flowering plants. *Proc Natl Acad Sci USA* **102**: 8369–8374.

Lahaye R, van der Bank M, Bogarin B *et al.* (2008) DNA barcoding the floras of biodiversity hotspots. *Proc Natl Acad Sci USA* **105**: 2923–2928.

Martin W, Stoebe B, Goremykin V *et al.* (1998) Gene transfer to the nucleus and the evolution of chloroplasts. *Nature* **393**: 162–165.

Martin W, Rujan T, Richly E *et al.* (2002) Evolutionary analysis of *Arabidopsis*, cyanobacterial, and chloroplast genomes reveals plastid phylogeny and thousands of cyanobacterial genes in the nucleus. *Proc Natl Acad Sci USA* **99**: 12246–12251.

McCoy SR, Kuehl JV, Boore JL, Raubeson LA. (2008) The complete plastid genome sequence of *Welwitschia mirabilis*: An unusually compact plastome with accelerated divergence rates. *BMC Evol Biol* **8**: 130.

McNeal JR, Kuehl JV, Boore JL, dePamphilis CW. (2007) Complete plastid genome sequences suggest strong selection for retention of photosynthetic genes in the parasitic plant genus *Cuscuta. BMC Plant Biol* **7**: 57.

Neyland R, Urbatsch LE. (1996) The *ndhF* chloroplast gene detected in all vascular plant divisions. *Planta* **200**: 273–277.

Palmer JD. (1985) Comparative organization of chloroplast genomes. *Annu Rev Genet* **19**: 325–354.

Palmer JD, Stein DB. (1986) Conservation of chloroplast genome structure among vascular plants. *Curr Genet* **10**: 823–833.

Plunkett GM, Downie SR. (2000) Expansion and contraction of the chloroplast inverted repeat in Apiaceae subfamily Apioideae. *Syst Bot* **25**: 648–667.

Raubeson LA, Jansen RK. (2005) Chloroplast genomes of plants. In: Henry RJ (ed.), Plant Diversity and Evolution: Genotypic and Phenotypic Variation in Higher Plants, pp. 45–68. CABI Publishing, Wallingford.

Revill MJ, Staney S, Hibberd JM. (2005) Plastid genome structure and loss of photosynthetic ability in the parasitic genus *Cuscuta. J Exp Bot* **56**: 2477–2486.

Sazanov LA, Burrows PA, Nixon PJ. (1998) The plastid *ndh* genes code for an NADH-specific dehydrogenase: Isolation of a complex I analogue from pea thylakoid membranes. *Proc Natl Acad Sci USA* **95**: 1319–1324.

Tsai CC, Huang SC, Huang PL *et al.* (2002) Phenetic relationship and identification of subtribe Oncidiinae genotypes by random amplified polymorphic DNA (RAPD) markers. *Sci Hort* **101**: 315–325.

Wakasugi T, Tsudzuki J, Ito S *et al.* (1994) Loss of all *ndh* genes as determined by sequencing the entire chloroplast genome of the black pin *Pinus thunbergii. Proc Natl Acad Sci USA* **91**: 9794–9798.

Webster, P. (1992) Oncidium subtribe. In: *The Orchid Genus Book.* Hunters Breeze, USA.

Wickett NJ, Zhang Y, Hansen SK *et al.* (2008) Functional gene losses occur with minimal size reduction in the plastid genome of the parasitic liverwort *Aneura mirabilis. Mol Biol Evol* **25**: 393–401.

Wu CS, Wang YN, Liu SM, Chaw S-M. (2007) Chloroplast genome (cpDNA) of *Cycas taitungensis* and 56 cp protein-coding genes of *Gnetum parvifolium*: Insights into cpDNA evolution and phylogeny of extant seed plants. *Mol Biol Evol* **24**: 1366–1379.

Wu FH, Kan DP, Lee SB *et al.* (2009a) Complete nucleotide sequence of *Dendrocalamus latiflorus* and *Bambusa oldhamii* chloroplast genomes. *Tree Physiol* **29**: 847–856.

Wu CS, Lai YT, Lin CP *et al.* (2009b) Evolution of reduced and compact chloroplast genomes (cpDNAs) in gnetophytes: Selection toward a lower-cost strategy. *Mol Phylog Evol* **52**: 115–124.

Wu FH, Chan MT, Liao DC *et al.* (2010) Complete chloroplast genome of *Oncidium* Gower Ramsey and evaluation of molecular markers for identification and breeding in Oncidiinae. *BMC Plant Biol* **10**: 68.

CHAPTER 14

Analysis of Chloroplast RNA Editing Sites in *Phalaenopsis aphrodite*

Ching-Chun Chang*,‡, Sin-Chung Liao† and Wun-Hong Zeng*

RNA editing sites were systematically examined for the transcripts of 74 known protein-coding genes in the chloroplasts of *P. aphrodite*. A total of 44 editing sites were identified in 24 transcripts, the highest number reported in seed plants to date. In addition, 21 editing sites were unique to *Phalaenopsis* as compared to other seed plants. All editing was C-to-U conversion, and 42 editing sites caused the change in amino acids. One of the remaining two editing sites occurred in transcripts of the *ndhB* pseudogene, and another in the 5′ untranslated region of *psbH* transcripts. However, RNA editing did not restore the continuous open reading frame in the frameshifted *ndh* genes, further confirming that they are pseudogenes.

14.1. Introduction

14.1.1. *RNA editing and regulation of gene function*

Chloroplast RNA editing, generally manifested as C-to-U conversion, but with U-to-C conversion also reported in a few cases, is one of the post-transcriptional regulation mechanisms of gene expression in land plants. Plastid RNA editing was first documented in maize *rpl2* transcripts for the creation of the initiation codon by ACG-to-AUG conversion, and since then instances of the generation and removal of the

*Institute of Biotechnology, National Cheng Kung University, Tainan 701, Taiwan.
†Department of Biological Science and Technology, Meiho Institute of Technology, Pingtung 912, Taiwan.
‡Corresponding author. E-mail: chingcc@mail.ncku.edu.tw

translational stop codon have also been identified (Bock, 2000; Fiebig *et al.*, 2004). However, editing sites generally reside at the internal positions in transcripts and most frequently affect the second codon positions. Sometimes they alter the first or third codon positions (Bock, 2000). Typically, the codon changes resulting from RNA editing restore the identity of conserved amino acids in plant phylogeny, suggesting that the conserved residues are important for the function of the protein. Previously, by assaying the activity of acetyl-CoA carboxylase expressed from unedited or edited *accD* mRNA in *E. coli*, it was suggested that editing is necessary for a functional enzyme (Sasaki *et al.*, 2001). In addition, transplastomic approaches using tobacco or *Chlamydomonas reinhardtii* have also demonstrated that the lack of specific RNA editing in *psbF* or *petB* transcripts, respectively, leads to a severe mutant phenotype (Bock *et al.*, 1994; Zito *et al.*, 1997). However, the editing frequency of chloroplast transcripts is dramatically affected by tissue types, developmental stages and environmental factors (Bock, 2000).

14.1.2. *Cis-elements surrounding the editing sites*

Chloroplast RNA editing has been systematically investigated for the protein-coding transcripts in the following plant species: dicot plants *Nicotiana tobacum* (Hirose *et al.*, 1999; Sasaki *et al.*, 2003, 2006), *Arabidopsis thaliana* (Lutz and Maliga, 2001; Tillich *et al.*, 2005), *Atropa belladonna* (Schmitz-Linneweber *et al.*, 2002), *Pisum sativum* (Inada *et al.*, 2004), *Solanum lycopersicum* (Kahlau *et al.*, 2006); monocot plants *Zea mays* (Maier *et al.*, 1995), *Oryza sativa* (Corneille *et al.*, 2000), *Saccharum officinarum* (Calsa Junior *et al.*, 2004); *Phalaenopsis aphrodite* (Zeng *et al.*, 2007); gymnosperm *Pinus thunbergii* (Wakasugi *et al.*, 1996); fern *Adiantum capillus-veneris* (Wolf *et al.*, 2004); and hornwort *Anthoceros formosae* (Kugita *et al.*, 2003). Genome-wide analysis of chloroplast transcripts has revealed that plants have undergone dramatic changes in both the levels and patterns of editing, from hornworts (1.2 percent conversions of all nucleotides examined) and ferns (0.38 percent) to seed plants (less than 0.05 percent) (Kugita *et al.*, 2003; Wolf *et al.*, 2004). In seed plants, a relatively constant number of editing sites, 21 to 44, have been identified in plastids (Tsudzuki *et al.*, 2001; Sasaki *et al.*, 2003, 2006; Kahlau *et al.*, 2006; Zeng *et al.*, 2007). On comparing editing sites among dicots (tobacco) and monocots (maize and rice), 12 common editing sites between tobacco and monocot plants, and 20 common sites between the two monocots were identified (Tsudzuki *et al.*, 2001). However, when looking at more closely related taxa, the number of shared sites increases. For instance, maize, rice, and sugarcane in the Poaceae family share at least 23 editing sites (Calsa Junior *et al.*, 2004), while tobacco, tomato, and *Atropa* in the Solanaceae family share 30 (Schmitz-Linneweber *et al.*, 2002; Kahlau *et al.*, 2006). Furthermore, 31 conserved sites out of a total of 35 editing sites were identified among three species of tobacco, *N. tobacum*, *N. sylvestris*, and *N. tomentosiformis* (Sasaki *et al.*, 2003). Moreover, at a subspecies level, three different ecotypes of *Arabidopsis thaliana* have all 28 editing sites in common, though the consequences of RNA editing differ at one position between the ecotypes (Tillich *et al.*, 2005). These studies suggest that determining the distribution and pattern of editing sites across taxa and across the entire chloroplast genome is an important step in investigating the evolutionary process of RNA editing in angiosperms. However, these plants still represent a poor sample of major clades in the phylogeny of seed plants.

Although plastids of seed plants have a relatively constant number of editing sites (21 to 44), apparently conserved *cis*-elements surrounding plastid editing sites are scarce. The regions respectively about 30 and 10 nucleotides immediately upstream and downstream of the editing sites have been mapped for selection of the correct editing target and for editing efficiency by transplastomic and *in vitro* approaches (Bock, 2000; Chateigner-Boutin and Hanson, 2003).

14.1.3. *Protein factors involved in RNA editing*

The *cis*-elements are recognized by nuclear-encoded *trans*-acting factors that are believed to be either site-specific or bind to small clusters of related sites (Hirose and Sugiura, 2001; Chateigner-Boutin and Hanson, 2003; Miyamoto *et al.*, 2004). A biochemical approach to

UV cross-linking RNA template with chloroplast extracts *in vitro* has identified several proteins, e.g., 25-, 56- and 70-kD proteins, which are potentially involved in site-specific RNA editing and can bind to specific *cis*-elements (Hirose and Sugiura, 2001; Miyamoto *et al.*, 2002). Using a genetic approach focusing on NADH dehydrogenase activity, two pentatricopeptide repeat proteins, (PPR)-CRR4 and CRR21, acting as a required site-specific recognition factor for editing at the translational initiation codon and a second site in *ndhD* transcripts, respectively, were identified in *Arabidopsis* (Okuda *et al.*, 2006, 2007). Furthermore, using a reverse genetic approach to screen T-DNA insertion mutants, six PPR proteins that account for nine editing sites in the chloroplasts of *Arabidopsis* were identified (Hammani *et al.*, 2009). The PPR family, characterized by tandem arrays of the 35-amino-acid motif, with more than 450 members in *Arabidopsis*, is involved in post-transcriptional processes such as RNA splicing, stability, editing and translation in plastid and mitochondria (Schmitz-Linneweber and Small, 2008). Those PPR proteins involved in plastid RNA editing belong to the plant specific E and DYW subgroups in the PLS subfamily (Okuda *et al.*, 2007; Hammani *et al.*, 2009).

The Orchidaceae, with approximately 30,000 species, is one of the largest families in flowering plants, and *Phalaenopsis aphrodite* subsp.*formosana* is the first species in which the chloroplast genome has been completely determined (Chang *et al.*, 2006). Previously, the ACG rather than an ATG codon at the translation initiation sites was observed in the *rpl2* and *ndhD* genes of *P. aphrodite* (Chang *et al.*, 2006). To confirm the presence of the RNA editing system, we study the RNA editing patterns for the plastid transcripts of all known protein-coding genes in *P. aphrodite* and compare them with those of other seed plants.

14.2. Determination of Chloroplast RNA Editing Sites in *P. aphrodite*

14.2.1. *RNA extraction and RT-PCR*

Leaves of *P. aphrodite* subsp.*formosana* were obtained from seedlings at the two-leaf stage of development, from which total cellular RNA was isolated according to the method described (Gehrig *et al.*, 2000). The RNA samples were then treated with DNase I (Promega, USA) for 30 min at 37°C to eliminate DNA contamination. To demonstrate the absence of DNA in the RNA preparation after DNase I digestion, the RNA quality was further checked by polymerase chain reaction (PCR) with at least three pairs of primers from plastid genes of the *Phalaenopsis* orchid. Primers were designed based on 74 known protein-coding genes encoded by the chloroplast genome of *P. aphrodite* (AY916449) using Vector NTI Suite software (InforMax, USA). Reverse transcriptase (Promega, USA) was used to synthesize cDNA from total RNA using a reverse primer for each gene at 42°C for 1 h. We multiplexed primers for up to 11 genes within a single reverse transcription (RT) reaction, and at least two independent RT reactions were performed. PCR was then applied to amplify cDNA with a primer pair of both forward and reverse primers for each gene separately. Genomic DNA was used as a template in a positive control to ensure the primer pairs were effective for PCR. The PCR reaction contained a final concentration of 200 nM gene-specific primers, 200 nM of each dNTP, three units of *Taq* DNA polymerase, and 5 μl of 10X*Taq* DNA polymerase buffer in a 50-μl reaction mixture. Amplification started with one 2-min cycle at 94°C, followed by 35–40 cycles of 1.5 min at 94°C, 2 min at 60°C, and 3 min at 72°C, and this was followed by one 5-min cycle at 72°C. Each PCR sample was electrophoresed on a 0.8 percent agarose gel and visualized by staining with ethidium bromide. PCR products were sliced from the gel and purified with a gel extraction kit (Viogene, Taiwan).

14.2.2. *Determination of RNA editing sites*

The cDNA fragments were either directly sequenced or first cloned into pGEM-T vector (Promega, USA), and then the propagated plasmid DNA was sequenced. The sequencing reaction was performed using the BigDye terminator cycle sequencing kit (Applied Biosystems, USA), according to the protocol recommended by the manufacturer. The DNA sequencer was an Applied Biosystems ABI 3700. To determine RNA editing sites, the cDNA sequences were then aligned to that of genomic DNA.

14.3. Chloroplast RNA Editing Patterns in *P. aphrodite*

14.3.1. *Overall properties of chloroplast RNA editing in P. aphrodite*

The chloroplast genome of *P. aphrodite* subsp.*formosana* contains 110 genes including 34 RNA genes, 74 known protein-coding genes and 2 conserved reading frames with unknown function (Chang *et al.* 2006). To extensively study the pattern of RNA editing, we sequenced 60,651 bp of cDNA representing the 74 known chloroplast protein-coding transcripts of *P. aphrodite* (Zeng *et al.*, 2007). A total of 44 editing sites were identified in the 24 transcripts of *P. aphrodite* chloroplast genes, which represented an average of 0.07 percent of the nucleotides examined (Table 14.1). This is the highest number of RNA editing sites reported in seed plants to date. All the RNA editing sites were of the C-to-U conversion type. Of the 42 sites that involved codons, four (9.5 percent) were in the first position, and 38 (90.5 percent) were in the second position; all resulted in the substitution of one amino acid for another (Table 14.1). This result is consistent with previous reports regarding the patterns of RNA editing across widely divergent taxa, which show a bias in favor of second codon position edits (Bock, 2000). The consequence of RNA editing in codon position for *Phalaenopsis* is to mostly restore the conservation of amino acids with other seed plants. The most frequently edited codon was Ser converted to Leu, followed by Ser to Phe, and Pro to Leu (Table 14.1). One of the remaining two editing sites occurred in transcripts of the *ndhB* pseudogene, and another in the 5′ untranslated region of *psbH* transcripts. Among 44 sites, seven partially edited sites were detected in the transcripts of *atpA* (site 2), *clpP* (site 2), *ndhB, psbF, rpoA* (sites 1, 3) and *rps8* genes in the *Phalaenopsis* orchid (Table 14.1). This is not surprising, since RNA editing efficiency has been reported to vary in different organs, developmental stages, and environmental conditions (Ruf and Kossel, 1997; Chateigner-Boutin and Hanson, 2003). In addition, there are no apparently conserved *cis*-elements surrounding the 44 editing sites in *Phalaenopsis*; however, a U_A context bias immediately before and after the editing site were observed (Fig. 14.1).

Table 14.1 RNA Editing Sites in Chloroplast Transcripts of the *Phalaenopsis* Orchid

Gene	Site	Nucleotide Position	Codon Position	Edited Codon	Amino Acid Change
accD	1^{φ}	1184	395	uCa	S → L
	2^{φ}	1412	471	cCa	P → L
	3^{φ}	1430	477	cCu	P → L
atpA	1	773	258	uCa	S → L
	2^{*}	1148	383	uCa	S → L
atpB	1^{φ}	1184	395	uCa	S → L
atpF	1	92	31	cCa	P → L
atpI	1^{φ}	428	143	cCu	P → L
	2^{φ}	629	210	uCa	S → L
clpP	1^{φ}	82	28	Cau	H → Y
	2^{*}	559	187	Cau	H → Y
matK	1^{φ}	533	178	uCu	S → F
	2^{φ}	718	240	Cau	H → Y
	3	1066	356	Cac	H → Y
ndhB$^{\#}$	1^{*}	1977			C → U
petB	1	611	204	cCa	P → L
petL	1	5	2	cCu	P → L
psaI	1^{φ}	80	27	uCu	S → F
psbF	1^{*}	77	26	uCu	S → F
psbH	1^{φ}	−30			C → U
rpl2	1	2	1	aCg	T → M
rpl23	1^{φ}	71	24	uCu	S → F
rpoA	$1^{*\varphi}$	200	67	uCu	S → F
	2^{φ}	368	123	uCa	S → L
	3^{*}	830	277	uCa	S → L
rpoB	1^{φ}	92	31	uCc	S → F
	2	401	134	uCu	S → F
	3	536	179	uCg	S → L
	4	614	205	uCa	S → L
	5	629	210	uCg	S → L
	6	686	229	cCg	P → L
	7	2489	830	uCa	S → L
rpoC1	1	62	21	uCa	S → L
	2^{φ}	203	68	uCu	S → F
	3^{φ}	509	170	uCa	S → L
	4^{φ}	638	213	uCg	S → L
rpoC2	1^{φ}	2846	949	uCu	S → F
rps2	1	134	45	aCa	T → I

(Continued)

14.3.3. *RNA editing in transcripts encoding for photosynthetic apparatuses*

In *P. aphrodite,* concerning the genes encoding subunits in the complexes (or assembly the complexes) of the photosynthetic electron transport chain, five of them, *psaI, psbF, psbH, petB* and *petL,* have transcripts with one RNA editing site, and *ycf3* transcripts have three sites. Three editing sites were unique to the *Phalaenopsis* orchid. The first was at the Ser^{27} (UCU) codon in *psaI* transcripts, and the second was at the Pro^{64} (CCA) codon in the *ycf3* transcripts. They changed to Phe^{27} (UUU) and Leu^{64} (CUA) in their respective transcripts, leading to amino acid conservation in plants. The third unique editing site resided in the −30 nucleotide (C) position in the upstream untranslated region (5′ UTR) of *psbH* transcripts, and this nucleotide position is located at the variable region when the 5′ UTR nucleotide sequences of *psbH* transcripts from eight species of seed plants were aligned (Fig. 14.2a). Previously, a C-to-U conversion identified at nucleotide position −10 of the *ndhG* 5′ UTR in monocot plants was predicted to modify the RNA secondary (stem/loop) structure (Drescher *et al.,* 2002). To find out if C-to-U conversion in the 5′ UTR of *psbH* transcripts in *Phalaenopsis* also affected the RNA secondary structure, the sequence extending 32 nt to each side of the editing site (−62 nt to +3 nt) was examined in its edited and unedited form with the RNA shapes prediction program (Giegerich *et al.,* 2004). The result suggests that the edited RNA sequence can form an energetically less stable secondary structure (Fig. 14.2b). Therefore, it is possible that editing in the *psbH* 5′ UTR might influence *psbH* expression. C-to-U conversions at the Ser^{26} (UCU) to Phe^{26} (UUU) and Pro^{204} (CCA) to Leu^{204} (CUA) codons were identified in the *psbF* and *petB* transcripts of *Phalaenopsis,* respectively. The functional importance of the corresponding editing events has been previously demonstrated in transplastomic tobacco and *Chlamydomonas reinhardtii,* respectively, in which the lack of RNA editing causes a severe mutant phenotype (Bock, 2000). Concerning the six plastid genes involved in the ATP synthase complex, four of them, *atpA, atpB, atpF* and *atpI,* have transcripts that are edited at six sites in *Phalaenopsis.* Three sites are unique to the orchid. One occurs at the Ser^{395}

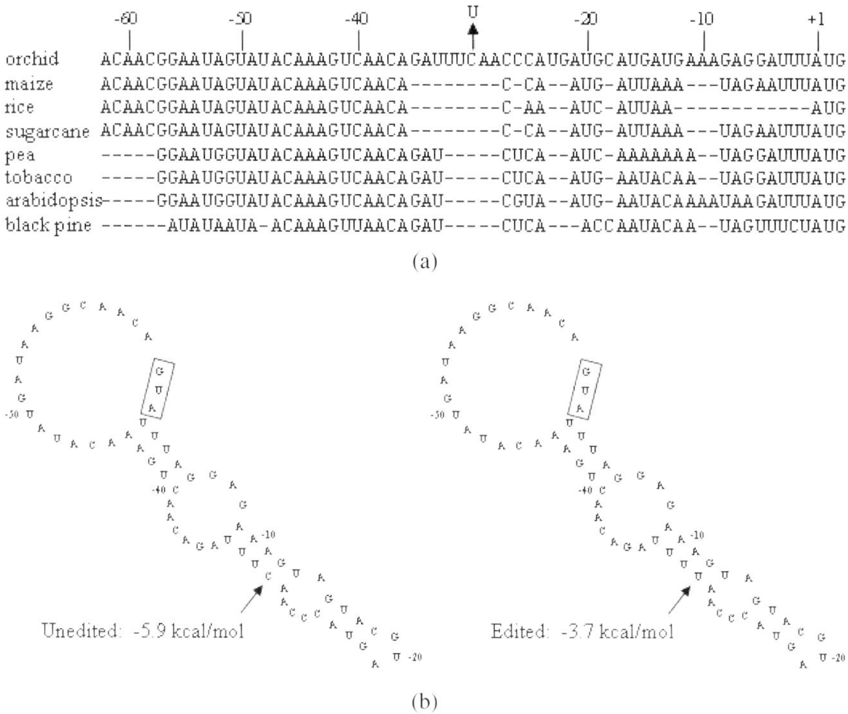

Fig. 14.2. RNA editing site of *psbH* transcripts in the chloroplasts of *P. aphrodite*. (a) The nucleotide sequences in the 5′ untranslated region of *psbH* transcripts from eight species of seed plants were aligned by GeneDoc. Position from translational start site A (+1) is indicated. RNA editing at position −30 is converted from C to U in the orchid as indicated by the arrow. Orchid: *P. aphrodite*; maize: *Zea mays*; rice: *Oryza sativa*; sugarcane: *Saccharum officinarum*; pea: *Pisum sativum*; tobacco: *Nicotiana tabacum*; arabidopsis: *Arabidopsis thaliana*; black pine: *Pinus thunbergii*. (b) Predicted RNA secondary structures formed by the unedited and edited *psbH* 5′UTR using the RNAshapes algorithm. The edited and unedited nucleotides are indicated by arrows. The *psbH* translational start codon is boxed.

(UCA) codon in *atpB* transcripts, while the other two occur at the Pro^{143} (CCU) and Ser^{210} (UCA) codons in *atpI* transcripts, and they convert to Leu^{395} (UUA), Leu^{143} (CUU), and Leu^{210} (UUA), respectively. Conversion of C to U at those sites leads to the restoration of codon conservation among seed plants (Zeng *et al.*, 2007).

14.3.4. *RNA editing in transcripts encoding for miscellaneous proteins*

Conversion of Ser^{267} or Pro^{267} to Leu^{267} by RNA editing was identified in *accD* transcripts of pea, soybean, canola, *Arabidopsis* and black pine, and resulted in amino acid conservation among seed plants (Sasaki *et al.*, 2001; Zeng *et al.*, 2007). RNA editing at this codon position was previously demonstrated to be required for functional acetyl-CoA carboxylase *in vitro* (Sasaki *et al.*, 2001). However, in the *Phalaenopsis* orchid, the corresponding codon is Phe^{266}, and RNA editing is not apparent (Zeng *et al.*, 2007). Considering the similarities in the hydrophobic properties of Phe and Leu, the *accD* gene is probably still functional in *Phalaenopsis* chloroplasts. On the other hand, three editing sites, which converted Ser^{395} (UCA) to Leu^{395} (UUA), Pro^{471} (CCA) to Leu^{471} (CUA), and Pro^{477} (CCU) to Leu^{477} (CUU), were uniquely observed in the *accD* transcripts of *Phalaenopsis*. Furthermore, three other unique editing sites were present in *clpP* and *matK* transcripts of *Phalaenopsis*. The former occurred at His^{28} (CAU) with conversion to Tyr^{28} (UAU), and the latter occurred at Ser^{178} (UCU) with conversion to Phe^{178} (UUU), and His^{240} (CAU) changed to Tyr^{240} (UAU). All of the above editing events tended to restore the codon conservation in seed plants (Zeng *et al.*, 2007).

Although *Ndh* genes encoding the subunits of the NADH dehydrogenase complex are involved in the cyclic electron flow of photosystem I and chlororespiration in tobacco, they are not essential for a plant's growth under normal conditions (Burrows *et al.*, 1998). All 11 subunits of *ndh* genes are present in the chloroplast genomes of photosynthetic vascular plants so far sequenced, with the exception of black pine, *Phalaenopsis* and *Oncidium* (Wakasugi *et al.*, 1994; Chang *et al.*, 2006; Wu *et al.*, 2010). RNA editing occurs frequently in the *ndh* transcripts, which account for more than 40 percent of the editing sites in the chloroplast transcripts of higher plants, except in *Phalaenopsis* and black pine (Zeng *et al.*, 2007). In particular, the *ndhB* transcripts of barley, tobacco, and *Arabidopsis* differ from the corresponding genomic sequence at nine sites, the highest number of editing events for a single chloroplast mRNA reported to date

(Tsudzuki *et al.*, 2001; Tillich *et al.*, 2005; Kahlau *et al.*, 2006; Zeng *et al.*, 2007). The "relative neutrality hypothesis" explains well why editing sites evolve more readily in those genes in which a transitory loss of function can be tolerated (Fiebig *et al.*, 2004). In *Phalaenopsis*, the *ndhA*, *ndhF* and *ndhH* genes are completely absent from the chloroplast genome. The other eight *ndh* genes (*ndhB, ndhC, ndhJ, ndhD, ndhE, ndhG, ndhI* and *ndhK*) have various degrees of nucleotide insertion/deletion as compared to tobacco, and they are all frame-shifted (Chang *et al.*, 2006). In addition, the plastid *ndhD* genes of *Phalaenopsis* have an ACG rather than an ATG codon at their translation initiation sites (Chang *et al.*, 2006). Previously, C-to-U conversion at the initiation codon of *ndhD* transcripts were reported in dicot plants (Tsudzuki *et al.*, 2001). Therefore, RNA editing was hypothesized to be required in repairing internal stop codons and/or initiation sites and thus restore the normal function of *ndh* genes in *Phalaenopsis*. However, from RT-PCR assays, no apparent RNA editing sites for *ndhC, ndhD, ndhE, ndhG, ndhI, ndhJ* or *ndhK* transcripts were detected. Only one partial editing site, corresponding to the maize codon 494, was identified in *ndhB* transcripts of the *Phalaenopsis* orchid, but it did not repair the internal stop codon of frame-shifted *ndhB* transcripts. Therefore, all the *ndh* genes are pseudogenes in the plastid genome of *Phalaenopsis*. The single remaining partial editing site is likely an evolutionary remnant from before the complete loss of plastid RNA editing sites for non-functional *ndh* genes.

14.4. Future Prospects

Determining the distribution and pattern of editing sites across taxa and across the entire chloroplast genome is an important step in investigating the evolutionary process of RNA editing in angiosperms. With the available information on the RNA editing patterns in *P. aphrodite*, the next step is to identify the protein factors as well as the editing enzymes involved in the RNA editing process in the chloroplasts of *P. aphrodite*. RNA interference (RNAi) will be a powerful tool for knocking out/down the potential PPR proteins involved in site-specific recognition of RNA editing in *P. aphrodite*.

Acknowledgments

This work was financially supported in part by grants from the National Science Council of Taiwan.

References

Bock R, Kossel H, Maliga P. (1994) Introduction of a heterologous editing site into the tobacco plastid genome the lack of RNA editing leads to a mutant phenotype *EMBOJ* **13**:4623–4628.

Bock R. (2000) Sense from nonsense: How the genetic information of chloroplasts is altered by RNA editing. *Biochimie* **82**(6–7): 549–557.

Burrows PA, Sazanov LA, Svab Z *et al.* (1998) Identification of a functional respiratory complex in chloroplasts through analysis of tobacco mutants containing disrupted plastid *ndh* genes. *EMBO J* **17**: 868–876.

Calsa Junior T, Carraro DM, Benatti MR *et al.* (2004) Structural features and transcript-editing analysis of sugarcane (*Saccharum officinarum* L.) chloroplast genome. *Curr Genet* **46**: 366–373.

Chang CC, Lin HC, Lin IP *et al.* (2006) The chloroplast genome of *Phalaenopsis aphrodite* (Orchidaceae): Comparative analysis of evolutionary rate with that of grasses and its phylogenetic implications. *Mol Biol Evol* **23**(2): 279–291.

Chateigner-Boutin AL, Hanson MR. (2003) Developmental co-variation of RNA editing extent of plastid editing sites exhibiting similar *cis*-elements. *Nucleic Acids Res* **31**(10): 2586–2594.

Corneille S, Lutz K, Maliga P. (2000) Conservation of RNA editing between rice and maize plastids: Are most editing events dispensable? *Mol Gen Genet* **264**(4): 419–424.

Drescher A, Hupfer H, Nickel C *et al.* (2002) C-to-U conversion in the intercistronic *ndhI/ndhG* RNA of plastids from monocot plants: Conventional editing in an unconventional small reading frame? *Mol Genet Genom* **267**(2): 262–269.

Fiebig A, Stegemann S, Bock R. (2004) Rapid evolution of RNA editing sites in a small non-essential plastid gene. *Nucleic Acids Res* **32**(12): 3615–3622.

Gehrig HH, Winter K, Cushman J *et al.* (2000) An improved RNA isolation method for succulent plant species rich in polyphenols and polysaccharides. *Plant Mol Biol Rep* **18**: 369–376.

Giegerich R, Voss B, Rehmsmeier M. (2004) Abstract shapes of RNA. *Nucleic Acids Res* **32**(16): 4843–4851.

Hammani K, Okuda K, Tanz SK *et al.* (2009) A study of new *Arabidopsis* chloroplast RNA editing mutants reveals general features of editing factors and their target sites. *Plant Cell* **21**(11): 3686–3699.

Hirose T, Kusumegi T, Tsudzuki T, Sugiura M. (1999) RNA editing sites in tobacco chloroplast transcripts: Editing as a possible regulator of chloroplast RNA polymerase activity. *Mol Gen Genet* **262**(3): 46–27.

Hirose T, Sugiura M. (2001) Involvement of a site-specific trans-acting factor and a common RNA-binding protein in the editing of chloroplast mRNAs: Development of a chloroplast *in vitro* RNA editing system. *EMBO J* **20**(5): 1144–1152.

Inada M, Sasaki T, Yukawa M *et al.* (2004) A systematic search for RNA editing sites in pea chloroplasts: An editing event causes diversification from the evolutionarily conserved amino acid sequence. *Plant Cell Physiol* **45**(11): 1615–1622.

Kahlau S, Aspinall S, Gray JC, Bock R. (2006) Sequence of the tomato chloroplast DNA and evolutionary comparison of solanaceous plastid genomes. *J Mol Evol* **63**(2): 194–207.

Kugita M, Yamamoto Y, Fujikawa T *et al.* (2003) RNA editing in hornwort chloroplasts makes more than half the genes functional. *Nucleic Acids Res* **31**: 2417–2423.

Lutz KA, Maliga P. (2001) Lack of conservation of editing sites in mRNAs that encode subunits of the NAD(P)H dehydrogenase complex in plastids and mitochondria of *Arabidopsis thaliana. Curr Genet* **40**(3): 241–249.

Maier RM, Neckermann K, Igloi GL, Kossel H. (1995) Complete sequence of the maize chloroplast genome: Gene content, hotspots of divergence and fine tuning of genetic information by transcript editing. *J Mol Biol* **251**: 614–628.

Miyamoto T, Obokata J, Sugiura M. (2002) Recognition of RNA editing sites is directed by unique proteins in chloroplasts: biochemical identification of cis-acting elements and trans-acting factors involved in RNA editing in tobacco and pea chloroplasts. *Mol Cell Biol* **22**(19):6726–6734.

Miyamoto T, Obokata J, Sugiura M. (2004) A site-specific factor interacts directly with its cognate RNA editing site in chloroplast transcripts. *Proc Natl Acad Sci USA* **101**(1): 48–52.

Okuda K, Nakamura T, Sugita M *et al.* (2006) A pentatricopeptide repeat protein is a site-recognition factor in chloroplast RNA editing. *J Biol Chem* **281**(49): 37661–37667.

Okuda K, Myouga F, Motohashi R *et al.* (2007) Conserved domain structure of pentatricopeptide repeat proteins involved in chloroplast RNA editing. *Proc Natl Acad Sci USA* **104**(19): 8178–8183.

Ruf S, Kossel H. (1997) Tissue-specific and differential editing of the two *ycf3* editing sites in maize plastids. *Curr Genet* **32**(1): 19–23.

Sasaki Y, Kozaki A, Ohmori A *et al.* (2001) Chloroplast RNA editing required for functional acetyl-CoA carboxylase in plants. *J Biol Chem* **276**(6): 3937–3940.

Sasaki T, Yukawa Y, Miyamoto T *et al.* (2003) Identification of RNA editing sites in chloroplast transcripts from the maternal and paternal progenitors of tobacco (*Nicotiana tabacum*): Comparative analysis shows the involvement of distinct trans-factors for *ndhB* editing. *Mol Biol Evol* **20**(7): 1028–1035.

Sasaki T, Yukawa Y, Wakasugi T *et al.* (2006) A simple *in vitro* RNA editing assay for chloroplast transcripts using fluorescent dideoxynucleotides: Distinct types of sequence elements required for editing of *ndh* transcripts. *Plant J* **47**: 802–810.

Schmitz-Linneweber C, Regel R, Du TG *et al.* (2002) The plastid chromosome of *Atropa belladonna* and its comparison with that of *Nicotiana tabacum*: The role of RNA editing in generating divergence in the process of plant speciation. *Mol Biol Evol* **19**: 1602–1612.

Schmitz-Linneweber C, Small I. (2008) Pentatricopeptide repeat proteins: A socket set for organelle gene expression. *Trends Plant Sci* **13**(12): 663–670.

Tillich M, Funk HT, Schmitz-Linneweber C *et al.* (2005) Editing of plastid RNA in *Arabidopsis thaliana* ecotypes. *Plant J* **43**(5): 708–715.

Tsudzuki T, Wakasugi T, Sugiura M. (2001) Comparative analysis of RNA editing sites in higher plant chloroplasts. *J Mol Evol* **53**: 327–332.

Wakasugi T, Tsudzuki J, Ito S *et al.* (1994) Loss of all *ndh* genes as determined by sequencing the entire chloroplast genome of the black pine *Pinus thunbergii*. *Proc Natl Acad Sci USA* **91**: 9794–9798.

Wakasugi T, Hirose T, Horihata M *et al.* (1996) Creation of a novel protein-coding region at the RNA level in black pine chloroplasts: The pattern of RNA editing in the gymnosperm chloroplast is different from that in angiosperms. *Proc Natl Acad Sci USA* **93**(16): 8766–8770.

Wolf PG, Rowe CA, Hasebe M. (2004) High levels of RNA editing in a vascular plant chloroplast genome: Analysis of transcripts from the fern *Adiantum capillus-veneris*. *Gene* **339**: 89–97.

Wu FH, Chan MT, Liao DC *et al.* (2010) Complete chloroplast genome of *Oncidium* Gower Ramsey and evaluation of molecular markers for identification and breeding in Oncidiinae. *BMC Plant Biol* **10**(1): 68.

Zeng WH, Liao SC, Chang CC. (2007) Identification of RNA editing sites in chloroplast transcripts of *Phalaenopsis aphrodite* and comparative analysis with those of other seed plants. *Plant Cell Physiol* **48**(2): 362–368.

Zito F, Kuras R, Choquet Y, Kossel H. (1997) Mutations of cytochrome b_6 in *Chlamydomonas reinhardtii* disclose the functional significance for a proline to leucine conversion by *per B* editing in maize and tobacco. *Plant Mol Biol* **33**(1): 79–86.

CHAPTER 15

Molecular Characterization of the Early Response of Orchid *Phalaenopsis amabilis* to *Erwinia Chrysanthemi* Infection

Shih-Feng Fu and Hao-Jen Huang*

Erwinia chrysanthemi is a devastating bacterial pathogen in orchid horticulture and causes soft-rot diseases by secretion of cell wall-degrading enzymes. However, the pathogenesis and resistance response to the pathogen in the orchid remain to be elucidated. In this chapter, the molecular events of the early host–pathogen interaction between *Phalaenopsis* orchids and *E. chrysanthemi* are described on the basis of gene expression profiles. At the early stage of pathogenesis, the defense mechanism, such as an alteration in reactive oxygen species level, was elicited upon infection. For the analysis of the signal transduction pathway triggered by *E. chrysanthemi* infection, a PCR-based strategy using degenerate primers was performed to identify the genes encoding protein kinases and phosphatases. The genes coding for *Phalaenopsis amabilis* calcium-dependent proteinkinase 1 (PaCDPK1) and tyrosine phosphatase (PaPTP1) were isolated from the orchid, and their expression in response to pathogen infection and wounding were characterized. To gain more insight into the mechanisms underlying the compatible interactions, genome-wide transcriptional regulation was investigated by a suppression subtractive hybridization strategy. A total of 170 clones were sequenced from the whole cDNA library to generate 65 nonredundant pathogen-responsive transcripts. Among the genes with assigned function, 37 percent were associated with metabolism, 30 percent with environmental interactions, 13 percent with energy, and 20 percent with others. One of the pathogen-responsive transcripts encoding trans-2-enoyl-CoA reductase was the most abundant representative of plant genes, comprising 15 percent of the total clones. A putative model of the early molecular events in the orchid in response to *E. chrysanthemi* infection was proposed. The chapter provides valuable information towards the understanding the orchid–*E. chrysanthemi* interaction at the molecular level.

*Department of Life Sciences, National Cheng Kung University, No. 1, University Road, Tainan 701, Taiwan. E-mail: haojen@mail.ncku.edu.tw

15.1. Introduction

Orchidaceae is one of the largest families of flowering plants. In tropical countries, orchids have become a major ornamental export crop, and the demand for their cut flowers has increased rapidly over the years. *Phalaenopsis amabilis* var. *formosa*, a monocotyledon ornamental crop belonging to the Orchidaceae, is indigenous to Taiwan. It is an epiphyte growing high in trees either in dense forests or sometimes, close to the ocean. *P. amabilis* is not only an important species for the floriculture industries but it is also extensively used in the breeding of *Phalaenopsis* hybrids (Law, 1995). However, bacterial diseases in *P. amabilis* and other orchid plant species, such as *Erwinia*, which cause soft rot diseases in a wide range of plant species including field crops and ornamental plants (Perombelon and Kelman, 1980; Lee, 2002), are major concerns among breeders due to their catastrophic effects on plant yield and market value. Although the Orchidaceae species have served as important plants to study flowering, very little is known about the signal transduction and host defense mechanisms with respect to the bacterial pathogen in these plant hosts.

Plants must deal with a wide range of biotic and abiotic environmental pressures by responding in an appropriate manner. Responses can be rapid and specific, and usually involve transcriptional expression of a set of genes whose products are tailored to the particular stress. In many of the best understood plant–pathogen interactions, the ability of the host plant to recognize the pathogen and activate defense responses is regulated in a specific gene-for-gene manner by the interaction between the products of a plant *R* gene and a pathogen *Avr* gene (Flor, 1971). Plants have also evolved various defense mechanisms to protect themselves from pathogen infection, including the generation of reactive oxygen species (ROS), the reinforcement of cell walls, and the activation of hypersensitive response (HR) (Jones and Takemoto, 2004). Salicylic acid (SA) is a messenger molecule in the activation pathways of systemic acquired resistance (SAR) (Wildermuth *et al.*, 2001), which results in the protection of the whole plant. Jasmonate (JA) and ethylene (ET) are also important signals in the induction of defense responses against necrotrophic pathogens that promote host cell death (Glazebrook, 2005). Previous studies indicated that JA had a central role in regulating plant defense gene expression against *Erwinia* (Norman-Setterblad *et al.*, 2000). Recently, studies on defense mechanisms have focused on several model plants, such as *Arabidopsis*, rice and tobacco. The knowledge obtained from these model plants has promoted the understanding of resistance mechanisms in other plants such as orchids; however, certain plants may have unique defense responses.

In this chapter, the molecular characteristics underlying the early response of orchid *P. amabilis* to *E. chrysanthemi* infection is described on the basis of gene expression profiles. During the pathogenesis of *E. chrysanthemi* in the orchid, the involvement of signal transduction pathways such as calcium-dependent protein kinase (CDPK) and classical tyrosine phosphatases (PTP) was investigated. The strategy for isolating and characterizing the protein kinases and phosphatases involved in these signal transduction pathways is also reported in this chapter. The use of degenerate oligonucleotides in the polymerase chain reaction (PCR) successfully isolated the genes encoding the protein kinases and phosphatase from the orchid. Further investigation on the protein kinases and phosphatases revealed the gene expression in response to *E. chrysanthemi* infection as well as wounding. In addition to the roles of protein kinases and phosphatases in the responsiveness to *E. chrysanthemi*, a genome-wide strategy was applied to help to gain a comprehensive understanding of the regulatory networks during early infection. The suppression subtractive hybridization (SSH) approach in conjunction with gene expression profiles facilitated the identification the pathogen-responsive factors that might be responsible for resistance to *E. chrysanthemi* infection. Finally, a putative molecular network model for the orchid was proposed to depict the host factors in response to early infection by *E. chrysanthemi*.

15.2. Pathogenesis of *E. chrysanthemi* in Orchids

The soft rot erwiniae, *E. carotovora* subsp. *Atroseptica*, *E. carotovora* subsp. *Carotovora* and *E. chrysanthemi* are major bacterial pathogens of potato and destructive to a wide variety of plants world-wide. This highly virulent necrotrophic pathogen attacks a variety of host tissues, including blossoms, actively growing shoots, and fruits. The pathogen is also capable of rapid systemic movement downward through stem tissue, resulting in infection and killing of rootstock crowns (Barras *et al.*, 1994; Toth *et al.*, 2003). Soft-rot disease caused by *E. chrysanthemi* is one of the most devastating diseases encountered in the economically important orchids (Liau *et al.*, 2003; Sjahril *et al.*, 2006). The outbreak of soft-rot in orchid floriculture is favored by high relative humidity and temperature during cultivation or transportation to markets. Its pathogenicity results from the secretion of extracellular enzymes responsible for the disorganization of the plant cell wall. These include pectin methyl esterases, pectin acetyl esterases, exo- and endopectate lyases, and exopolygalacturonases (Kazemi-Pour *et al.*, 2004). These enzymes cause maceration of plant tissues, cell separation and cell death, indicating that the pectinases are important virulence determinants (Bell *et al.*, 2004; Zhao *et al.*, 2005).

However, no genetically defined resistance to *Erwinia* has been described in orchids and other plant species, and the pathogen does not appear to contain typical *Avr* genes nor does it normally cause HR in plants. Resistance to this pathogen has been observed in *Solanum brevidens* (Austin *et al.*, 1988), a wild and nontuber-bearing potato species, but the genetic basis for this resistance remains elusive. Therefore, development of resistance to this pathogen appears to rely on generally induced defense mechanisms.

To study the early response of orchid *P. amabilis* to *E. chrysanthemi* infection, an experimental system was established for the inoculation of the orchid plants with the bacterial pathogen (Tsai *et al.*, 2007). The *E. chrysanthemi* was cultured overnight at 28°C in liquid Luria–Bertani medium. To evaluate bacterial growth *in planta*, plants were inoculated with 30 μl of the bacterial culture containing 8 to 9 $\times 10^7$ colony forming units (CFU). After inoculation, the plants were incubated for 24 h but in high humidity (80$\sim$100%) to facilitate bacterial growth. When the leaves of the orchid plants were infected with *E. chrysanthemi*, the typical symptoms of tissue maceration were observed 12 h post-inoculation (Fig. 15.1a). The time course of symptoms on the orchid leaves after inoculation was monitored and compared. It was found

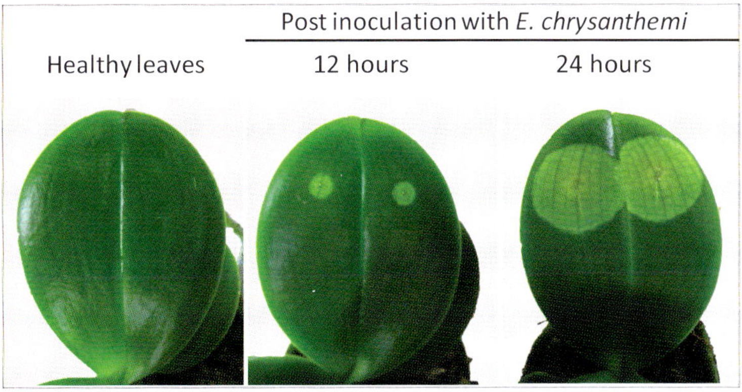

(a)

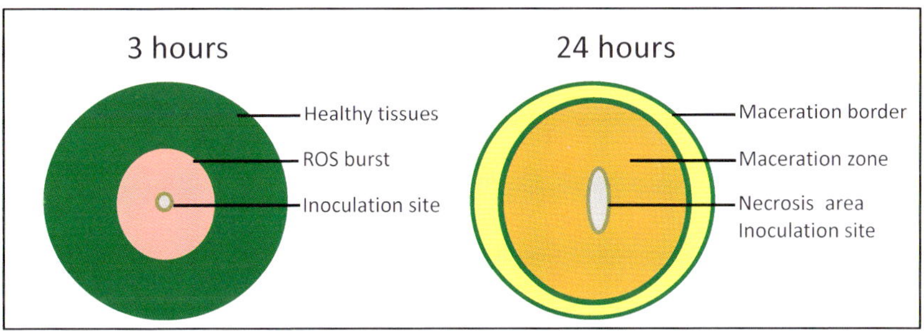

(b)

Fig. 15.1. Disease symptoms induced by *E. chrysanthemi* on *P. amabilis* leaves. (a) The leaves of *P. amabilis* plants (24 month old) were inoculated with an *E. chrysanthemi* suspension containing the bacteria at the concentration of 8 to 9×10^7 colony forming units (CFU). The inoculated plants were kept in a growth room at high humidity (80~100%) at 28°C for different periods of time. Typical tissue maceration developed 12 h post-inoculation. The symptoms progressed by 24 h, and the area of soft-tissue maceration increased in size, with severe necrosis occurring at the site of inoculation. The panel on the left represents the healthy control. (b) *E. chrysanthemi* caused an oxidative burst on the leaves of the orchid and reactive oxygen species (ROS) were detected. Leaves were stained with 2,7-dichlorofluorescein in order to visualize the ROS at 3 h post-inoculation. Left: an illustration of oxidative burst; right: a model showing typical symptoms after inoculation.

that at 24 h post-inoculation, a progressive disorganization of tissues was more visible (Fig. 15.1a). At 24 h post-inoculation, tissue necrosis at the initial site of inoculation was observed and more obvious. Therefore, the symptoms observed during the early stage of infection (24 h) included tissue maceration and necrosis. An illustration is shown in Fig. 15.1b.

Interestingly, it was found that the inoculated leaf exhibited a significantly greater elicited release of reactive oxygen species (ROS) in the surrounding areas adjacent to the site of inoculation than the mock-inoculated leaf (wounding treatment) (Fig. 15.1b). Therefore, the oxidative burst was the early event in response to *E. chrysanthemi* infection. In basal resistance to *E. chrysanthemi*, it was suggested that the increased level of ROS may link to cell wall formation and the assembly of barriers. In addition, the oxidative burst might induce a signal transduction pathway, which leads to subsequent defense responses in the local or systemic plant tissues.

The effects of cell wall-degrading enzymes secreted by *E. chrysanthemi* on the pathogenesis in the orchid were investigated by inoculation of the orchid leaves with culture filtrate (CF). The CF was collected from an *in vitro* grown *E. chrysanthemi* suspension after 24 h. The maceration symptoms spread following inoculation with *E. chrysanthemi* and CF at 6 h post-inoculation (Fig. 15.2). In the same period, only the wounded tissues were seen at the site with mock-inoculation (Fig. 15.2). Interestingly, the area with maceration symptoms increased in size following inoculation with *E. chrysanthemi* at 36 h but not in the leaves with CF- and mock-inoculation (Fig. 15.2). Thus, the CF containing the cell wall-degrading enzyme was one of the factors responsible for early pathogenesis of *E. chrysanthemi* in the orchids. The successful pathogenesis of *E. chrysanthemi* in the orchids mainly relied on the propagation of the bacterial pathogen. In addition, it was suggested that CF might be an elicitor that triggered the defense response to *E. chrysanthemi*. The initiation of basal resistance by CF might subsequently facilitate the plants in confining the pathogen to the site of inoculation. An illustration is shown on the basis of the experiments in orchids when inoculated with *E. chrysanthemi* and CF (Fig. 15.2). The results implied that the orchid might exhibit basal resistance to *E. chrysanthemi* and initiate defense responses against the pathogen.

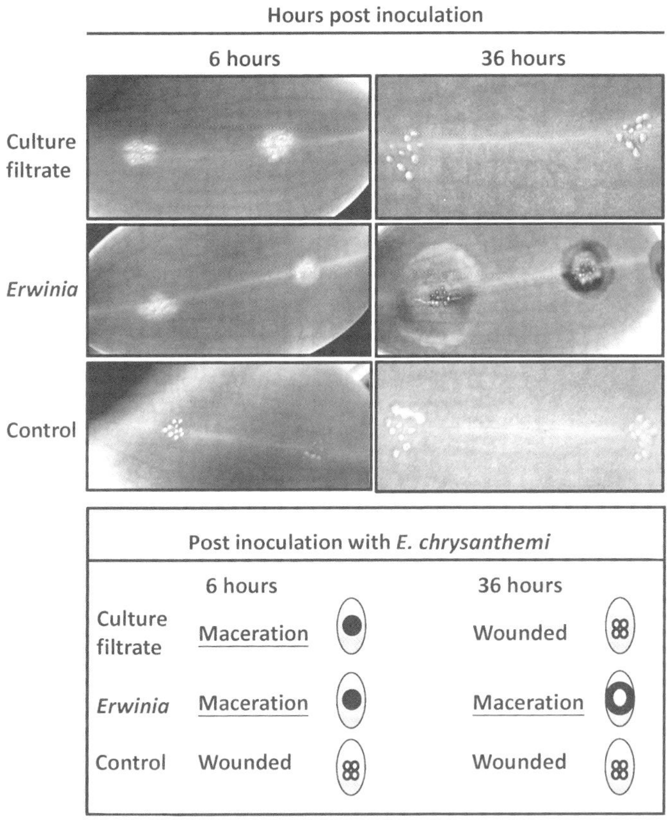

Fig. 15.2. Symptom development in response to culture filtrate (CF) treatment and inoculation with *E. chrysanthemi*. The leaves of *P. amabilis* plants (24 months old) were inoculated with CF and *E. chrysanthemi*. Mock-inoculated leaves served as controls. These photos of representative symptoms were taken at 6 h and 36 h post-inoculation. The bottom panel illustrates the various symptoms observed on the plant tissues inoculated with CF and *E. chrysanthemi*.

15.3. Signaling Transduction in Orchids During Early Infection with *E. chrysanthemi*

A common feature of plant defense responses is the transcriptional activation of a large number of defense-related genes upon pathogen infection including the accumulation of proteins with antimicrobial activity, such as pathogenesis-related proteins (PR), and the enzymes involved in the biosynthesis of antimicrobial compounds, such as phytoalexins. In previous reports, tobacco leaves treated with *E. carotovora* or its elicitor systemically induced a set of PR genes. The activation of different signal transduction pathways connected to protein kinases, such as mitogen-activated protein kinases (MAPK) and calcium-dependent protein kinases (CDPK), and transcription factors regulating the activity of defense gene promoters is crucial for defeating pathogens (Glazebrook *et al.*, 2003). It has been suggested that CDPKs have different functions and participate in multiple diverse signaling pathways. The evidence suggests that CDPKs show specific calcium activations and substrate specificities as well as the typical inducement of plant hormones (Abo-el-Saad and Wu, 1995; Martin and Busconi, 2000). In addition, CDPKs play essential roles in biotic stress signal transduction in plants. For example, exposure to non-specific elicitors and wounding have been reported to cause an increased expression of the *NtCDPK1* gene in tobacco (Yoon *et al.*, 1999).

In eukaryotes, the protein tyrosine phosphatase (PTP) family is very diverse in domain structure and substrate preference with some family enzymes now recognized to dephosphorylate complex carbohydrates, mRNA and phosphoinositides (Tonks, 2006). The PTPs are further categorized into two large groups based on phosphoamino acid specificity: the phosphotyrosine-specific phosphatases (classical PTPs) and dual-specificity phosphatases (DsPTPs) (Luan, 2003). The classical PTPs consist both of the transmembrane receptor-like PTPs and intracellular PTPs. These classical PTPs are characterized by a single conserved catalytic domain and highly variable N- and C-terminal extensions (Tonks, 2006; Gandhi *et al.*, 2005). The DsPTPs cannot only dephosphorylate tyrosine residues but also show phosphatase activities toward serine and threonine residues (Tonks, 2006).

Accumulating evidence now suggests that tyrosine phosphorylation and dephosphorylation serve functions in plants as they do in animals. Changes in tyrosine phosphorylation levels have been correlated with somatic embryogenesis (Barizza *et al.*, 1999), petiole movement (Kameyama *et al.*, 2000) as well as disease-resistance signaling in plants (van Bentem and Hirt, 2009). For instance, kinase cascades of the mitogen-activated protein kinase (MAPK) class play a remarkably important role in the establishment of resistance to plant pathogens (Pitzschke *et al.*, 2009). Tyrosine phosphorylation is essential for activation of MAP kinase proteins, which modulate downstream gene expression (Pitzschke *et al.*, 2009). In addition, tyrosine dephosphorylation is also involved in the disease-signaling pathway during pathogenesis. The plant pathogen *Pseudomonas syringae* injects the virulence factor HopPtoD2, which is a PTP, into its host cytoplasm. Its catalytic activity suppresses programmed cell death, just like that of another *P. syringae* virulence factor, HopAO1, which is also predicted to be a PTP. Host MAPKs are potential targets of these virulence factors encoding tyrosine phosphatases (van Bentem and Hirt, 2009). However, very little is known about the roles of tyrosine phosphorylation in orchids during the early infection with *E. chrysanthemi*. In the next sections, the isolation and characterization of genes involved in the disease-mediated signal transduction in the orchid are described. Their roles in the early response to *E. chrysanthemi* are investigated.

15.3.1. *Strategy for isolating genes encoding protein kinases and phosphatases*

In order to isolate orchid genes encoding kinases and phosphatases in response to infection with *E. chrysanthemi*, a reverse transcription (RT)-PCR-based cloning strategy using degenerate primers was employed to isolate these genes from *P. amabilis*. Total RNA was isolated from orchids inoculated with the pathogen or from healthy control plants. The RNA samples from the pathogen-infected plants or healthy plants were reverse-transcribed into cDNA. Degenerate primers corresponding to the conserved domains of known plant kinase or phosphatase protein sequences were designed and used in a RT-PCR reaction. A cDNA clones with 300~1,000 bp in length containing partial sequences of kinase or phosphatase genes were obtained. The candidate clones with homology to kinases or phosphatases were subjected to further DNA sequencing.

A full-length cDNA, including 5′ and 3′ untranslated region (UTR), from the orchid was subsequently isolated using the 5′ and 3′ rapid amplification of cDNA ends (RACE) approach. In a RACE reaction, primary PCR amplification reactions are used to obtain individual 5′ or 3′ sequences. The PCR products were cloned and verified by DNA sequencing. Secondary PCR amplifications were performed to combine 5′ and 3′ sequences and generate full-length cDNAs. The full-length cDNA of the kinases or phosphatases were amplified by using primers designed from the extreme 5′ and 3′ ends.

The strategy described above is based on the design of degenerate oligonucleotide primers using regions of amino acid and DNA sequence homology, and allows the isolation of related sequences. Many genes encoding kinases and phosphatases are members of families or super-families, and therefore, this technique has wide applications in the field of molecular characterization of signal transduction genes in response to pathogen infection in orchids.

15.3.2. *Calcium-dependent protein kinases*

15.3.2.1. *Sequence analysis of PaCDPK1 from P. amabilis*

The *PaCDPK1* (accession number EF555574) cDNA was 2,285-bp long, including 3′ UTR, an open reading frame (ORF) and part of 5′ UTR. The *PaCDPK1* cDNA contained a 1,779-bp ORF which encoded a 593 amino acid polypeptide with a predicted molecular mass of 65.6 kDa. The deduced protein sequence of PaCDPK1 was highly homologous to other plant CDPKs, including *Arabidopsis* AtCPK1 (accession number At5g04870, 71 percent identity), *Arabidopsis* AtCPK2 (accession number At3g10660, 67 percent identity), maize ZmCDPK9 (accession number D85039, 61 percent identity), and tomato LeCDPK1 (accession number AF363784, 59 percent identity). The sequence alignment indicated that the PaCDPK1 protein possessed all of the structural characteristics of CDPKs (Tsai *et al.*, 2007). The CDPKs have a variable N-terminal domain, a central kinase domain, an auto-inhibitory domain, and a C-terminal calmodulin like domain. The kinase domain was a typical serine/threonine kinase containing the 11 sub-domains. The calmodulin-like region contained four EF-hand Ca^{2+}-binding motifs at its C-terminus. In addition, the normally well-conserved auto-inhibitory domain connecting the kinase domain to the calmodulin-like domain in CDPKs was degenerated in PaCDPK1 (Tsai *et al.*, 2007).

15.3.2.2. *Expression of the PaCDPK1 gene in response to E. chrysanthemi*

To elucidate the roles of *PaCDPK1* in biotic stress responses, we investigated whether *PaCDPK1* could be induced by infection with *E. chrysanthemi*. The effects of wounding treatment on the expression of *PaCDPK1* were also studied. The apex of the *P. amabilis* leaf was inoculated with *E. chrysanthemi* by needle. The undamaged leaf tissue adjacent to the inoculated site was defined as P (pathogen) 1, P2, P3 and P4, respectively. The four regions were illustrated and used as plant samples (Fig. 15.3a). The level of *PaCDPK1* transcript in leaves was significantly higher in the P2 region than that of the controls after inoculation with *E. chrysanthemi* (Fig. 15.3a). In addition, the gene expression of *PaPTP1* in response to wounding treatment was examined, and the collection of wounded samples (W1, W2, W3 and W4) was similar to that of pathogen inoculation. The gene expression of *PaCDPK1* was also induced significantly in the W4 and W3 regions after wounding treatment (Fig. 15.3b).

The possible involvement of *PaCDPK1* in response to other environmental constraints such as low and high temperature was also studied. The orchid plants were grown at 28°C for two weeks and then either kept at 28°C (control plants) or moved to 4°C (cold-treated plants) or 37°C (heat-treated plants). It was found that the *PaCDPK1* transcript levels increased 3 h after low temperature treatment (4°C) but not after 24 h at 37°C (Tsai *et al.*, 2007). In conclusion, the expression of the *PaCDPK1* gene was not only induced after infection with *E. chrysanthemi* but also in response to environmental stresses.

15.3.2.3. *Promoter activity of the PaCDPK1 gene in response to E. chrysanthemi*

To facilitate functional analysis of the promoter region of the orchid *PaCDPK1* gene, the 1,301-bp promoter fragment was amplified, resulting in transcriptional fusion with the *GUS* gene. Transgenic

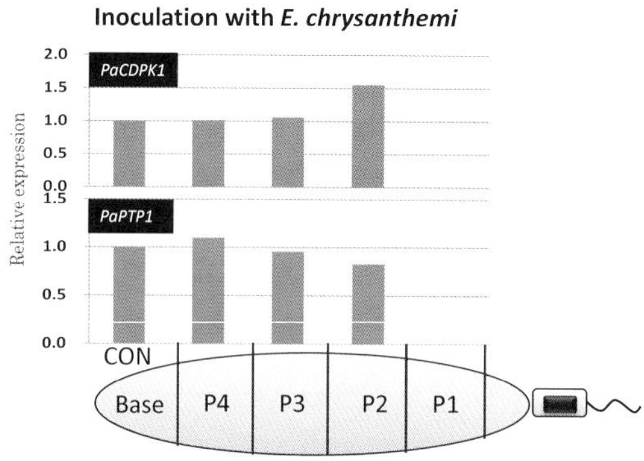

(a)

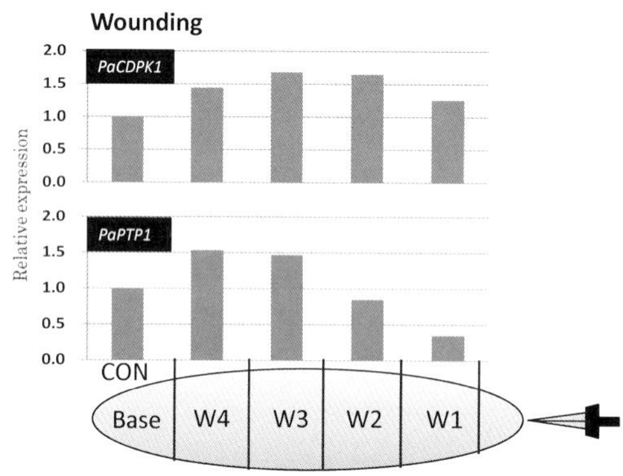

(b)

Fig. 15.3. Analysis of *PaCDPK1* and *PaPTP1* gene expression in response to *E. chrysanthemi* infection and mechanical wounding. (a) For infection with *E. chrysanthemi*, the apical end of the orchid leaf was inoculated and the surrounding area with 1 cm interval was defined as region P (pathogen) 1. The adjacent regions distal to the P1 site were designated P2, P3 and P4, respectively. RNA samples were collected from the different regions (P1–P4) at 48 h after infection with the bacterial pathogen. The leaf tissues from healthy plants served as controls (CON). (b) The collection and analysis of the samples after wounding treatment was similar to that of the inoculation experiment described in (a). The RNA samples were denatured and subjected to northern blot analysis with a gene-specific probe. The transcript level of ribosomal RNA (*rRNA*) served as equal loading control. Hybridization signals were quantified with Adobe Photoshop version 7.0 to develop histograms of signal intensity. The values were normalized with respect to ribosomal RNA content. The signal of the first sample on the panel was defined as 1.0 (arbitrary units), and other abundances are expressed relative to that value.

Arabidopsis plants harboring the *PaCDPK1::GUS* construction were generated, and the GUS activity in *Arabidopsis* seedlings was analyzed under biotic and abiotic environmental stresses. When the transgenic *PaCDPK1/GUS Arabidopsis* seedlings were subjected to the pathogen infection, the transgenic plants carrying the *GUS* reporter gene exhibited deep blue GUS staining (Tsai *et al.*, 2007). In addition, the GUS activity was significantly activated in the transgenic *Arabidopsis* seedlings after treatment with low temperature and wounding. This suggests a responsiveness of the *PaCDPK1* promoter to infection with *E. chrysanthemi* as well as to low temperature and wounding (Tsai *et al.*, 2007).

Our findings revealed that the *PaCDPK1* gene from the orchid could be regulated by environmental stresses at the transcriptional level. We also demonstrated that the cloned *P. amabilis CDPK1* promoter is activated by low temperature, wounding, and pathogen challenge in transgenic *Arabidopsis*. The study presented here is promising, but it would be interesting to elucidate the potential tissue specific nature of the *PaCDPK1* promoter in the orchid. Furthermore, this activity of the inducible promoter in the orchid may be used as a biotechnological tool and applied in the floriculture industries to cope with pathogen attack and environmental constraints.

15.4. Protein Tyrosine Phosphatases

15.4.1. *Sequence analysis of PaPTP1 from P. amabilis*

A full-length cDNA clone 1,588-nucleotide bp in length was isolated from *P. amabilis* and subsequently termed *PaPTP1*. The nucleotide sequence of the *PaPTP1* cDNA contained an ORF that would encode a protein of 346 amino acids, with an ATG codon at the 215th nucleotide, and the poly(A)$^+$ tail starting at the 1,253rd nucleotide downstream from the stop codon. The molecular mass of the predicted polypeptide is 38.7 kDa. The PaPTP1 protein sequence showed significant homology to *Arabidopsis* AtPTP1, *Pisum sativum* PsPTP1 and *Glycine max* GmPTP1 with 57 percent, 54 percent and 53 percent identities, respectively (Fordham-Skelton *et al.*, 1999; Xu *et al.*, 1998). The gene sequence encoding the PaPTP1 was deposited in GenBank under the accession number of GU119901. A phylogenetic tree of the PaPTP1 protein sequence with a number of PTPs from eukaryotes was constructed (Fu *et al.*, 2010). The phylogenetic relationship suggested that PaPTP1 is a novel member of the classical PTP family in plants. In addition, the sequence alignments of PaPTP1 with PTPs from other organisms were generated (Fu *et al.*, 2010). The predicted catalytic domain, containing sequence motifs (I/V)HCXAGXXR(S/T)G (Luan, 2003) characteristic of tyrosine-specific protein phosphatase, was also conserved in the PaPTP1 amino acid sequence.

15.4.2. *Expression of the PaPTP1 gene in response to wounding and E. chrysanthemi*

To gain further insight into the possible function of *PaPTP1* in *P. amabilis,* the expression of the *PaPTP1* gene in response to *E. chrysanthemi* as well as wounding was investigated. The four regions of infected tissues (P1, P2, P3 and P4) were illustrated and used as plant samples (Fig. 15.3a). The steady-state level of *PaPTP1* transcripts remained unchanged in the regions (P4, P2 and P3) near the site infected with *E. chrysanthemi* (Fig. 15.3a). These results indicated that expression of the *PaPTP1* gene was not induced by infection with *E. chrysanthemi.*

In addition, the gene expression of *PaPTP1* in response to wounding treatment was examined, and the collection of wounded samples (W1, W2, W3 and W4) was similar to that of pathogen infection (Fig. 15.3b). The wound-induced accumulation of *PaPTP1* transcripts was detected after 48 h in regions near the wound (W1 and W2) and gradually increased in quantity with distance distal to the wound site (W3 and W4). The maximum accumulation of *PaPTP1* transcripts was observed in the W4 region when compared to that of leaf tissues from healthy plants (Fig. 15.3b). This indicated that expression of the *PaPTP1* gene was induced by mechanical wounding.

15.4.3. *The PaPTP1 gene encoded a functional PTP*

To understand whether the *PaPTP1* gene encodes a functional PTP, the recombinant PaPTP1 fused to glutathione-S-transferase (GST-PaPTP1) was created and its enzyme properties were characterized. The GST-PaPTP1 protein was overexpressed in *Escherichia coli* culture in the presence of isopropylthio-β-galactoside (IPTG) (Fu *et al.*, 2010). Subsequently, the recombinant GST-PaPTP1 protein was purified and analyzed by SDS-PAGE. A purified protein with predicted molecular weight of 64 kDa corresponding to recombinant GST-PaPTP1 was detected and subjected to protein phosphatase assay. The phosphatase activity assay was performed by using the generic phosphatase substrate *p*-nitrophenolphosphate (*p*NPP). The PaPTP1 protein hydrolyzed *p*NPP rapidly, and the activity was almost linear during the first 120 min of the reaction (Fu *et al.*, 2010). With the increased concentration of PaPTP1 protein in the reaction, *p*NPP hydrolysis was enhanced proportionally (Fu *et al.*, 2010). In contrast, the reaction mixture containing purified GST alone exhibited no phosphatase activity, and this served as a negative control. These results demonstrated that PaPTP1 protein was a functional PTP (Fu *et al.*, 2010).

In summary, the *PaPTP1* gene from the economically important orchids encodes a functional classical PTP. The expression of the *PaPTP1* gene is mainly localized in floral organs and in response to wounding. Generation of loss-of-function orchid plants by virus-induced gene silencing has been described and is achievable (Lu *et al.*, 2007). Further functional analysis using *PaPTP1*-silenced orchid plants should provide novel information on the role of reversible tyrosine phosphorylation in developmental biology and stress responses.

15.5. Genome-Wide Responses of Orchid Plants to *E. chrysanthemi* Infection

15.5.1. *Strategy for studying transcriptional regulation during the pathogenesis*

To study the molecular basis of the orchid response during the early stage of *E. chrysanthemi* infection, the pathogen-induced genes were isolated through the construction of a subtractive cDNA library. The time course-dependent development of symptoms on the orchid leaves were monitored after inoculation, showing that at

24 h post-inoculation, a progressive disorganization of tissues called maceration was visible.

The major objective of this study was to identify the genes that are differentially expressed in the orchid with *E. chrysanthemi* infection. A pathogen-induced cDNA library was constructed using the SSH strategy. The mRNA from non-infected plants and plants 24 h post-inoculation with *E. chrysanthemi* was prepared as driver and tester RNA samples, respectively. After two PCR and differential hybridization experiments, the subtracted DNA sample was analyzed by agarose gel electrophoresis. The distinct DNA bands ranged from 500 bp to 1,400 bp, with a medial size of approximately 800 bp being observed in the enriched DNA. The insert DNA of the three distinct regions were cloned into vectors and stored. The EST clones were subsequently sequenced and defined with a unique number in the pathogen-induced cDNA library. Sequence similarity searching was performed against the NCBI (National Center for Biotechnology Information, USA) database using the ExPASy Proteomics Server, which compared translated nucleotide sequences with protein sequences. The gene expression of the pathogen-responsive ESTs was further verified by semi-quantitative RT-PCR analysis. Thus, a combination of SSH and expression profile strategies facilitated our study of the genome-wide expression in orchids during the pathogenesis. The approach described here successfully identified novel *E. chrysanthemi*-responsive transcripts from the orchid.

15.5.2. *Pathogen-responsive expressed sequence tags (ESTs) in orchids*

The nucleotide sequences of the 170 randomly selected clones were analyzed and their putative functions identified by carrying out NCBI BLAST searches (*E*-value cut-off at 10^{-5}) (Table 15.1). A summary of the pathogen-responsive ESTs was obtained according to functional categorization (Fig. 15.4). As a result of homology searches, 156 clones were matched to previously known genes (Fig. 15.4 and Table 15.1). They shared significant similarity to a broad range of gene families encoding metabolic enzymes, structural proteins, regulatory factors,

Table 15.1 Summary of EST Clones from the PaEI Library

	Number	Percentage (%)
Total clones	170	—
BLAST match clones	156	91.8
Cluster	20	—
Redundancy	103	60.5
Singlet	53	31.2
BLAST no-match clones	14	8.2
Cluster	1	—
Redundancy	2	1.2
Singlet	12	7.1

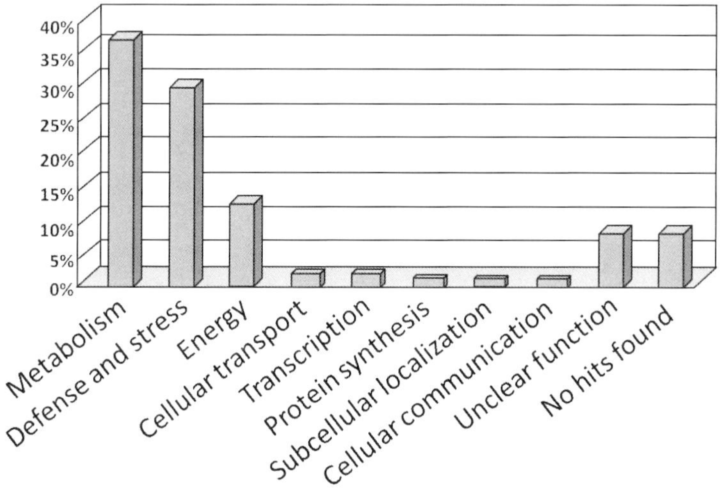

Fig. 15.4. Gene ontology analysis of the EST clones from the PaEI Library. The EST clones were subjected to gene ontology analysis, and grouped into ten functional categories. The genes with functions related to metabolism, defense and stress and energy were predominant among the PaEI library.

transporter or protective proteins (Fig. 15.4). The total EST clones in the library included 65 unique clones of cDNA fragments, which in turn consisted of 105 clones assembled in 21 clusters (Table 15.1). Transcript redundancy is a measure of the abundance of the corresponding cDNA in non-normalized libraries. In other words, information on randomly picked cDNAs represents the relative repression levels of genes in a particular library.

15.5.3. *Functional significance of the pathogen-responsive ESTs*

The SSH strategy was used to identify and characterize genes that were differentially expressed in the orchid following *E. chrysanthemi* inoculation. This study focused on the early stages of pathogen stress in leaf tissues due to the important roles of early response genes in mediating the effects of *E. chrysanthemi* infection. A total of 170 EST clones after *E. chrysanthemi* inoculation for 24 h were isolated (Fig. 15.4 and Table 15.1). The individual EST clones with the functional significance related to defense responses was described (Table 15.2), and their expression patterns in response to biotic and abiotic stresses were further verified (Table 15.3). Our results may provide important insights into the regulation of orchid defense-related genes.

Chalcone synthase (CHS) catalyzes the production of naringenin chalcone from malonyl-CoA (Abe *et al.*, 2001). This was of more importance in our results because CHS (EST clone 1), another key enzyme involved in flavonoid synthesis, was also shown as an enhanced expression coincidentally (Tables 15.2 and 15.3). Similar results were observed in studies carried out on barley leaves upon infection with the

Table 15.2 EST Clones Related to Defense Responses from the PaEI Library

Clone	Gene Name	Description
1	CHS	Chalcone synthase Flavonoid biosynthesis
2	Putative PDR-like ABC transporter	Putative PDR (pleiotropic drug resistance) — like ATP-binding cassette (ABC) transporter
3	GST	Glutathione transferase Scavenger of toxic oxidation products
4	ACC	Acetyl-CoA carboxylase Malonyl-CoA biosynthesis
5	PR10	Pathogenesis-related protein 10 Response to pathogens or other stresses

Table 15.3 Verification of the Expression of EST Clones from the PaEI Library

	RT-PCR Summary of Ten Infection Time-Course					RT-PCR Summary of Ten Level Spatial Treatment				
Clone	**Control**	**6 h**	**12 h**	**24 h**	**48 h**	**Control**	**Wounding**	**Me-JA**	**Pathogen**	**CF**
1 CHS	+	++	++	+++	++	–	++	+	+++	+
2 ABC	+	+++	+++	+++	+++	++	++	++	+++	+
3 GST	+	++	++	++	++	+	+	++	++	+
4 ACC	+	++	++	++	+++	+	+	++	+++	+
5 PR10	+	+++	+++	+++	++	+	+	++	+++	+

fungus *Blumeria graminis* (Christensen *et al.*, 1998). Its expression has also been demonstrated in citrus (Moriguchi *et al.*, 2001) and in cotton (Cui *et al.*, 2000). Thus, considering the function of these genes, and also integrating our expression, it is reasonable to propose that enhancement of flavonoid synthesis may increase defense responses.

ATP-binding cassette (ABC) transporters function as ATP-driven efflux pumps that export a wide variety of compounds (Sanchez-Fernandez *et al.*, 2001). The pleiotropic drug resistance (PDR) subfamily of plant ABC transporters has been implicated in plant defense. Various microbial elicitors induce tobacco *PDR* (*NtPDR1*) gene expression (Sasabe *et al.*, 2002), which has been demonstrated to be involved in pathogen resistance (Stukkens *et al.*, 2005). Recently, it was demonstrated that *Arabidopsis* AtPDR8 was a key factor controlling the extent of cell death in defense response, and it was suggested that AtPDR8 transported some substance which was closely related to the response of plants to pathogens (Kobae *et al.*, 2006). We also found that the putative PDR-like ABC transporter (EST clone 2) had higher expression during *E. chrysanthemi* inoculation, indicating that it may play a major role in orchid defense against pathogens (Tables 15.2 and 15.3).

During early pathogenesis of *E. chrysanthemi* in the orchid, ROS were produced quickly and might play an important role in triggering defense responses. Plants have a network of genes controlling ROS metabolisms including catalase, glutathione reductase, and glutathione transferase. In the pathogen-induced cDNA library, glutathione transferase (GST) (EST clone 3), correlated with cell protection from oxidative stress damage, was concurrently expressed at a higher level in leaves inoculated with *E. chrysanthemi* (Tables 15.2 and 15.3). Toxic oxidation products and other toxic compounds produced in the plant for pathogen defense were detoxified by GST-catalyzed conjugation of reduced glutathione (GSH). The GSTs were abundant in plants, representing up to 1 percent of soluble proteins (Frova, 2006). GST in many plant species (Lamb and Dixon, 1997) is believed to protect cells against damage from the oxidative burst triggered by pathogen invasion.

Acetyl-CoA carboxylase (ACCase) catalyzed the ATP-dependent formation of malonyl-CoA from acetyl-CoA and bicarbonate. In plant cells, cytosolic malonyl-CoA was required for flavonoid synthesis (Roesler *et al.*, 1994). Flavonoids have frequently been proposed to participate in plant response to pathogen attack and the corresponding enzymes involved are induced by pathogens. Previous studies have revealed a significant role for acetyl-CoA carboxylase (ACCase) in the synthesis of flavonoids (Roesler *et al.*, 2006). This implies that the increasing of ACCase (EST clone 4) expression may at least partially account for the role of the ACCase gene in response to pathogen infection (Tables 15.2 and 15.3). Moreover, the previous study suggested that the promoter activity of the *PvACCase* gene was enhanced by infection with *Pseudomonas syringae* (Figueroa-Balderas *et al.*, 2006).

Similar results were obtained upon RT-PCR analysis with pathogenesis-related protein 10 (PR10) (EST clone 5), which indicated that earlier stages after inoculation in the orchid. The expression of PR10 was observed at 6 h and high levels were maintained throughout the course of the experiment (Tables 15.2 and 15.3). Previous studies have indicated that pathogenesis-related (PR) proteins in plants are one of the crucial components in the induction of the plant's self-defense mechanism (Jwa *et al.*, 2001). Based on sequence homology and biological activity, the PR proteins have been grouped into 14 families (Kitajima and Sato, 1999). It has been reported that antimicrobial defense-related proteins such as PR10 may contribute to the establishment of SAR (Lee and Hwang, 2005). For seedlings incubated with sap containing Tobacco mosaic virus, the transcript level of *SsPR10* from *Solanum surattense* was increased significantly, suggesting that SsPR10 was pathogen-inducible (Liu *et al.*, 2006).

Taken together, by using the SSH strategy, we have identified a number of differentially expressed genes in the orchid when infected with *E. chrysanthemi*. A putative molecular model underlying early responses of orchid plants to the pathogen is proposed (Fig. 15.5). Further analysis combined with the results from others suggests that the improvement of defense-related genes perhaps accounts for the orchid's increasing resistance against pathogens. This study, showing the characterization and expression pattern of genes under pathogen stress in the orchid (Fig. 15.5), together with other related reports from microarray databases, importantly complements our understanding

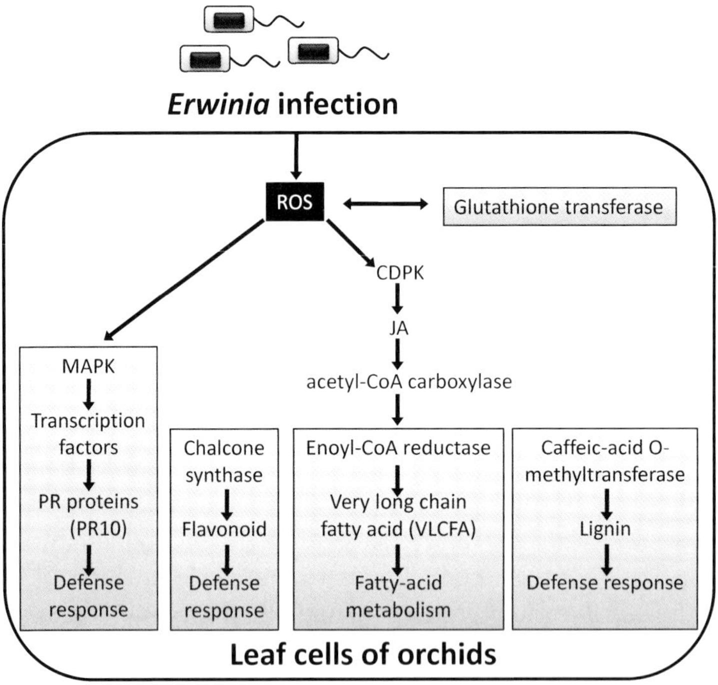

Fig. 15.5. A putative model of the early molecular characteristics of the orchid plants in response to *E. chrysanthemi* infection. A summary of the molecular events, including increase in reactive oxygen species (ROS), signal transduction via kinases, regulation by transcription factors and downstream defense responses, was the basis of the study described in this chapter. The arrows indicate a direct or indirect interaction of the molecular components.

of the host–pathogen interaction between *Phalaenopsis* orchids and *E. chrysanthemi*.

15.6. Conclusions and Future Prospects

Studies on model plants such as *Arabidopsis* and potatoes have provided a wealth of information on pathogen recognition, signal transduction and the activation of defense responses. However, very little is known about the infection and defense responses of the orchid to the well-studied phytopathogen *E. chrysanthemi*. In this chapter, we have demonstrated that *E. chrysanthemi* and its elicitors induced a defense response in *P. amabilis*, as evidenced by enhanced expression of conserved signal transduction and plant defense-related genes. In addition, our study isolated novel pathogen-responsive factors in orchids that have potential roles in resistance to *E. chrysanthemi*. This work thus establishes *P. amabilis* as an experimental host for the investigation of plant–bacterial pathogen interactions. The two strategies described in this chapter, degenerate primer-based PCR and SSH, are promising tools for understanding the mechanisms involved in the pathogen-induced symptoms and defense responses in orchids.

In conclusion, the fact that most of these gene expressions have already been reported in other plant–pathogen interactions should be underscored. However, these first results must be augmented because only a few clones from the SSH library have been sequenced and studied for their expression patterns. From the data obtained here, it is clear that there were many unknown components in the pathogen stress responses that required detailed characterization. In addition, the extent to which these genes and the associated proteins contribute to defense response is an important focus of ongoing studies. Experiments are underway at present to determine whether over- or under-expression of these genes in transgenic orchids could yield instructive or useful changes in pathogen defense.

Acknowledgments

The study described in this chapter was funded by grants from the Taiwan National Science Council (project number NSC 96-2317-B-006-005) and the Council of Agriculture (project number 91-3.1.3-Z3). This work was also supported by a grant from the Ministry of Education, Taiwan ("Landmark Project Grant" for National Cheng Kung University's "Top-University Project", B024).

References

Abe I, Takahashi Y, Morita H, Noguchi H. (2001) Benzalacetone synthase — A novel polyketide synthase that plays a crucial role in the biosynthesis of phenylbutanones in *Rheum palmatum*. *Eur J Biochem* **268**: 3354–3359.

Abo-el-Saad M, Wu R. (1995) A rice membrane calcium-dependent protein kinase is induced by gibberellin. *Plant Physiol* **108**: 787–793.

Austin S, Lojkowska E, Ehlenfeldt MK *et al.* (1988) Fertile interspecific somatic hybrids of *Solanum* — A novel source of resistance to *Erwinia* soft rot. *Phytopathology* **78**: 1216–1220.

Barizza E, Lo Schiavo F, Terzi M, Filippini F. (1999) Evidence suggesting protein tyrosine phosphorylation in plants depends on the developmental conditions. *FEBS Lett* **447**: 191–194.

Barras F, Vangijsegem F, Chatterjee AK. (1994) Extracellular enzymes and pathogenesis of soft-rot *Erwinia. Annu Rev Phytopathol* **32**: 201–234.

Bell KS *et al.* (2004) Genome sequence of the enterobacterial phytopathogen *Erwinia carotovora* subsp. atroseptica and characterization of virulence factors. *Proc Natl Acad Sci USA* **101**: 11105–11110.

Christensen AB, Gregersen PL, Schroder J, Collinge DB. (1998) A chalcone synthase with an unusual substrate preference is expressed in barley leaves in response to UV light and pathogen attack. *Plant Mol Biol* **37**: 849–857.

Cui YX, Bell AA, Joost O, Magill C. (2000) Expression of potential defense response genes in cotton. *Physiol Mol Plant Pathol* **56**: 25–31.

Figueroa-Balderas RE, Garcia-Ponce B, Rocha-Sosa M. (2006) Hormonal and stress induction of the gene encoding common bean acetyl-coenzyme A carboxylase. *Plant Physiol* **142**: 609–619.

Flor HH. (1971) Current status of gene-for-gene concept. *Annu Rev Phytopathol* **9**: 275.

Fordham-Skelton AP, Skipsey M, Eveans IM *et al.* (1999) Higher plant tyrosine-specific protein phosphatases (PTPs) contain novel amino-terminal domains: Expression during embryogenesis. *Plant Mol Biol* **39**: 593–605.

Frova C. (2006) Glutathione transferases in the genomics era: New insights and perspectives. *Biomol Eng* **23**: 149–169.

Fu SF, Lin CW, Kao TW *et al.* (2010) PaPTP1, a gene encoding protein tyrosine phosphatase from orchid, *Phalaenopsis amabilis,* is regulated during floral development and induced by wounding. *Plant Mol Biol Rep.* DOI:10.1007/s11105-010-0216-y.

Gandhi TK *et al.* (2005) A bioinformatics analysis of protein tyrosine phosphatases in humans. *DNA Res* **12**: 79–89.

Glazebrook J *et al.* (2003) Topology of the network integrating salicylate and jasmonate signal transduction derived from global expression phenotyping. *Plant J* **34**: 217–228.

Glazebrook J. (2005) Contrasting mechanisms of defense against biotrophic and necrotrophic pathogens. *Annu Rev Phytopathol* **43**: 205–227.

Jones DA, Takemoto D. (2004) Plant innate immunity — Direct and indirect recognition of general and specific pathogen-associated molecules. *Curr Opin Immunol* **16**: 48–62.

Jwa NS, Kumar Agrawal G, Rakwal R *et al.* (2001) Molecular cloning and characterization of a novel jasmonate inducible pathogenesis-related class 10 protein gene, JIOsPR10, from rice (*Oryza sativa* L.) seedling leaves. *Biochem Biophys Res Commun* **286**: 973–983.

Kameyama K *et al.* (2000) Tyrosine phosphorylation in plant bending. *Nature* **407**: 37.

Kazemi-Pour N, Condemine G, Hugouvieux-Cotte-Pattat N. (2004) The secretome of the plant pathogenic bacterium *Erwinia chrysanthemi*. *Proteomics* **4**: 3177–3186.

Kitajima S, Sato F. (1999) Plant pathogenesis-related proteins: Molecular mechanisms of gene expression and protein function. *J Biochem* **125**: 1–8.

Kobae Y *et al.* (2006) Loss of AtPDR8, a plasma membrane ABC transporter of *Arabidopsis thaliana*, causes hypersensitive cell death upon pathogen infection. *Plant Cell Physiol* **47**: 309–318.

Lamb C, Dixon RA. (1997) The oxidative burst in plant disease resistance. *Annu Rev Plant Physiol Plant Mol Biol* **48**: 251–275.

Law N. (1995) Cut orchids in the world market. *Flor Cult Int* **5**: 12–15.

Lee YA, Chen KP, Chang YC. (2002) First report of bacterial soft rot of white flowered calla lily caused by *Erwinia chrysanthemi* in Taiwan. *Plant Dis* **86**: 1273.

Lee SC, Hwang BK. (2005) Induction of some defense-related genes and oxidative burst is required for the establishment of systemic acquired resistance in *Capsicum annuum*. *Planta* **221**: 790–800.

Liau CH *et al.* (2003) The sweet pepper ferredoxin-like protein (pflp) conferred resistance against soft rot disease in *Oncidium* orchid. *Transgenic Res* **12**: 329–336.

Liu X *et al.* (2006) A novel pathogenesis-related protein (SsPR10) from *Solanum surattense* with ribonucleolytic and antimicrobial activity is stress- and pathogen-inducible. *J Plant Physiol* **163**: 546–556.

Lu HC *et al.* (2007) Strategies for functional validation of genes involved in reproductive stages of orchids. *Plant Physiol* **143**: 558–569.

Luan, S. (2003) Protein phosphatases in plants. *Annu Rev Plant Biol* **54**: 63–92.

Martin ML, Busconi L. (2000) Membrane localization of a rice calcium-dependent protein kinase (CDPK) is mediated by myristoylation and palmitoylation. *Plant J* **24**: 429–435.

Moriguchi T, Kita M, Tomono Y *et al.* (2001) Gene expression in flavonoid biosynthesis: Correlation with flavonoid accumulation in developing citrus fruit. *Physiol Plantarum* **111**: 66–74.

Norman-Setterblad C, Vidal S, Palva ET. (2000) Interacting signal pathways control defense gene expression in *Arabidopsis* in response to cell wall-degrading enzymes from *Erwinia carotovora*. *Mol Plant Microbe In* **13**: 430–438.

Perombelon MCM, Kelman A. (1980) Ecology of the soft rot *Erwinias*. *Annu Rev Phytopathol* **18**: 361–387.

Pitzschke A, Schikora A, Hirt H. (2009) MAPK cascade signalling networks in plant defence. *Curr Opin Plant Biol* **12**: 421–426.

Roesler KR, Shorrosh BS, Ohlrogge JB. (1994) Structure and expression of an *Arabidopsis* acetyl-coenzyme A carboxylase gene. *Plant Physiol* **105**: 611–617.

Sanchez-Fernandez R, Davies TG, Coleman JO, Rea PA. (2001) The *Arabidopsis thaliana* ABC protein superfamily, a complete inventory. *J Biol Chem* **276**: 30231–30244.

Sasabe M, Toyoda K, Shiraishi T *et al.* (2002) cDNA cloning and characterization of tobacco ABC transporter: NtPDR1 is a novel elicitor-responsive gene. *FEBS Lett* **518**: 164–168.

Sjahril R, Chin DP, Khan RS *et al.* (2006) Transgenic Phalaenopsis plants with resistance to Erwinia carotovora produced by introducing wasabi defensin gene using Agrobacterium method. *Plant Biotechnol* **23**: 191–194.

Stukkens Y *et al.* (2005) NpPDR1, a pleiotropic drug resistance-type ATP-binding cassette transporter from *Nicotiana plumbaginifolia*, plays a major role in plant pathogen defense. *Plant Physiol* **139**: 341–352.

Tonks NK. (2006) Protein tyrosine phosphatases: From genes, to function, to disease. *Nat Rev Mol Cell Bio* **7**: 833–846.

Toth IK, Bell KS, Holeva MC, Birch PRJ, (2003) Soft rot *Erwiniae*: From genes to genomes. *Mol Plant Pathol* **4**: 17–30.

Tsai TM *et al.* (2007) PaCDPK1, a gene encoding calcium-dependent protein kinase from orchid, *Phalaenopsis amabilis*, is induced by cold, wounding, and pathogen challenge. *Plant Cell Rep* **26**: 1899–1908.

van Bentem SD, Hirt H. (2009) Protein tyrosine phosphorylation in plants: More abundant than expected? *Trends Plant Sci* **14**: 71–76.

Wildermuth MC, Dewdney J, Wu G, Ausubel FM. (2001) Isochorismate synthase is required to synthesize salicylic acid for plant defence. *Nature* **414**: 562–565 [Erratum: *Nature* **417**: 571–571 (2002)].

Xu Q, Fu HF, Gupta R, Luan S. (1998) Molecular characterization of a tyrosine-specific protein phosphatase encoded by a stress-responsive gene in *Arabidopsis Plant Cell* **10**: 1769–1769.

Yoon GM, Cho HS, Ha HJ *et al.* (1999) Characterization of NtCDPK1, a calcium-dependent protein kinase gene in *Nicotiana tabacum*, and the activity of its encoded protein. *Plant Mol Biol* **39**: 991–1001.

Zhao YF, Blumer SE, Sundin GW. (2005) Identification of *Erwinia amylovora* genes induced during infection of immature pear tissue. *J Bacteriol* **187**: 8088–8103.

CHAPTER 16

Identification and Characterization of New *Phalaenopsis* Orchid-Infecting Viruses

You-Xiu Zheng* and Fuh-Jyh Jan*,†

More than 30 orchid-infecting viruses have been reported. However, only three viruses, *Cymbidium mosaic virus* (CymMV), *Odontoglossum ringspot virus* (ORSV) and *Cucumber mosaic virus* (CMV), have previously been reported to infect *Phalaenopsis* orchids in Taiwan. In the past several years, we have identified and characterized the causal agents of four newly emerging *Phalaenopsis* orchid viral diseases that occurred in Taiwan. *Phalaenopsis* plants bearing virus-like symptoms of chlorotic ringspots and chlorotic spots with centric necrosis have been observed in Taiwan for more than a decade. The causal agent has been described as the "Taiwan virus" since these symptoms were frequently observed in orchids from Taiwan. Recently, we identified and characterized the "Taiwan virus" as an isolate of Capsicum chlorosis virus (CaCV-Ph), which reportedly infects capsicum and tomato plants in Australia and Thailand, respectively. Another symptom, chlorotic spots on *Phalaenopsis* leaves, was found to be caused by a virus with long filamentous particles. This virus reacted with a commercialized potyvirus-specific monoclonal antibody and was confirmed to be a new potyvirus based on the results of molecular characterization. Thus, the orchid-infecting potyvirus was designated as Phalaenopsis chlorotic spot virus (PhCSV). The third viral disease of *Phalaenopsis* orchids exhibited symptoms of chlorotic rings from which isometric particles were observed. The genomic sequence of this orchid virus revealed that it is an isolate of *Carnation mottle virus* (CarMV) and was designated as CarMV-Ph. Furthermore, *Phalaenopsis* plants with white flowers bearing symptoms of flower crinkle have been observed in Taiwan, China and Japan since 2006. Crinkled *Phalaenopsis* flowers generally suffered shorter flower longevity. In Japan, the symptom of flower crinkle is called "Hanafuke", meaning flower senescence. Our results demonstrate that ORSV was the sole causal virus of flower crinkle in *Phalaenopsis* orchids. The specific antisera against these four

*Department of Plant Pathology, National Chung Hsing University, Taichung, Taiwan.
†Corresponding author. E-mail: fjjan@nchu.edu.tw

viruses and the specific primers of RT-PCR have been developed and used successfully for detecting these viruses. The occurrence of these viruses in *Phaelenopsis* orchids has direct implications for the economically important nursery and floral industry worldwide.

16.1. Orchid-Infecting Viruses

16.1.1. *Orchid-infecting viruses*

A plant virus is a pathogen that infects plants and replicates in the infected cells. The virus can move from the infected cells to the neighboring cells via plasmodesmata and to the whole plant via the vascular system. When a plant virus replicates its genomic RNA or DNA, expresses viral proteins and packages viral particles, the biosynthesis and biometabolism of the host plant are interfered and symptoms appear gradually on leaves, stems, flowers and fruits. Orchids are also the hosts of plant viruses and various symptoms are observed on leaves and/or flowers when infected.

Odontoglossum ringspot virus (ORSV, *Tobamovirus*) is the first identified virus infecting orchids (Jensen and Gold, 1951). Since then, more than 31 viruses have been reported to infect plants in the *Orchidaceae* family (Lawson and Hsu, 1995; Zettler *et al.*, 1990; Zheng, 2007). Orchid-infecting viruses with definite classification are *Cymbidium mosaic virus* (CymMV, *Potexvirus*) (Francki, 1970), *Cucumber mosaic virus* (CMV, *Cucumovirus*) (Inouye, 1969), *Bean yellow mosaic virus* (BYMV, *Potyvirus*) (Lesemann and Koenig, 1985), *Colombian datura virus* (CDV, *Potyvirus*) (Fry *et al.*, 2004), *Calanthe mild mosaic virus* (CalMMV, *Potyvirus*) (Gara *et al.*, 1998), *Dasheen mosaic virus* (DsMV, *Potyvirus*) (Jordan *et al.*, 2002), Spiranthes mosaic virus 2 (SpiMV 2, *Potyvirus*), *Spiranthes mosaic virus* 3 (SpiMV 3, *Potyvirus*) (Guaragna *et al.*, 2006), Dendrobium mosaic virus (DeMV, *Potyvirus*) (Hu *et al.*, 1995), *Clover yellow vein virus* (ClYVV, *Potyvirus*) (Inouye *et al.*, 1988), *Turnip mosaic virus* (TuMV, *Potyvirus*) (Lesemann and Vetten, 1985), Vanilla mosaic virus (VanMV, *Potyvirus*) (Wisler *et al.*, 1987), *Ceratobium mosaic virus* (CerMV, *Potyvirus*) (Mackenzie *et al.*, 1998), *Sarcochilus virus Y* (SVY, *Potyvirus*) (Gibbs *et al.*, 2000), Habenaria mosaic virus (HaMV, *Potyvirus*) (Inouye *et al.*,

1998), Pecteilis mosaic virus (PcMV, *Potyvirus*) (Yora *et al.*, 1983), Vanilla necrosis virus (VNV, *Potyvirus*) (Pearson and Pone, 1988), Diurus virus Y (DVY, *Potyvirus*) (Gibbs *et al.*, 2000), Dendrobium vein necrosis virus (DVNV, *Closterovirus*) (Lesemann, 1977), *Tobacco rattle virus* (TRV, *Tobravirus*) (Lesemann and Vetten, 1985), *Cymbidium ringspot virus* (CymRSV, *Tombusvirus*) (Hollings *et al.*, 1977), *Tomato ringspot virus* (TomRSV, *Nepovirus*) (Goff and Corbett, 1977), Orchid fleck virus (OFV, *Rhabdovirus*) (Chang *et al.*, 1976), *Tomato spotted wilt virus* (TSWV, *Tospovirus*) (Hu *et al.*, 1993), *Impatiens necrotic spot virus* (INSV, *Tospovirus*) (Lawson and Hsu, 1995), Calanthe mosaic virus (CalMV, *Carlavirus*) (Yamamoto and Ishii, 1981), Capsicum chlorosis virus (CaCV, *Tospovirus*) (Zheng *et al.*, 2008c), Phalaenopsis chlorotic spot virus (PhCSV, *Potyvirus*) (Zheng *et al.*, 2008a) and *Carnation mottle virus* (CarMV, *Carmovirus*) (Zheng *et al.*, 2008b; Zheng *et al.*, 2011).

In the past, virus classification was mainly based on serological relationship, virus morphology and host range. However, it was difficult to classify a virus because of limited research information. Therefore, besides the 30 classified orchid-infecting viruses, there were some unclassified viruses such as Cymbidium mild mosaic virus (CyMMV, isometric virions) (Doi and Yora, 1978), Phalaenopsis necrosis virus (bacilliform) (Lesemann and Begtrup, 1971), Dendrobium leaf spot virus (bacilliform) (Peters, 1974) and Laelia red leafspot virus (LRLV, bacilliform) (Lawson and Hsu, 1995). Phalaenopsis necrosis virus had bacilliform virions and was considered to be an isolate of OFV. Today, it is easier to identify a virus with globally available research information, sequencing techniques and sequence databases.

16.1.2. Phalaenopsis *orchid-infecting viruses*

Among the orchid-infecting viruses, CymMV, ORSV (Wey, 1988), CMV (Lawson and Hsu, 1995; Ko, 1988), OFV (Lawson and Brannigan, 1986), DVNV (Lesemann, 1977), CaCV (Zheng *et al.*, 2008c), PhCSV (Zheng *et al.*, 2008a) and CarMV (Zheng *et al.*, 2008b; Zheng *et al.*, 2011) have been reported to infect *Phalaenopsis* orchids. All but

OFV and DVNV can reportedly infect *Phalaenopsis* orchids in Taiwan. CymMV-infected *Phalaenopsis* orchids show symptoms of black necrotic spots, line pattern, mosaic and strip on leaves. ORSV causes mosaic symptoms on *Phalaenopsis* leaves or is symptomless. *Phalaenopsis* orchids develop chlorosis or chlorotic streaks on leaves when infected by CMV. The co-infection of CymMV and ORSV often causes severe symptoms of necrosis and dark sunken spots on orchid leaves.

Orchid-infecting viruses are easily transmitted to new orchid plants via tissue propagations, which have a domino effect on transmission. CymMV and ORSV are very stable via mechanical transmission and have spread worldwide via tissue propagation and world trade. *Phalaenopsis* orchids infected by CymMV and/or ORSV show various symptoms on leaves and subsequently lose their commercial value.

16.1.3. *Challenges of the identification of new* Phalaenopsis *orchid-infecting viruses*

Besides the symptoms caused by the identified orchid-infecting viruses, there were some observed symptoms in *Phalaenopsis* orchids caused by unidentified viruses. Various limitations have hindered investigation of the unidentified viruses that cause disease in *Phalaenopsis* orchids. For example, a plant virus can cause different symptoms in various plants, and different viruses might cause similar symptoms in a single plant. Furthermore, plants might be co-infected by multiple viruses and exhibit complex symptoms. Hence, it is difficult to identify a virus based on symptoms. For preliminary diagnosis of viral diseases, the antisera against plant viruses are often used for quick screening. The antisera against orchid viruses are, however, not easy to obtain. Isolating a virus from orchids is difficult and back-inoculation of an isolated virus is a challenge. However, if the isolation and back-inoculation of a new orchid virus are not done, that virus cannot be claimed as the causal agent of a particular symptom. Recently, four new viral diseases of *Phalaenopsis* orchids were identified by our group via virus isolation, genome sequencing and back-inoculation.

16.2. *Capsicum Chlorosis Virus* (CaCV-Ph)

16.2.1. *Virus isolation and serological test*

Phalaenopsis plants bearing virus-like symptoms of chlorotic ringspots and necrotic spots have been observed in central and southern Taiwan for more than a decade (Chen *et al.*, 1998). A tospovirus-like culture has been isolated from symptomatic orchids. However, back-inoculation of this virus culture onto *Phalaenopsis* to confirm the culture as the causal agent was not successful and the sequence of the viral gene for virus taxonomy could not be determined (Chen *et al.*, 1998). As the causal agent of this orchid disease was unclear, the virus was described as the "Taiwan virus" since these symptoms were often observed in orchids from Taiwan.

In November 2002, one *Phalaenopsis* plant bearing chlorotic spots with necrosis in the central region (Fig. 16.1A) and another plant bearing chlorotic ringspots (Fig. 16.1B) were tested by enzyme-linked immunosorbent assay (ELISA) with antisera against 15 viruses, including four tospoviruses. The symptomatic leaves reacted positively only with the antiserum of *Watermelon silver mottle virus* (WSMoV, *Tospovirus*). Thereafter, a virus culture (91-orchid-1) was collected from *Phalaenopsis* showing symptoms of chlorotic ringspots on leaves and was established in *Nicotiana benthamiana* and *Chenopodium quinoa* through triple single-lesion isolation (Zheng *et al.*, 2008c).

Both antisera produced against the nucleocapsid proteins (NP) of 91-orchid-1 and WSMoV reacted strongly with the crude sap of 91-orchid-1 and WSMoV in indirect ELISA tests. A protein band measuring about 30 kDa in the crude sap of 91-orchid-1-infected *C. quinoa* was observed in western blotting using the antiserum against WSMoV or 91-orchid-1. The results of the serological test indicated that 91-orchid-1 is an isolate of a tospovirus belonging to the WSMoV serogroup.

16.2.2. *Virus identification*

For the identification of the *Phalaenopsis*-infecting virus, 91-orchid-1, the conserved region of L RNA was amplified by RT-PCR using

Fig. 16.1. Viral symptoms in *Phalaenopsis* orchids: (A) chlorotic spots with necrosis in the central region and (B) chlorotic ringspots caused by Capasicum chlorosis virus isolate *Phalaenopsis* (CaCV-Ph); (C) Phalaenopsis chlorotic spot virus (PhCSV)-caused chlorotic spots on leaves; (D) *Carnation mottle virus* (CarMV)-infected *Phalaenopsis* orchid with chlorotic rings; (E) white flower-bearing *Phalaenopsis* orchid infected by *Odontoglossum ringspot virus* (ORSV) with partially crinkled regions on petals; (F) as the symptoms progress, the crinkled regions gradually enlarge and fuse, and the petals become noticeably transparent.

primers gL3637 and gL4435c (Chu *et al.*, 2001). The 0.8-kbp conserved region of L RNA of 91-orchid-1 shares 86.9, 82.5, and 83.8 percent nucleotide identity and 98.8, 97.7, and 96.5 percent amino acid identity with those of CaCV-AIT (Knierim *et al.*, 2006), WSMoV, and *Groundnut* (GBNV), respectively (Zheng *et al.*, 2008c). The sequence comparison of partial L RNAs indicated that 91-orchid-1 is a member of the genus *Tospovirus*, closely related to CaCV, WSMoV and GBNV. Following the amplification of the nucleocapsid gene (N gene) with primers of five tospoviruses, the N gene of 91-orchid-1 was only amplified by CaCV primers (FJJ2003-8 and FJJ2003-10). After sequencing the 0.9-kbp RT-PCR product, the N gene of 91-orchid-1 was found to be 825 bp in length and to encode a predicted protein of 275 amino acids with a calculated molecular mass of 30.6 kDa. The nucleotide and predicted amino acid sequences of the 91-orchid-1 N gene share identities of 85.9–96.1 percent and 93.1–97.5 percent, respectively, with those of CaCV isolates available in GenBank (AY036057, NC_008301, DQ355974, AF059578 and AY312061). Based on the sequence analyses of the N gene, this *Phalaenopsis* orchid-infecting virus (91-orchid-1) belonged to the species CaCV in genus *Tospovirus* and was designated as CaCV-Ph.

16.2.3. *Back-inoculation*

To determine whether CaCV-Ph was the causal agent for the disorder observed in the field, the virus was back-inoculated onto several cultivars of *Phalaenopsis* orchid seedlings by mechanical inoculation. All of the cultivars were susceptible to the virus and showed symptoms of chlorotic ringspots and spots similar to those occurring in nature at 10–14 days post-inoculation. The chlorotic spots gradually necrotized and enlarged to cover the whole leaf. Finally, defoliation of the infected leaves occurred. The virus induced local infection on the inoculated leaves of *Phalaenopsis* plants, and no systemic symptoms were observed on new leaves of the inoculated *Phalaenopsis* plants during the 60-day observation period. The results of back-inoculation demonstrated that the virus causing chlorotic ringspots and spots with centric necrosis on *Phalaenopsis* orchids was indeed CaCV-Ph (Zheng *et al.*, 2008c).

16.2.4. *Significance of CaCV-Ph*

CaCV-Ph belongs to the genus *Tospovirus,* which has a wide host range including tobacco, tomato, peanut, cucurbits, vegetables and ornamentals (Prins and Goldbach, 1998; Jan *et al.*, 2003). In laboratory experiments, CaCV-Ph infected not only several *Phalaenopsis* cultivars but also systemically infected tomato, tobacco and pepper plants, which are economically important crops worldwide. CaCV-Ph initially caused severe symptoms of chlorotic ringspots and spots on the leaves of these susceptible crops which later developed into necrosis in the central region of the spots. Moreover, CaCV has been found to infect gloxinia (Lin *et al.*, 1999), calla lily (Chen *et al.*, 2007a), amaryllis and blood lily (Chen *et al.*, 2009) in Taiwan and has caused loss of tomato, capsicum and peanut crops in Australia (McMichael *et al.*, 2002), Thailand (Pongsapich and Chiemsombat, 2002) and China (Chen *et al.*, 2007b). Recently, necrotic spots on tomato leaves in Taiwan was reportedly caused by CaCV (Huang *et al.*, 2010). These reports indicated that CaCV infects not only ornamentals but also vegetables and other crops and has spread to other countries in Asia and Australia.

Besides being transmitted via mechanical transmission, tospoviruses are also transmitted by thrips such as *Frankliniella occidentalis, F. schultzei, F. fusca, F. intonsa, F. bisponsa, Thrips tabaci, T. setosus* and *T. palmi* (Jan *et al.*, 2003). *T. palmi* transmits most tospoviruses belonging to the WSMoV serogroup and is the potential vector of CaCV (Chiemsombat *et al.*, 2008; Persley *et al.*, 2006), which was reportedly transmitted by tomato thrips (*Ceratothripoides claratris*) (Premachandra *et al.*, 2005). *T. palmi* has not yet been found to transmit CaCV-Ph in Taiwan but this insect has been the major vector of WSMoV in Taiwan. Thus, *T. palmi* could potentially transmit CaCV-Ph in the near future.

Unlike CymMV and ORSV, which possess a narrow host range with no known insect vectors, CaCV-Ph infects many crops and can be transmitted by thrips and tissue propagation. With these properties,

CaCV-Ph, as a member of the tospoviruses, can be efficiently spread via orchid tissue cultures and insect transmission. The geographic distribution of CaCV reveals that this virus is a worldwide plant pathogen, and CaCV-Ph can potentially be spread to America and Europe via the trade in *Phalaenopsis* orchids and to other crops via thrips transmission. Thus, CaCV-Ph should be considered an economically impactful plant virus throughout the world.

16.3. *Phalaenopsis Chlorotic Spot Virus* (PhCSV)

16.3.1. *Virus isolation and serological test*

In October 2004, *Phalaenopsis* plants bearing virus-like symptoms of chlorotic spots on leaves (Fig. 16.1C) were observed in central Taiwan (Chen *et al.*, 2006b). After using 18 anti-viral antibodies to test the symptomatic tissues via indirect ELISA, the crude saps of the collected leaves reacted positively only to the POTY monoclonal antibody (MAb) (ADI, LLC., Arkansas, USA), which is widely used for the detection of potyviruses. A potyvirus-like culture (isolate 7-2) was then isolated from the symptomatic orchids and maintained in the systemic hosts, *N. benthamiana* and *C. quinoa*, by mechanical inoculation (Zheng *et al.*, 2008a). One structural protein of 33–34 kDa was present in purified virus preparations in SDS-PAGE. Electron microscopic examination of ultra-thin sections of isolate 7-2-infected *N. benthamiana* plant tissues revealed the presence of pinwheel inclusion bodies typical of potyviruses. Long flexuous filament particles about 12–15 by 750–800 nm in size were also observed in the crude sap of 7-2-infected *C. quinoa* leaves. A serological test and viral morphology indicated that the orchid virus 7-2 might belong to the genus *Potyvirus*.

16.3.2. *Virus identification*

To classify and characterize the orchid virus 7-2, the conserved region around 1.2 kbp including the 3′-end of NIb and the 5′-end of coat protein (CP) genes of the potyvirus genome was amplified using the degenerate primers Pot 1 and Pot 2 (Colinet and Kummert,

1993). When BLAST analysis against the GenBank database was performed, the 1,189-bp DNA fragment from isolate 7-2 shared 60.6 percent, 63.1 percent, 58.9 percent and 56.6 percent nucleotide identity and 55.6 percent, 65.1 percent, 52.5 percent and 44.8 percent amino acid identity with those of BYMV, *Beet mosaic virus* (BtMV), *Bean common mosaic virus* (BCMV) and TuMV, respectively, indicating they are found to be closely homologous. The sequence comparison of the conserved region indicated that isolate 7-2 should be a member of the potyviruses. For further classification of this virus, the nucleotide and amino acid sequences of the CP gene, the 3′ terminal region of viral genomic RNA was amplified from isolate 7-2 using primers Pot 2 and olig d(T). The 2,018 nt of the 3′ terminal region of isolate 7-2 RNA encodes a predicted polypeptide of 546 amino acids, and the CP of 280 amino acids is encoded at C-terminal region of this polypeptide. When BLAST analysis against the GenBank database was performed, the CP gene of isolate 7-2 shared low nucleotide and predicted amino acid identities of 57.9–67.3 percent and 53.5–70.3 percent, respectively, with those of the 18 potyviruses that share the highest sequence identities (Zheng *et al.*, 2008a). These results demonstrated that isolate 7-2 that caused chlorotic spots on *Phalaenopsis* orchids was a new potyvirus, and was designated as Phalaenopsis chlorotic spot virus (PhCSV) (Chen *et al.*, 2006b; Zheng *et al.*, 2008a).

16.3.3. *Back-inoculation*

To determine whether the chlorotic spots on *Phalaenopsis* were caused by isolate 7-2, orchids were mechanically inoculated with inocula prepared from isolate 7-2-infected *N. benthamiana*. Initially, the chlorotic spots were not obvious on the leaves, but the inoculated *Phalaenopsis* seedlings did eventually produce chlorotic spots similar to those found on naturally infected plants (Zheng *et al.*, 2008a).

16.3.4. *Significance of PhCSV*

The genus *Potyvirus* consists of a large group of plant viruses comprising 132 species and more than 90 tentative species that together cause serious diseases in major crops around the world. Potyviruses are transmitted by mechanical inoculation, grafting, seeds, pollen and insect vectors (aphids) and infect more than 59 families of plants. To date, 17 potyviruses have been reported to infect orchid plants. Among them, BYMV, BCMV, DsMV, TuMV and VNV [a member of the *Watermelon mosaic viruses* (WMV)] are responsible for a severe loss of fruits, vegetables and ornamentals worldwide. Under experimental conditions, PhCSV also infects *Solanum lycopersicum* (*Lycopersicum esculentum*, tomato), *Capsicum anmuum* var. *grossum* (sweet pepper), *Lisianthus russellianus* (*Eustoma grandiflorum*, lisianthus), *Vigna angularis* (adzuki bean), *Cucurbita moschata* (pumpkin), *Celosia argentea* (cockscomb), *Gomphrena globosa* (globe amaranth), and *Amaranthus viridis* (slender amaranth) (Zheng *et al.*, 2008a). Besides naturally infecting *Phalaenopsis* orchids, *Basella rugose mosaic virus* (BaRMV), a synonym virus of PhCSV, was also found to infect malabar spinach in Taiwan (Huang and Chang, 2006). Peace lily mosaic virus (PeLMV), a recently reported potyvirus infecting the peace lily in Vietnam (Ha *et al.*, 2008), was found to be a member of the PhCSV group based on sequence homology. All the above strongly suggests that PhCSV can infect vegetables and ornamentals, and might potentially cause disease in more crops via aphid transmission. Like other orchid-infecting viruses, PhCSV might be transmitted to other crops and wild plants via aphids in fields and be spread globally via orchid trading.

16.4. Carnation Mottle Virus (CarMV)

16.4.1. *Virus isolation and serological test*

In 2003, *Phalaenopsis* plants bearing previously unreported virus-like symptoms of chlorotic rings (Fig. 16.1D) were observed in central Taiwan (Zheng *et al.*, 2006). A virus culture (92-orchid-1) was collected from these infected *Phalaenopsis* leaves and maintained in *C. quinoa* by mechanical inoculation. Electron microscopic examination of ultra-thin sections of naturally infected *Phalaenopsis* orchids revealed the presence of isometric virions about 32 nm in diameter. Similar size particles were also observed in the crude sap of 92-orchid-1-infected *C. quinoa* plants. The crude saps of the symptomatic leaves from *Phalaenopsis* and infected *C. quinoa* were tested by ELISA with antisera against CymMV, ORSV, CMV, CaCV-Ph, PhCSV and some isometric viruses. Both crude saps reacted positively with CarMV antiserum. In western blotting tests using CarMV antiserum, a protein of 40 kDa was detected in purified viral particles and the leaf extracts of infected *C. quinoa*. These results indicated that the virus isolate, 92-orchid-1, might be a carmovirus close to CarMV (Zheng, 2007; Zheng *et al.*, 2008b; Zheng *et al.*, 2011).

16.4.2. *Virus identification*

To determine the taxonomic relationships and to characterize the genomic sequence of 92-orchid-1, one cDNA fragment around 0.9 kbp was amplified from the total RNAs extracted from 92-orchid-1-infected *C. quinoa* plants by RT-PCR using the degenerate primers, FJJ2003-17 and FJJ2003-18, of the conserved region located in the ORF1RT gene of carmoviruses (Zheng, 2007). This conserved region of the 92-orchid-1 ORF1RT gene shared 97.1 percent nucleotide identity and 98.3 percent amino acid identity with that of CarMV-B (Guilley *et al.*, 1985), and low sequence identities (less than 70 percent) with those of other carmoviruses. The sequence comparison indicated that 92-orchid-1 shared high homology in the conserved region of RNA-dependent RNA polymerase (RdRp) with CarMV. In order to classify this orchid virus, the CP gene of 92-orchid-1 was amplified using the specific primers of the CarMV CP gene and found to share 95.8–98.8 percent nucleotide identity and 96.8–98.9 percent amino acid identity with those of other CarMV isolates available in GenBank. Hence, 92-orchid-1 was considered to be an isolate of CarMV and was designated as CarMV-Ph (Zheng, 2007; Zheng *et al.*, 2008b; Zheng *et al.*, 2011).

16.4.3. *Significance of CarMV*

CarMV is a major viral pathogen of carnations and causes significant losses (Canizares *et al.*, 2001; Singh *et al.*, 2005). CarMV infects begonias (Paludan and Begtrup, 1985), daphnes (Lee *et al.*, 2006) and calla lily (Chen *et al.*, 2003a) naturally, and cow herb (*Saponaria vaccaria*)

(Hakkaart and van Olphen, 1971) and rainbow pink (*Dianthus chinensis*) (Chen *et al.*, 2003b) mechanically. Like CymMV and ORSV, CarMV is not transmitted by insect vectors but has highly stable virions. Thus, CarMV is a highly infectious virus and often arises in crops around the world as a result of simple transmission via contact and cultivation. In Taiwan, CarMV has been reported to infect carnation (Chen *et al.*, 2003b), calla lily (Chen *et al.*, 2003a) and *Phalaenopsis* orchids (Zheng *et al.*, 2008b; Zheng *et al.*, 2011). Because CarMV readily infects ornamental plants and causes severe symptoms, it is possible that this virus will infect other orchids and ornamental plants in Taiwan.

16.5. The Virus Causing a New Disease with Flower Crinkle in *Phalaenopsis*

16.5.1. *Virus isolation and serological test*

Most orchid viruses cause symptoms on leaves. However, since 2006 a new disease has been observed with symptoms of flower crinkle on white flower-bearing *Phalaenopsis* hybrids, especially on *Phal.* Sogo Yukidian "V3" (*Phal.* Yukimai × *Phal.* Taisuco Kochdian) (Fig. 16.1E), in Taiwan, China and Japan. Crinkled *Phalaenopsis* flowers generally had shorter flower longevity. In Japan, the symptom of flower crinkle is called "Hanafuke", meaning flower senescence (personal communication with Dr. Kazuo Yoneda, Nihon University). Because the cause of flower crinkle in *Phalaenopsis* orchids was unclear and no obvious symptom was shown in *Phalaenopsis* leaves, for a long time the disorder was considered a result of physiological dysfunction.

Five plants of *Phal.* Sogo Yukidian "V3" with flower crinkle symptoms were collected and detected by indirect ELISA using antisera against ORSV, CymMV, CaCV-Ph, PhCSV, CarMV-Ph and 13 other viruses. The leaves and flowers from one symptomatic plant reacted positively only to the antiserum against ORSV, while those from the other four plants reacted positively to antisera against ORSV and CymMV (Zheng *et al.*, 2010). RT-PCR confirmed the results of the ELISA tests for the presence of the viruses. The rod-shaped particles were found in ORSV-infected but not healthy tissues when examined under an electron microscope. Based on these results, it was hypothesized that flower crinkle is not a physiological problem and is probably caused by ORSV. Thereafter, five ORSV-like virus cultures, one each from the flowers of the five symptomatic *Phalaenopsis* orchids exhibiting flower crinkle symptoms, were established in *C. quinoa*. One CymMV-like virus culture was isolated from the flowers of one of the ORSV/CymMV mix-infected *Phalaenopsis* orchids.

16.5.2. *Virus identification*

In order to identify that the six isolated virus cultures were indeed ORSV or CymMV, the specific primers for ORSV and CymMV were used for amplifying the CP genes of these isolated viruses. Sequence analysis showed that the CP genes of five ORSV-like virus cultures were identical and shared 97.3–100 percent nucleotide identity and 96.2–100 percent amino acid identity with those of 41 ORSV isolates available in GenBank. The CP genes of the CymMV-like virus culture shared 89.3–97.2 percent nucleotide identity and 95.1–99.6 percent amino acid identity with those of 10 CymMV isolates. The results of sequence comparison demonstrated that these isolated viruses are indeed ORSV or CymMV (Zheng *et al.*, 2010).

16.5.3. *Back-inoculation*

In the ELISA test, one symptomatic *Phalaenopsis* orchid appeared to be infected only by ORSV. Therefore, ORSV was suspected as the causal agent of flower crinkle. To confirm this hypothesis, plants of *Phal.* Sogo Yukidian "V3" and *Doritaenopsis* Join Angel "274-1" (*Dtps.* Casablanca Joy × *Phal.* Taida inlong "KC1932") bearing white flowers were inoculated with ORSV and CymMV under several conditions. No matter where the inoculated sites were (leaf or flower stalk), the flowers of *Phalaenopsis* plants infected by ORSV showed the crinkle symptoms (Zheng *et al.*, 2010). The *Phalaenopsis* plants also exhibited flower crinkle when doubly inoculated with ORSV and CymMV, regardless of the order of the inoculation. However, none of the *Phalaenopsis* plants, either naturally or experimentally infected by CymMV alone, exhibited flower crinkle. At the same time, the naturally CymMV-infected plants bore crinkled flowers when mechanically inoculated with ORSV. The results of back-inoculation revealed that ORSV is the single causal agent of flower crinkle in *Phalaenopsis*.

Initially, the *Phalaenopsis* orchids showed partially crinkled regions on the petals (Fig. 16.1E), and then the crinkled regions became transparent. As the symptoms continued to develop, the crinkled regions gradually enlarged and fused. At that stage the petals became noticeably transparent (Fig. 16.1F). In addition, the blooming period of diseased orchids was shortened as compared to that of asymptomatic plants. The symptoms were then observed on new flowers systemically, and some flower buds shriveled and dried, resulting in a drastic reduction in flower numbers (Zheng *et al.*, 2010).

16.5.4. *Significance of the virus*

ORSV-infected *Phalaenopsis* orchids showed symptoms of mosaic and leaf-yellowing (Yamane *et al.*, 2008) or unapparent symptoms (Chen *et al.*, 2006a) on leaves. ORSV also causes the symptoms of ringspots, chlorotic mottle and mild mosaic in *Odontoglossum* (Jensen and Gold, 1951), *Calanthe* (Matsumoto *et al.*, 1995) and *Cymbidium* (Kondo *et al.*, 1992) orchids, respectively, and the symptoms of color break in *Cattleya* flowers (Corbett, 1967) and color break and deformation in *Phalaenopsis* flowers (Inouye, 1990). ORSV is easily transmitted via tissue propagation, this partly explains why most orchids are ORSV infected. *Phalaenopsis* hybrids bearing large white flowers, such as *Phal.* Sogo Yukidian "V3", are popular potted flowers in Japan, America and Europe. However, crinkle symptoms with obvious transparency often arises in *Phalaenopsis* orchids bearing white flowers, and the reduction in visual appeal has caused a drastic loss of market value. Appropriate management of *Phalaenopsis* culturing and planting is necessary to minimize the losses caused by ORSV and other orchid viruses.

16.6. Conclusion and Future Prospects

Containing more than 25,000 species, *Orchidaceae* is one of the largest families in the plant kingdom. Comparing with the viral pathogens associated with plants in *Solanaceae, Asteraceae, Leguminosae, Cucurbitaceae and Brassicaceae,* orchids are reported to be infected by only 30 classified and several unclassified viruses. The above-mentioned difficulties and challenges in the identification of new orchid viral diseases could well explain the low numbers of orchid-infecting viruses identified to date. Indeed, many virus-induced diseases without definite classification have been observed in orchids.

In order to identify more unknown virus-like diseases, we concentrated on studying viral pathogens of *Phalaenopsis* orchids from the past years. We successfully demonstrated that chlorotic ringspots caused by the so-called "Taiwan virus" were indeed attributable to CaCV-Ph. In addition, we showed that chlorotic spots were caused by PhCSV, chlorotic rings by CarMV, and flower crinkle by ORSV. These four viruses are major viral pathogens of ornamentals and vegetables; it is necessary to monitor the diseases caused by these viruses to prevent infection from spreading to other crops.

The approaches to the control and prevention of viral diseases in orchids include the use of virus-free tissues for orchid propagation, virus quarantine during culturing and planting, insect management, and removal of inoculum sources. Some systems have been developed to regenerate virus-free orchids (chemical and thermal therapies) and to detect orchid viruses (ELISA and RT-PCR). However, the identification of orchid-infecting viruses remains extremely vital for developing strategic approaches to viral disease management.

Acknowledgments

The authors are grateful to Dr. Chung-Jan Chang, Professor of Plant Pathology at the University of Georgia, Griffin Campus, for his critical review of the manuscript for journal publication and for publication in this volume.

References

Canizares MC, Marcos JF, Pallás V. (2001) Molecular variability of twenty-one geographically distinct isolates of *Carnation mottle virus* (CarMV) and phylogenetic relationships within the *Tombusviridae* family. *Arch Virol* **146**: 2039–2051.

Chang MU, Arai K, Doi Y, Yora K. (1976) Morphology and intracellular appearance of orchid fleck virus. *Ann Phytopathol Soc Japan* **42**: 156–157.

Chen CC, Chen YK, Ko WF, Yeh ST. (2006a) Symptomatological and cytopathological studies of *Odontoglossum ringspot virus* and*Cymbidium mosaic virus* infections on *Phalaenopsis* orchids in Taiwan. *Bull Taichung Dist Agric Improv Stn* **93**: 15–27.

Chen CC, Huang CH, Chen TC *et al.* (2007a) First report of Capsicum chlorosis virus causing yellow stripes on calla lilies in Taiwan. *Plant Dis* **91**: 1201.

Chen CC, Huang CH, Cheng YH *et al.* (2009) First report of Capsicum chlorosis virus infecting amaryllis and blood lily in Taiwan. *Plant Dis* **93**: 1346.

Chen CC, Ko WF, Lin CY *et al.* (2003a) First report of *Carnation mottle virus* in calla lily (*Zantedeschia* spp.) *Plant Dis* **87**: 1539.

Chen CC, Lin CY, Ko WF, Jan FJ. (2003b) Isolation and characterization of *Carnation mottle virus* from carnation. *Plant Pathol Bull* **12**: 199–208.

Chen CC, Zheng YX, Chen YK, Jan FJ. (2006b) Isolation and characterization of a new potyvirus causing chlorotic spots on moth orchids (*Phalaenopsis* spp.) Phytopathology **96**: S22.

Chen K, Xu Z, Yan L, Wang G. (2007b) Characterization of a new strain of Capsicum chlorosis virus from peanut (*Arachis hypogaea* L.) in China. *J Phytopathol* **155**: 178–181.

Chen TC, Hsu HT, Yeh SD. (1998) A new tospovirus-like virus isolated from orchid. In: *IV International Symposium on Tospoviruses and Thrips in Floral and Vegetable Crops*, p. 94. May 2–6, Wageningen, Netherlands.

Chiemsombat P, Gajanandana O, Warin N *et al.* (2008) Biological and molecular characterization of tospoviruses in Thailand. *Arch Virol* **153**: 571–577.

Chu FH, Chao CH, Chung MH *et al.* (2001) Completion of the genome sequence of *Watermelon silver mottle virus* and utilization of degenerate primers for detecting tospoviruses in five serogroups. *Phytopathology* **91**: 361–368.

Colinet D, Kummert J. (1993) Identification of a sweetpotato feathery mottle virus isolate from China (SPFMVCH) by the polymerase chain reaction with degenerate primers. *J Virol Methods* **45**: 149–159.

Corbett MK. (1967) Some distinguishing characteristics of the orchid strain of tobacco mosaic virus. *Phytopathology* **57**: 164–172.

Doi Y, Yora K. (1978) Cymbidium mild mosaic virus, isolation of cymbidium mild mosaic virus. *Korean J Plant Prot* **17**: 131–138.

Francki RIB. (1970) *Cymbidium Mosaic Virus.* CMI/AAB Descriptions of Plant Viruses No. 27.

Fry CR, Zimmerman MT, Scott SW. (2004) Occurrence of *Colombian datura virus* in the terrestrial orchid, *Spiranthes cernua. J Phytopathol* **152**: 200–203.

Gara IW, Kondo H, Maeda T *et al.* (1998) Calanthe mild mosaic virus, a new potyvirus causing a mild mosaic disease of *Calanthe* orchid in Japan. *J Phytopathol* **146**: 357–363.

Gibbs A, Mackenzie A, Blanchfield A *et al.* (2000) Viruses of orchids in Australia: Their identification, biology and control. *Aust Orchid Rev* **65**: 10–21.

Goff LM, Corbett MK. (1977) Association of tomato ringspot virus with a chlorotic leaf streak of *Cymbidium* orchids. *Phytopathology* **67**: 1096–1100.

Guaragna MA, Ndum O, Jordan R. (2006) Detection and characterization of two previously undescribed potyviruses in the terrestrial orchid *Spiranthes cernua. Acta Hortic* **722**: 209–218.

Guilley H, Carrington JC, Balazs E *et al.* (1985) Nucleotide sequence and genome organization of carnation mottle virus RNA. *Nucleic Acids Res* **13**: 6663–6677.

Ha C, Coombs S, Revill PA *et al.* (2008) Design and application of two novel degenerate primer pairs for the detection and complete genomic characterization of potyviruses. *Arch Virol* **153**: 25–36.

Hakkaart FA, van Olphen M. (1971) *Saponaria vaccaria* "Pink Beauty", a new test plant for carnation mottle virus. *Eur J Plant Pathol* **77**: 127–133.

Hollings M, Stone OM, Barton RJ. (1977) Pathology, soil transmission and characterization of cymbidium ringspot, a virus from *Cymbidium* orchids and white clover (*Trifolium repens*) *Ann Appl Biol* **85**: 233–248.

Hu JS, Ferreria S, Wang M *et al.* (1995) Purification, host range, serology and partial sequencing of dendrobium mosaic potyvirus, a new member of the bean common mosaic virus subgroup. *Phytopathology* **85**: 542–546.

Hu JS, Ferreira S, Wang M, Xu MQ. (1993) Detection of *Cymbidium mosaic virus, Odontoglossum ringspot virus,* tomato spotted wilt virus, and potyviruses infecting orchids in Hawaii. *Plant Dis* **77**: 464–468.

Huang CH, Chang YC. (2006) Basella rugose mosaic virus, a new potyvirus infecting *Basella rubra. Plant Pathol* **55**: 819.

Huang CH, Zhang YX, Cheng YH, Lee WS, Jan FJ. (2010) First report of Capsicum *chlorosis* virus infecting tomato in Taiwan. *Plant Dis* **94**:1263.

Inouye N. (1969) Cucumber mosaic virus isolated from *Dendrobium. Agric Res* **53**: 49–60.

Inouye N. (1990) Virus, fungal and bacterial diseases of orchids. *Plant Prot* **44**: 177–181.

Inouye N, Maeda T, Mistsuhata K. (1988) A strain of clover yellow vein virus isolated from *Calanthe* sp. *Acta Hortic* **234**: 61–68.

Inouye N, Maeda T, Mitsuhata K, Gara IW. (1998) Habenaria mosaic virus, a new member of potyvirus from *Habenaria radiata* orchid. *Bull Res Inst Bioresour Okayama Univ* **5**: 155–168.

Jan FJ, Chen TC, Yeh SD. (2003) Occurrence, importance, taxonomy and control of thrips-borne tospoviruses. In: Huang H, Acharya SN (eds.), *Advances in Plant Disease Management*, pp. 391–411. Research Signpost, Kerala.

Jensen DD, Gold HA. (1951) A virus ring spot of *Odontoglossum* orchid: Symptoms, transmission and electron microscopy. *Phytopathology* **41**: 648–653.

Jordan R, Guaragna MA, Kinard G, Lynn S. (2002) Detection and first report of *Dasheen mosaic virus* and a second potyvirus infecting the terrestrial orchid *Spiranthes cernua. Acta Hortic* **568**: 253–260.

Knierim D, Blawid R, Maiss E. (2006) The complete nucleotide sequence of a capsicum chlorosis virus isolate from *Lycopersicum esculentum* in Thailand. *Arch Virol* **151**: 1761–1782.

Ko NJ. (1988) Cytological identification of *Cucumber mosaic virus* infecting *Phalaenopsis*. *Proc Natl Sci Counc Repub China* **12**: 46–51.

Kondo H, Maeda T, Inouye N. (1992) Host range and properties of odontoglossum ringspot tobamovirus (ORSV) isolated from miniature *Cymbidium* in Japan. *Bull Res Inst Bioresour Okayama Univ* **1**: 21–34.

Lawson RH, Brannigan M. (1986) Virus disease of orchids. In: *Handbook on Orchid Pests and Diseases*, pp. 2–49 American Orchid Society, Florida.

Lawson RH, Hsu HT. (1995) Orchid. In: Loebenstein G, Lawson RH, Brunt AA (eds.), Virus and Virus-Like Diseases of Bulb and Flower Crops, pp. 409–420. John Wiley & Sons, West Sussex.

Lee BY, Ha JH, Ryu KH. (2006) Daphne-infecting viruses in Korea and properties of *Daphne virus S*. *Acta Hortic* **722**: 55–58.

Lesemann DE. (1977) Long, filamentous virus-like particles associated with vein necrosis of *Dendrobium phalaenopsis*. *Phytopathol Z* **89**: 330–339.

Lesemann DE, Begtrup J. (1971) Electron microscopic demonstration of a bacilliform virus in *Phalaenopsis*. *Phytopathol Z* **71**: 257–269.

Lesemann DE, Koenig R. (1985) Identification of bean yellow mosaic virus in *Masdevallia*. *Acta Hortic* **164**: 347–354.

Lesemann DE, Vetten HJ. (1985) The occurrence of tobacco rattle and turnip mosaic viruses in *Orchis* sp, and of an unidentified potyvirus in *Cypripedium calceolus*. *Acta Hortic* **164**: 45–54.

Lin JL, Chang CA, Chen CC. (1999) Cloning and sequencing the nucleocapsid protein gene of a tospovirus isolate from gloxinia. *Plant Pathol Bull* **8**: 182.

Mackenzie AM, Nolan M, Wei KJ *et al.* (1998) Ceratobium mosaic potyvirus: Another virus from orchids. *Arch Virol* **143**: 903–914.

Matsumoto JI, Takanori M, Inouye N. (1995) *Odontoglossum ringspot virus* isolated from *Calanthe* sp. in Japan. *Bull Res Inst Bioresour Okayama Univ* **3**: 163–174.

McMichael LA, Persley DM, Thomas JE. (2002) A new tospovirus serogroup IV species infecting capsicum and tomato in Queensland, Australia. *Austral Plant Pathol* **31**: 231–239.

Paludan N, Begtrup J. (1985) *Carnation mottle virus* demonstrated in *Begonia elatior* and *Begonia* × *cheimantha* showing vein clearing, leaf curl and flower break. *Acta Hortic* **164**: 33–39.

Pearson MN, Pone SP. (1988) Viruses of *Vanilla* in the Kingdom of Tonga. *Austral Plant Pathol* **17**: 59–60.

Persley DM, Thomas JE, Sharman M. (2006) Tospoviruses — an Australian perspective. *Austral Plant Pathol* **35**: 161–180.

Peters KR. (1974) Charakterisierung schwer nachweisbarer phytopathogener Viren mit physikalischen und elektronenmikroskopischen Methoden. Dissertation, University of Hamburg, Hamburg, 93 pp.

Pongsapich P, Chiemsombat P. (2002) Characterization of tospovirus infecting tomatoes in Thailand revealed the presence of serogroup IV-tospovirus but not serogroup I-tomato spotted wilt virus. In: The *1st International Conference on Tropical and Subtropical Plant Diseases*, November 5–8, Chiang Mai, Thailand.

Premachandra WTSD, Borgemeister C, Maiss E *et al.* (2005) *Ceratothripoides claratris*, a new vector of a Capsicum chlorosis virus isolate infecting tomato in Thailand. *Phytopathology* **95**: 659–663.

Prins M, Goldbach R. (1998) The emerging problem of tospovirus infection and nonconventional methods of control. *Trends Microbiol* **6**: 31–35.

Singh HP, Hallan V, Raikhy G *et al.* (2005) Characterization of an Indian isolate of *Carnation mottle virus* infecting carnations. *Curr Sci* **88**: 594–601.

Wey GC. (1988) Occurrence and investigation of important diseases on *Phalaenopsis* in Taiwan. *Rep Taiwan Sugar Res Inst* **122**: 31–41.

Wisler GC, Zettler FW, Mu L. (1987) Virus infections of *Vanilla* and other orchids in French Polynesia. *Plant Dis* **71**: 1125–1129.

Yamamoto T, Ishii M. (1981) Mosaic disease of *Calanthe discolor* Lind. *Proc Assoc Plant Prot Shikoku* **16**: 75–79.

Yamane K, Oyama K, Iuchi E *et al.* (2008) RT-PCR detection of *Odontoglossum ringspot virus*, *Cymbidium mosaic virus* and tospoviruses and association of infections with leaf-yellowing symptoms in *Phalaenopsis*. *J Phytopathol* **156**: 268–273.

Yora K, Saito Y, Doi Y *et al.* (1983) *Handbook of Plant Viruses*. Asakura Shoten, Tokyo.

Zettler FW, Ko NJ, Wisler GC *et al.* (1990) Viruses of orchids and their control. *Plant Dis* **74**: 621–625.

Zheng YX. (2007) Identification and characterization of new *Phalaenopsis* orchid-infecting viruses and developing new transgenic approach for broad-spectrum resistance to viruses by expressing designed siRNA-generating synthetic nucleotides. PhD Dissertation, National Chung Hsing University, Taichung, Taiwan.

Zheng YX, Chen CC, Chen CC, Jan FJ. (2006) Identification and characterization of an isolate of *Carnation mottle virus* from *Phaelenopsis* orchids in Taiwan. *Plant Pathol Bull* **15**: 316.

Zheng YX, Chen CC, Chen YK, Jan FJ. (2008a) Identification and characterization of a potyvirus causing chlorotic spots on *Phalaenopsis* orchids. *Eur J Plant Pathol* **121**: 87–95.

Zheng YX, Chen CC, Jan FJ. (2008b) *Phalaenopsis* orchids showing chlorotic rings, a new disease caused by *Carnation mottle virus*. *J Plant Pathol* **90**: 192 (S2)

Zheng YX, Chen CC, Jan FJ (2011) First report of *Carnation mottle virus* in *Phalaenopsis* orchids. Plants Dis **95**: 354.

Zheng YX, Chen CC, Yang CJ *et al.* (2008c) Identification and characterization of a tospovirus causing chlorotic ringspots on *Phalaenopsis* orchids. *Eur J Plant Pathol* **120**: 199–209.

Zheng YX, Shen BN, Chen CC, Jan FJ. (2010) *Odontoglossum ringspot virus* causing flower crinkle in *Phalaenopsis* hybrids. *Eur J Plant Pathol* **128**: 1–5.

CHAPTER 17

What We have Learned from Studies of the Differential Movement Mechanism of *Cymbidium Mosaic Virus*

Hsiang-Chia Lu*, Hsiang-Iu Wang† and Hsin-Hung Yeh*,‡

Cymbidium mosaic virus (CymMV) is one of the most important viruses that infect orchids. Previous observations indicated that although CymMV was mostly restricted to infecting the Orchidaceae, experimental inoculation indicated that CymMV can infect several families of plant hosts. In addition, CymMV has never been isolated from wild orchids. We found two CymMV isolates that can move differentially between hosts. Both can infect *Phalaenopsis* orchids, but only one can systemically infect *Nicotiana benthamiana* (the model plant used by plant virologists). Genetic studies revealed two control modes involved in CymMV movement, with the coat protein (CP) playing a dominant role. Biochemistry studies revealed that the movement ability of CymMV is associated with the binding affinity of the CP and viral RNA, and the important amino acids for CymMV movement in *N. benthamiana* were identified. Sequence analysis of worldwide CymMV CPs suggests that CymMV may have once had a broader host range, and only some isolates evolved to infect orchids only. A remaining issue is identifying the natural host of CymMV, so important for disease management in the orchid industry.

*Department of Plant Pathology and Microbiology, National Taiwan University, Taipei 106, Taiwan.
†Department of Computer Science, National Tsing Hua University, 101, Section 2, Kuang-Fu Road, Hsinchu 30013, Taiwan.
‡Corresponding author. E-mail: hyeh@ntu.edu.tw

17.1. Introduction

Cymbidium mosaic virus (CymMV) is one of the important prevalent viruses that infect orchids and a major threat to the industry. CymMV is mainly spread through mechanical transmission (Jensen, 1951; Zettler *et al.*, 1987). It can cause chlorotic or necrotic sunken-patch symptoms on leaves and flowers, which results in economic loss for the orchid industry (Zettler *et al.*, 1990). No known natural vectors have been reported for the spread of CymMV. The management of disease with CymMV infection has been studied with RT-PCR, ELISA, DIG-labeled cRNA probes (Hu *et al.*, 1998), real-time RT-PCR (Eun *et al.*, 2000), LC- and/or MALDI-mass spectrometry (Tan *et al.*, 2000) and quartz crystal microbalance-based DNA biosensors (Eun *et al.*, 2002). A recent breakthrough found CymMV-resistant transgenic *Dendrobium* and *Phalaenopsis* expressing the coat protein (CP) gene of CymMV (Chang *et al.*, 2005; Chan *et al.*, 2005). In addition, we recently developed a gene-silencing vector induced by CymMV that can help overcome several challenges of functional genomics studies of orchids (including large genome size, long life cycle and low transformation efficiency) (Lu *et al.*, 2007).

Viruses usually have a defined, restricted host range that varies among different viruses and sometimes even different isolates of the same virus species. One mystery about CymMV is that in general, the host range of CymMV is considered restricted to the Orchidaceae, although CymMV has never been isolated from wild orchids (Zettler *et al.*, 1990). Recently, our studies of naturally collected variants of CymMV helped us better understand the functional diversity that CymMV generates for adaptation (Lu *et al.*, 2009). Our results also support a new hypothesis about the origin of CymM and may help in the management of CymMV infection.

17.2. Two Distinct Isolates of CymMV Move Differentially Between Hosts

We previously collected several isolates of CymMV to construct virus-induced gene-silencing vectors for orchid functional genomics studies

(Lu *et al.*, 2007, 2009). To biologically characterize these isolates, we inoculated the virus in plants maintained in the laboratory for biological differentiation. Interestingly, we found two distinct types of CymMV (Lu *et al.*, 2009). Both M1 and M2 can infect *Phalaenopsis* orchids, but only M1 can systemically infect *Nicotiana benthamiana* (the model plant used by plant virologists). However, even with M1 infection of *N. benthamiana*, no visible symptoms were observed. This finding may explain why *N. benthamiana* is not recorded as a host of CymMV (ICTVdB: The Universal Virus Database of the International Committee on Taxonomy of Viruses, *www.ictvdb.iacr.ac.uk/Ictv/fr-index.htm*). The M1 and M2 isolates of CymMV were cloned and completely sequenced. Our whole-mount *in situ* hybridization revealed that both isolates can replicate in *N. benthamiana* protoplasts, but M2 is restricted to the initially infected cells (Lu *et al.*, 2009). Therefore, M2 cannot move cell to cell. Because both isolates share 97 percent sequence identity, a subtle sequence variation contributes to the differences in host-specific movement — the virus-encoded movement protein(s) may function in one plant but not another.

17.3. Two Control Modes are Involved in CymMV Movement and CP Plays a Dominant Role in Movement

CymMV belongs to the genus *Potexvirus* in the family Flexiviridae (Adams *et al.*, 2005). Previous studies of potexviruses revealed that triple gene block proteins (TGBps) and CPs are required for viral movement between cells (Bayne *et al.*, 2005; Chapman *et al.*, 1992; Krishnamurthy *et al.*, 2002, 2003; Lough *et al.*, 2000; Morozov and Solovyev, 2003; Tamai and Meshi, 2001; Verchot-Lubicz, 2005). We used the natural mutants of CymMV, M1 and M2, for detailed study of the movement of CymMV. Interestingly, genome exchange between these two isolates revealed two distinct control modes (dominant and matching) involved in CymMV movement (Fig. 17.1). Our studies revealed that among the components involved in CymMV movement, the CP plays a dominant role, but a surprisingly specific combination (we termed "matching mode") of certain

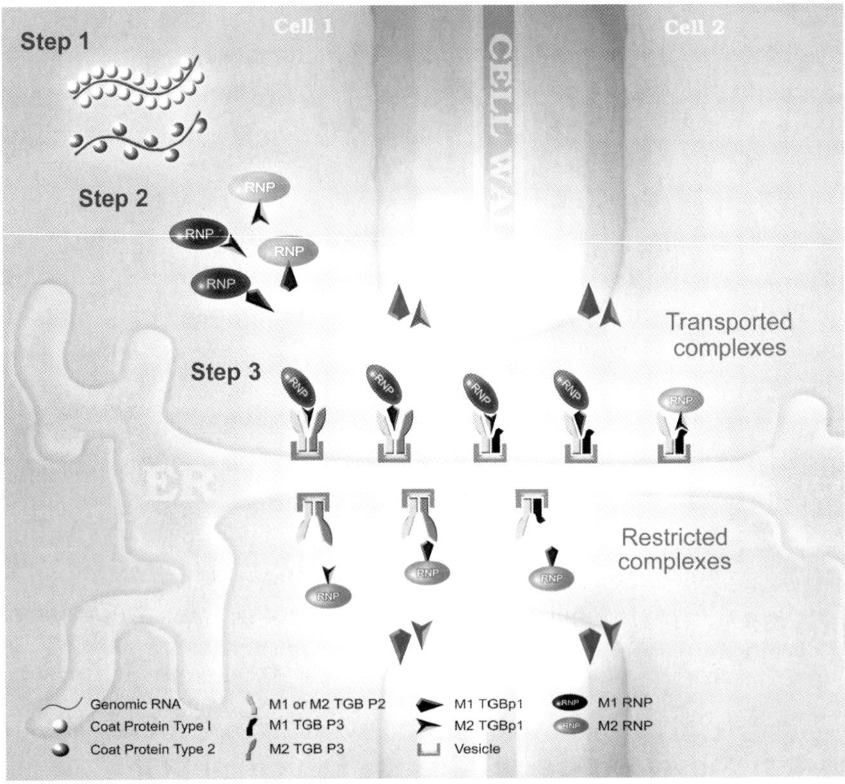

Fig. 17.1. Schematic representation of transported complexes and restricted complexes in *Cymbidium mosaic virus* (CymMV) movement between cells. In most transported complexes, the CP is dominant in controlling the movement between cells. However, another control mode that requires correct matching of triple gene block protein 1 (TGBp1), TGBp3 and CP allows for the movement of CymMV between cells. The transported complexes and restricted complexes are indicated. The RNAs of CymMV isolates M1 and M2, TGBp1s, TGBp2s, TGBp3s, CP, and ribonucleoprotein (RNP) are indicated.

movement-related components also allows the CymMV to move cell to cell in *N. benthamiana* when the CP is not functional. The combination includes the M2 CP and 5′ RNA sequence (1–4,333, including the whole RNA-dependent RNA polymerase encoding region) but also the TGB open reading frames of M1 (Fig. 17.2).

	RdRp	TGB1	TGB2	TGB3	CP	Local Infection	Systemic Movement
1. pCymMV-M1						+	+
2. pCymMV-M2						+	–
3. pCymMV-M1/RdRp-M2						+	+
4.pCymMV-M2/RdRp-M1						+	–
5.pCymMV-M1/TGB-M2						+	+
6.pCymMV-M2/TGB-M1						+	+
7.pCymMV-M1/CP-M2						+	–
8.pCymMV-M2/CP-M1						+	+
9.pCymMV-M2/TGBp1,3-M1						+	+

Fig. 17.2. Schematic representation of genome organization and infectivity assay of the parents of CymMV M1 and M2 isolates and the derived chimeric constructs. Genome organization of CymMV is shown at the top. Rectangles represent open reading frames encoded by CymMV genomic RNA, RNA-dependent RNA polymerase (RdRp), TGB open reading frames 1, 2, and 3, and coat protein (CP). Sequences corresponding to pCymMV-M1 and pCymMV-M2 are indicated by gray and white rectangles, respectively. Clones with competent systemic infection in *N. benthamiana* are indicated by plus symbols.

The results are intriguing because although the movement-required components for potexviruses have been previously determined, the hierarchic roles of the components in CymMV were previously unknown.

17.4. The Movement Ability of CymMV is Associated with the Binding Affinity of CP and Viral RNA

Although M1 and M2 CPs determine the movement of CymMV in *N. benthamiana*, the CPs contain only four amino acid differences. Thus, we could identify the important amino acids involved in CymMV movement by site-directed mutagenesis. We found two CP amino acid changes, in G82A and L89P, that were both needed for M2 to

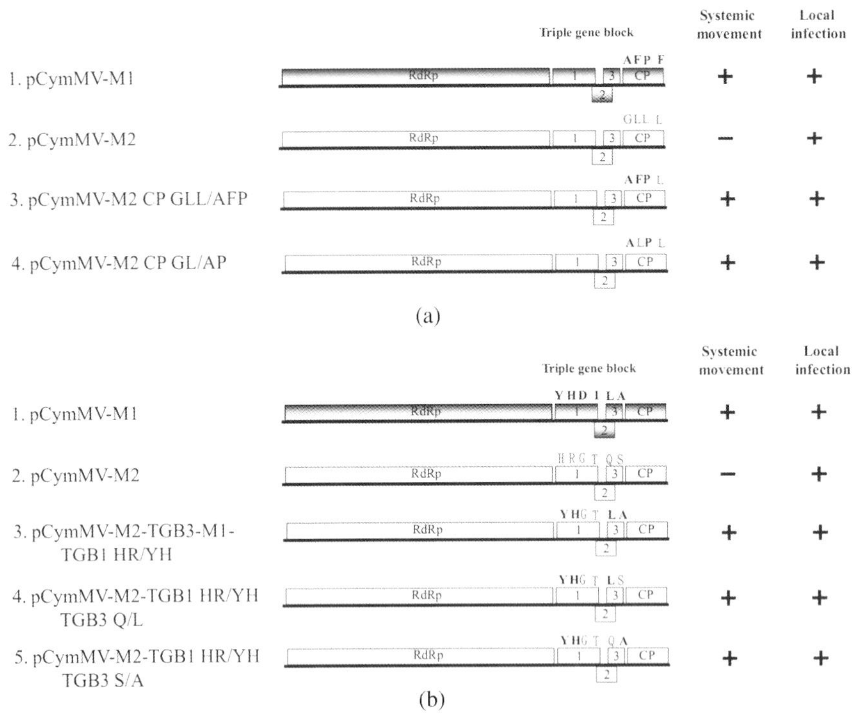

Fig. 17.3. Schematic representation of coat protein (CP) and TGBp1,3 mutant infectivity assays. Rectangles represent open reading frames (ORFs) encoded by CymMV genomic RNA, RNA-dependent RNA polymerase (RdRp), TGB ORFs 1, 2, and 3 and CP. Black and gray letters in TGBp1 and TGBp3 and CP indicate different amino acids between pCymMV-M1 and pCymMV-M2, respectively. Chimeric viruses with competent systemic infection in *N. benthamiana*plants are indicated by plus symbols. Systemic infection was detected two weeks post-inoculation by RT-PCR.

systemically infect *N. benthamiana* (Fig. 17.3a, 17.4). Interestingly, these amino acids are located within the previously predicted RNA binding domain of the CP (Abouhaidar and Lai, 1989). Our further biochemical RNA–protein binding results indicated that the CP for M1 has a higher binding affinity than does the CP for M2. The CP for M2 with amino acid changes in G82A and L89P has a similar RNA binding affinity, as compared to M1 CP.

Although the movement of potexviruses has been extensively studied, the nature of the CymMV that moves between cells remains unresolved. Two forms of movement are proposed: one is the linear viral ribonucleoprotein complex (vRNP) formed by TGBp1, viral RNA (vRNA) and CP (Lough *et al.*, 2000), and another is the virion or altered virion with a single-tailed particle comprising TGBp1, vRNA and CP (Cruz *et al.*, 1998; Karpova *et al.*, 2006). However, in both types, potexvirus CP and vRNA are both involved in forming the vRNP that moves between cells. Our data suggest that the movement of CymMV in different hosts may be affected by the binding between CP and vRNA that alters the nature of the vRNP.

We have also mapped the minimum amino acid changes required in the CymMV TGBp1 and TGBp3 to allow CymMV M2 to move systemically in *N. benthamiana* (Fig. 17.3). All the amino acids are not located in the conserved motifs essential for potexvirus movement (Kadare and Haenni, 1997; Kalinina *et al.*, 2002; Leshchiner *et al.*, 2006). Subtle changes in these regions may only slightly alter protein–protein or protein–RNA interaction, which may not completely block the movement of CymMV but may modify the movement ability between hosts. This observation may explain why M1 and M2 can both systemically infect *Phalaenopsis* orchids, but only M1 can systematically infect *N. benthamiana*.

17.5. Sequence Analysis Revealed that Orchids May Not be the Natural Hosts of CymMV

Because we have determined the important amino acids involved in CymMV movement, we performed sequence alignment to examine the composition of the amino acids in certain positions among all published CymMV isolates (Fig. 17.4). Surprisingly, the CP sequences of most of the worldwide CymMV isolates (including seven of our collected isolates) were identical to that of the CymMV M1-encoded CP, which suggests that most CymMV strains in nature can systemically infect *N. benthamiana*. We inoculated *N. benthamiana* with the seven collected isolates, and all could systemically infect *N. benthamiana*. In addition, all the inoculated CymMV did not induce visible symptoms

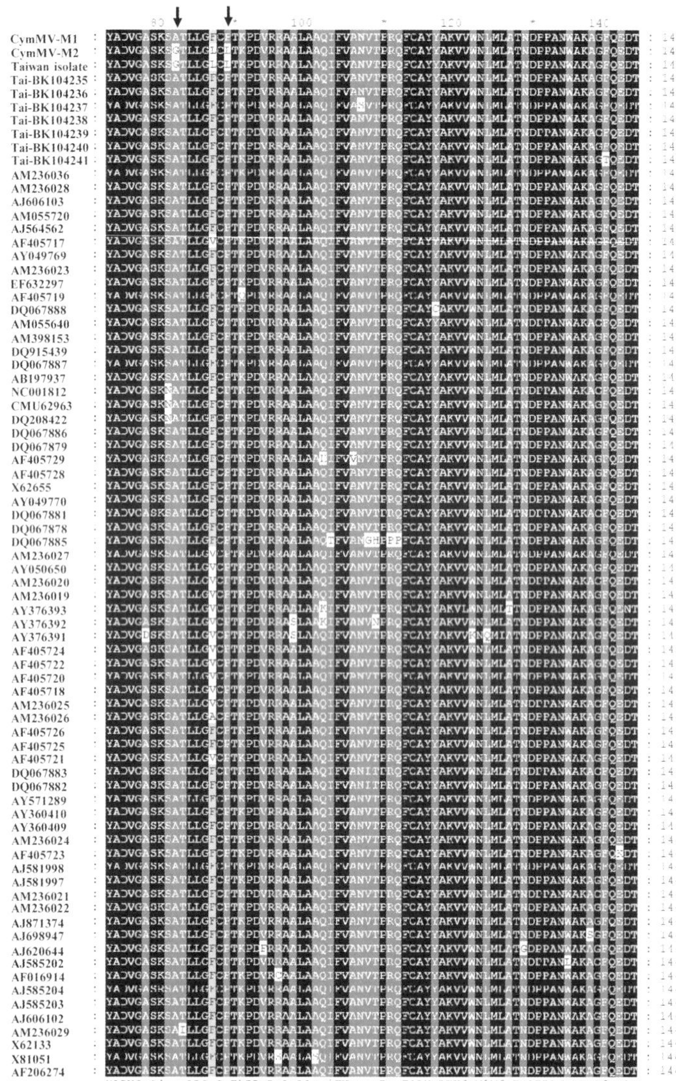

Fig. 17.4. Sequence alignment of the coat protein (CP) amino acids of Cymbidium mosaic virus isolates. pCymMV-M1, -M2 and other CymMV isolates were collected. The sequences of the remaining CymMV isolates were obtained from GenBank, and all accession numbers are indicated. The arrows indicate the amino acid positions important for pCymMV-M2 to systemically infect *N. benthamiana*.

on *N. benthamiana*. Therefore, the M2 type most likely evolved from the M1 type. If this is true, the M2 type may have evolved from CymMV isolates with a broader host range through viral serial passage on orchids because of human cultivation and thus gradually lost the ability to systemically infect *N. benthamiana*. A similar situation has been reported for Hibiscus Chlorotic Ringspot Virus (HCRSV), which lost the ability to infect its experimental systemic host *Hibiscus cannabinus* L. (kenaf) because of serial passage in a local lesion host, *Chenopodium quinoa* (Liang *et al*., 2002). In view of the observation that CymMV has never been isolated from wild orchids, we speculate that the host range of CymMV is broader than current records, and orchids are not the prevalent natural hosts of CymMV. CymMV may be introduced to orchids through their mass cultivation. Thus, the natural hosts of CymMV must be identified.

17.6. Functional Diversity Generated by Potexviruses Through a Combination of Mutated Proteins from CymMV Quasi-Species

Because viral replicase is prone to error, individuals of a viral population within its host are all considered to carry different mutations and are termed "quasi-species". Our analysis revealed that when CP is not functional, another movement control mode (matching), which requires exact matching of particular components of mutated CymMV, allows CymMV to move between cells (Fig. 17.1). This phenomenon is intriguing because it suggests that the virus may adapt to environmental change through mutation in a single virus and through the combination of mutated proteins or RNA from quasi-species to generate more functional diversity. Because the cellular conditions and morphology differ among hosts or among different stages of a host, potexviruses may need different strategies to move between cells. Adjusting movement accessory components could be an efficient way to generate the functional diversity to adapt to changes in cellular condition. This flexibility may allow the virus to move between different hosts or different conditions in a host (Fig. 17.5).

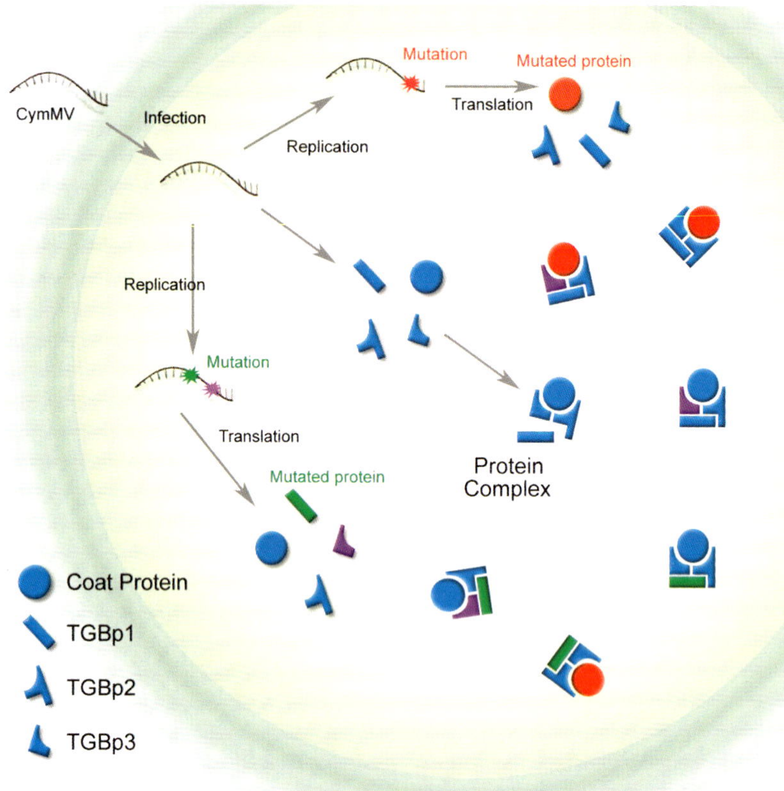

Fig. 17.5. Model of accumulation of virus-encoded protein complexes in Cymbidium mosaic virus (CymMV)-infected cells. CymMV infects and replicates in host cells, and progeny viruses form quasi-species. Then, the genomes of the viral quasi-species translate slightly different proteins, and these proteins make up the numerous combinations of protein complexes to increase the functional diversity of the protein complex. The blue icons indicate the original virus-encoded protein, and other colors indicate the mutated proteins. The triple gene block protein 1 (TGBp1), TGBp2, TGBp3 and coat protein are indicated.

17.7. Conclusions

Our studies revealed several detailed mechanisms of CymMV movement between hosts. By identifying the important amino acids involved in CymMV movement, we could perform sequence analysis to resolve the possible evolution direction of CymMV. CymMV originally may have had a broader host range, and only some isolates evolved to infect orchids only. Our analysis is supported by previous observations: although CymMV was found restricted to the Orchidaceae, experimental inoculation indicated that CymMV can infect several families of plant hosts, and CymMV has never been isolated from wild orchids. Therefore, identifying the natural host of CymMV has important implications for disease management in the orchid industry.

References

Abouhaidar MG, Lai R. (1989) Nucleotide sequence of the 3-terminal region of clover yellow mosaic virus RNA. *J Gen Virol* **70**: 1871–1875.

Adams MJ, Accotto GP, Agranovsky AA *et al.* (2005) *Virus Taxonomy: Eighth Report of the International Committee on Taxonomy of Viruses.* Elsevier Academic Press, San Diego, CA, pp. 1101–1124.

Bayne EH, Rakitina DV, Morozov SY, Baulcombe DC. (2005) Cell-to-cell movement of potato potexvirus X is dependent on suppression of RNA silencing. *Plant J* **44**: 471–482.

Chan YL, Lin KH, Sanjaya *et al.* (2005) Gene stacking in *Phalaenopsis* orchid enhances dual tolerance to pathogen attack. *Transgenic Res* **14**: 279–288.

Chang C, Chen YC, Hsu YH *et al.* (2005) Transgenic resistance to *Cymbidium mosaic* virus in *Dendrobium* expressing the viral capsid protein gene. *Transgenic Res* **14**: 41–46.

Chapman S, Hills G, Watts J, Baulcombe DC. (1992) Mutational analysis of the coat protein gene of potato virus X: Effects on virion morphology and viral pathogenicity. *Virology* **191**: 223–230.

Cruz SS, Roberts AG, Prior DA *et al.* (1998) Cell-to-cell and phloem-mediated transport of potato virus X. The role of virions. *Plant Cell* **10**: 495–510.

Eun AJ, Huang L, Chew FT *et al.* (2002) Detection of two orchid viruses using quartz crystal microbalance (QCM) immunosensers. *J Virol Methods* **99**: 71–79.

Eun AJ, Seoh M, Wong S. (2000) Simultaneous quantitation of two orchid viruses by the TaqMan real-time RT-PCR. *J Virol Methods* **87**: 151–160.

Hu WW, Wong SM. (1998) The use of DIG-labelled cRNA probes for the detection of *cymbidium mosaic potexvirus* (CymMV) and *odontoglossum ringspot tobamovirus* (ORSV) in orchids. *J Virol Methods* **70**: 193–199.

Jensen DD. (1951) Mosaic or black streak disease of *cymbidium* orchids. *Phytopathology* **41**: 401–414.

Kadare G, Haenni AL. (1997) Virus-encoded RNA helicases. *J Virol* **71**: 2583–2590.

Karpova OV, Zayakina OV, Arkhipenko MV *et al.* (2006) Potato virus X RNA-mediated assembly of single-tailed ternary 'coat protein–RNA–movement protein' complexes. *J Gen Virol* **87**: 2731–2740.

Kalinina NO, Rakitina DV, Solovyev AG *et al.* (2002) RNA helicase activity of the plant virus movement proteins encoded by the first gene of the triple gene block. *Virology* **296**: 321–329.

Krishnamurthy K, Mitra R, Payton ME, Verchot-Lubicz J. (2002) Cell-to-cell movement of the PVX1 2K, 8K, or coat proteins may depend on the host, leaf developmental stage, and the PVX 25K protein. *Virology* **300**: 269–281.

Krishnamurthy K, Heppler M, Mitra R *et al.* (2003) The potato virus X TGBp3 protein associates with the ER network for virus cell-to-cell movement. *Virology* **309**: 135–151.

Leshchiner AD, Solovyev AG, Morozov SY, Kalinina NO. (2006) A minimal region in the NTPase/helicase domain of the TGBp1 plant virus movement proteins responsible for ATPase activity and cooperative RNA binding. *J Gen Virol* **87**: 3087–3095.

Liang XZ, Lee BT, Wong SM. (2002) Covariation in the capsid protein of hibiscus chlorotic ringspot virus induced by serial passaging in a host that restricts movement leads to avirulence in its systemic host. *J virol* **76**: 12320–12324.

Lu HC, Chen HH, Tsai WC *et al.* (2007) Strategies for functional validation of genes involved in reproductive stage of orchids. *Plant Physiol* **143**: 558–569.

Lu HC, Chen CN, Tsai MH *et al.* (2009) Cymbidium mosaic potexvirus isolate-dependent host movement system reveal two movement control determinants and the coat protein is the dominant. *Virology* **388**: 147–159.

Lough TJ, Netzler NE, Emerson SJ. (2000) Cell-to-cell movement of potexviruses: Evidence for a ribonucleoprotein complex involving the coat protein and first triple gene block protein. *Mol Plant Microbe Interact* **13**: 962–974.

Morozov SYu, Solovyev AG. (2003) Triple gene block: Modular design of a multifunctional machine for plant virus movement. *J Gen Virol* **84**: 1351–1366.

Tamai A, Meshi T. (2001) Cell-to-cell movement of potato virus X: The role of p12 and p8 encoded by the second and third open reading frames of the triple gene block. *Mol Plant Microbe Interact* **14**: 1158–1167.

Tan SW, Wong SM, Kini RM. (2000) Rapid simultaneous detection of two orchid viruses using LC-and/or MALDI-mass spectrometry. *J Virol Methods* **85**: 93–99.

Verchot-Lubicz J. (2005) A new cell-to-cell transport model for potexviruses. *Mol Plant Microbe Interact* **18**: 283–290.

Zettler FW, Ko NJ, Wisler GC *et al.* (1990) Viruses of orchids and their control. *Plant Dis* **74**: 621–626.

Zettler FW, Wisler GC, Elliott MS, Ko NJ. (1987) Some new potentially significant viruses of orchids and their probable means of transmission. *Am Orchid Soc Bull* **56**: 1044–1051.

CHAPTER 18

Transgenic Protection Against Virus Infection in Orchids

Ting-Yu Chen* and Na-Sheng Lin*,†

Almost all cultivated orchids are susceptible to virus infection. Cymbidium mosaic virus (CymMV) and Odontoglossum ringspot virus (ORSV) are the two most prevalent and important viruses and cause great economic loss worldwide. To date, the generation of resistant orchids through genetic engineering is the most efficient approach to controlling virus diseases. Gene transfer by *Agrobacterium*-mediated transformation or particle bombardment has been successful in orchids. Via the concept of pathogen-derived resistance, transgenic *Phalaenopsis, Dendrobium* and *Oncidium* expressing the coat protein (CP) gene of CymMV showed a substantial level of resistance through both protein- and RNA-mediated mechanisms. The development of transgenic orchids is slow and time-consuming, but we expect to obtain transgenic orchids with resistance to CymMV or both ORSV and CymMV in the near future. The stability of engineered protection and the environmental impact should be tested in the field before the commercial release of transgenic orchids.

18.1. Introduction

18.1.1. *Viruses occurring naturally in orchids*

The Orchidaceae, the largest family of monocotyledons, contains flowering plants comprising 900 genera and more than 25,000 species (Stewart *et al.*, 1995). Plant viruses are the major problems in the cultivation of orchids throughout the world. Approximately 30 different viruses have been reported to infect orchids, including Cymbidium

*Institute of Plant and Microbial Biology Academia Sinica, Taipei 11529, Taiwan.
†Corresponding author. E-mail: nslin@sinica.edu.tw

mosaic virus (CymMV) (Navalinskienë *et al.*, 2005; Zettler *et al.*, 1990), Odontoglossum ringspot virus (ORSV) (Jensen *et al.*, 1951; Zettler *et al.*, 1990), Cucumber mosaic virus (CMV) (Hadley, 1987), Orchid fleck virus (OFV) (Hideki *et al.*, 2006) and Tomato spotted wilt virus (TSWV) (Hu, 1990). Of these, CymMV and ORSV are the two most prevalent and economically important orchid viruses worldwide (Zettler *et al.*, 1990). They can reduce orchid vigor, lower flower quality and lead to economic losses. Because these two viruses are very stable and are frequently transmitted together, mixed infections are common, often leading to greater disease severity and economic loss.

CymMV, a member of the *Potexvirus* group, is the most important virus that affects the quality and quantity of flower production in orchids. The symptoms on leaves of CymMV-infected orchids such as *Cymbidium, Oncidium, Phalaenopsis, Cattleya* and *Vanda* include yellow spots, chlorotic mosaic, vein chlorosis and ringspots with sunken necrotic lines. The virus is highly virulent in susceptible hosts, systematically infecting all parts of the orchid (Wong *et al.*, 1996; Yamane *et al.*, 2008; Zettler *et al.*, 1990). CymMV has a genome consisting of single-stranded, positive-sense RNA with five open reading frames (ORFs), which encode a 160-kDa RNA-dependent RNA polymerase (RdRp) for replication, a triple gene block for cell-to-cell movement, and a 24-kDa capsid protein (CP) for viral encapsidation (Wong *et al.*, 1997). The particle morphology consists of a flexuous rod approximately about 500 nm long and 15 nm wide (Navalinskienë *et al.*, 2005).

ORSV, a member of the *Tobamovirus* group, is a rigid-rod particle about 300 nm long and 18 nm wide (Navalinskienë *et al.*, 2005). The genome of ORSV consists of single-stranded, positive-sense RNA with four ORFs: 126- and 183-kDa RNA-dependent RNA polymerase and readthrough protein, respectively; movement protein (MP) and CP. The virus infects more than 20 genera of the Orchidaceae and produces different disease symptoms on different genera or species. The ORSV induces ringspot symptoms, originally found on *Odontoglossum grande* in the United States (Jensen and Gold, 1951). Other disease symptoms are mosaic, mottle, chlorotic streak, breaking and necrosis. However, disease symptoms on ORSV-infected *Phalaenopsis* and *Oncidium* are not obvious, and the regular growth and flower structure are normal. ORSV and CymMV co-infected *Phalaenopsis* and *Oncidium* show enhanced symptoms (Navalinskienë *et al.*, 2005).

18.1.2. *Isolates of CymMV and ORSV*

The complete nucleotide sequences of CymMV (Vaughan *et al.*, 2008; Wong *et al.*, 1997) and ORSV (Chng *et al.*, 1996) have been reported. Genomic sequences vary among the isolates. In 2002, Ajjikuttira reported that the homology of the CP gene among CymMV isolates was 89.1–99.7 percent and 93.2–100 percent at the nucleotide and amino acid levels, respectively, whereas the homology of the CP gene among ORSV isolates was 95.5–100 percent and 93–100 percent, respectively. In deduced amino acid sequences, the N-terminus was more conserved than the C-terminus in both CymMV and ORSV (Ajjikuttira *et al.*, 2002).

18.1.3. *Control of orchid viruses*

The control of virus disease is essential in the orchid industry. To achieve this goal, growers must be aware of the modes of virus spread and adopt handling practices that limit the possibility of virus transmission. When a new seedling of orchids is brought into the garden, it should be isolated from other plants for at least two months under low light intensity conditions. In general, orchids infected with viruses should show symptomatic leaves or flowers. However, some infected orchids may not show symptoms. The most reliable methods for detecting virus infection are enzyme-linked immunosorbant assay (ELISA), nucleic acid hybridization and RT-PCR (Zettler *et al.*, 1990).

CymMV and ORSV can infect orchids by mechanical transmission. Transmission of these viruses by specific insects or through seeds is not known. Control of virus diseases in orchids is possible only with good management practices. To propagate the orchids, the equipment for dividing the plants should be sterilized. Clamps, knives, blades should be dipped in alkaline solution and sterilized by flame. Mentholated spirits, hypochlorite, and formalin may be used. As well, benches should also be washed thoroughly with soap or sodium hypochlorite. Because orchids cannot be cured of virus infection, they should be destroyed. Ridding the infection by meristem cultures should be attempted only with a valuable breeding plant or valuable unique clone (Retheesh and Bhat, 2010). Growers may control the spread of the viruses by the treatments mentioned above. However, limited measures are available to resist virus infection in natural orchids.

18.1.4. *Transgenic resistance to CymMV in orchids*

Although good horticultural practices can almost curb the spread of CymMV and ORSV in plant collections, methods to improve traits of virus resistance are still needed. In 1986, engineered cross-protection against virus infection through the expression of viral CP in transgenic plants was first demonstrated with the CP of Tobacco mosaic virus (TMV) (Powell-Abel *et al.*, 1986). With control by pathogen-derived resistance (PDR) (Sanford and Johnson, 1985), the expression of the viral CP gene in transgenic plants conferred protection against many other virus infections (Gottula and Fuchs, 2009). CP-mediated protection has been widely used and is successful in producing a broad range of virus-resistant crops (Miller and Hemenway, 1998). This protection depends on sequence similarity between the transgene and the infecting viruses (Lomonosoff, 1995).

Transformation of the CymMV CP gene into *Dendrobium* protocorms through particle bombardment has been successful in generating a virus-resistant orchid variety (Chang *et al.*, 2005). The transgenic plants exhibited milder or no symptoms as compared with the non-transgenic orchids on challenge with CymMV virions. Because the CymMV-CS isolate from infected *Cymbidium sinesis* Wild shares high sequence homology with other geographically distinct CymMV isolates, transgenic *Dendrobium* expressing this CP gene has great potential to be resistant to infection by various CymMV isolates (Chang *et al.*, 2005). Similarly, transgenic *Phalaenopsis* expressing the CP gene showed enhanced tolerance to CymMV infection (Chan *et al.*, 2005). Monitoring the RdRp gene by RT-PCR in transgenic orchids challenged with CymMV revealed no virus, which suggests that CymMV could not efficiently replicate in these plants. Moreover, the detection of CymMV-specific small interfering RNAs rather than the transgene suggests that the resistance of transgenic plants is via a RNA-mediated mechanism (Chan *et al.*, 2005).

The ORSV CP gene derived from ORSV-infected *Epidendrum* was also transformed into the rhizomes of *Cymbidium niveo-marginatum*. The presence of the ORSV CP gene was confirmed in transgenic shoots by PCR analysis, and the expression of ORSV CP was confirmed by ELISA (Chen *et al.*, 2006). To date, no reports are available for exist of transgenic orchids resistant to ORSV or to both CymMV and ORSV.

18.2. Generating Transgenic Orchids

The hypersensitive response is commonly used as an active defense against viruses. This response is preceded by a specific recognition of the virus, which is based on matching gene products of plants (dominant resistance gene, R genes) and virus (avirulence genes) (Goldbach *et al.*, 2003). Some R genes are involved in virus resistance in hosts (e.g., to trigger the immune response to invading viruses). One example is the tomato *Sw-5* gene conferring resistance to a number of topoviruses. The R genes are available for the working mechanism of the defense systems in plants. Another way to build up virus resistance is to follow the principle of PDR (Sanford and Johnson, 1985). One of the successful and widely applied strategies is the exploitation of the viral CP gene. This method has been used with a number of plant viruses: transgenic plants expressing the CP gene showed high levels of operational resistance (Baulcombe, 1996; Beachy, 1993; Wilson, 1993). As well, the transformation of host plants with modified MP (Obsuwan *et al.*, 2009) or replicase gene was successful for PDR (Anderson *et al.*, 1992; Lomonosoff, 1995). Post-transcriptional gene silencing is one of the true mechanisms of PDR. The engineered antiviral strategies have been used in a variety of crops but have rarely been applied in orchids (Lin *et al.*, 2007).

18.2.1. *Preparation of transgene*

Virus-derived genes or fragments can be inserted into the appropriate vector for gene transformation. The insertion can be a full-length CP gene, modified MP, RdRp or any untranslated region, which may be truncated sense constructs, antisense constructs, constructs containing ribozymes or inverted repeats, regardless of whether the insertion is designed or expected to be expressed in the transgenic plants. Plasmids harboring the viral sequence for transformation usually comprise several elements, such as promoters (regulatory sequences for transcription initiation; i.e., nopaline synthase or cauliflower mosaic virus 35S), transcriptional regulatory elements (enhancers), protein targeting polypeptide signals, multiple cloning sites (restriction sites), fusion tags (epitopes, reporters, affinity, cleavage sites), transcriptional termination signals (polyadenylation), antibiotic selection markers (i.e., hygromycin, kanamycin resistant genes), and origin of replication for high copy number of plasmids.

18.2.2. *Transformation of orchids*

Two different methodologies for gene transfer are commonly used for orchids. The first is *Agrobacteriun*-mediated transformation (Fig. 18.1). The plasmid containing any cDNA fragment of viral sequence can be transformed into *Agrobacterium tumefaciens* for transformation into the explants, such as protocorm-like bodies (PLBs), leaves or protocorms of orchids. These explants are cultured on selection medium for several months. The transgenic lines survived on the selection medium can be induced to develop new shoots and healthy roots. Putative transformants are tested by PCR to confirm the presence of the transgene.

The second method is the transformation of orchids by particle bombardment (Fig. 18.2). The plasmid containing the transgene is coated on gold particles for bombardment. The bombardment involves the use of a particle delivery system (PDS). After bombardment, the explants are transferred onto the selection medium for several months for continued subculture. When transgenic shoots are induced, the plantlets are tested by PCR or other molecular analysis. After molecular

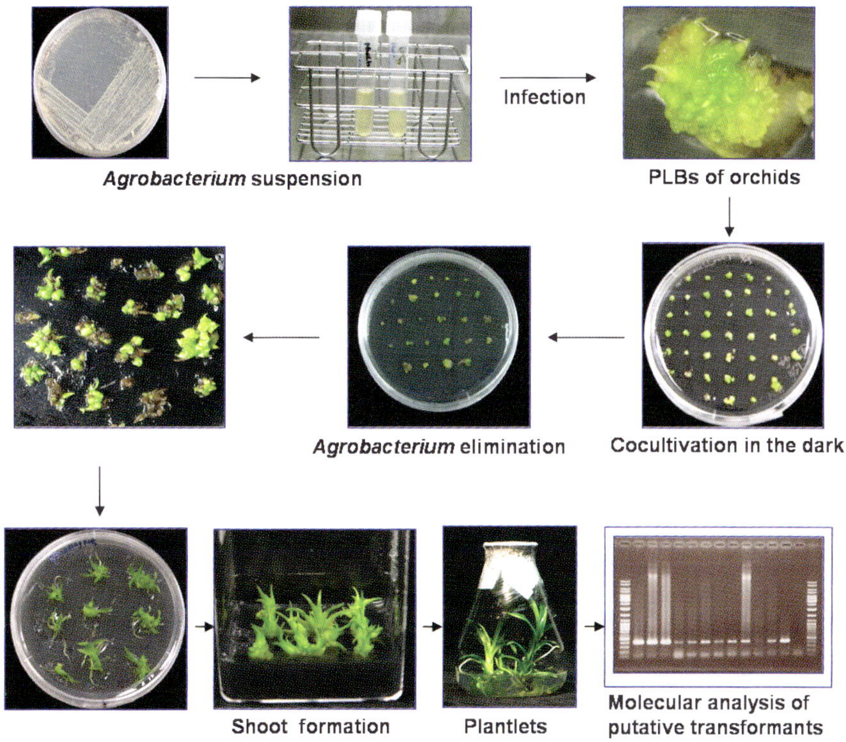

Fig. 18.1. *Agrobacterium*-mediated transformation of protocorm-like bodies (PLBs) in orchids.

characterization, the transgenic plants are transferred to a growth chamber for viral resistance tests.

18.3. Molecular Characterization of Transgene

Any putative transformants grown on the selection medium are confirmed by molecular characterization. DNA analysis can be conducted for introns, leader sequences, terminators, and enhancers of plant-expressible cassettes. DNA analyses may be in the form of PCR, southern analyses and DNA sequencing, for example. As well, to demonstrate whether the inserted gene is expressed just as expected, confirmation at the RNA or protein level is important. For transgenic

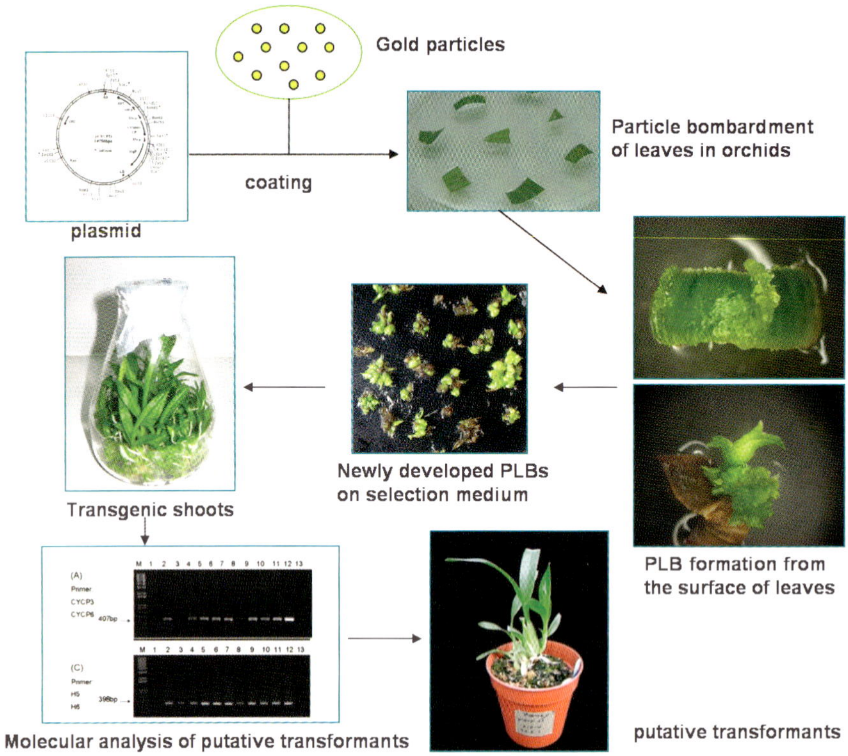

Fig. 18.2. Transformation of orchid leaves by particle bombardment.

lines in which the transgenes are derived from a viral coding region, the RNA and protein expression of the transgenes must be determined. The expression of the transgenes can be routinely determined by RT-PCR, northern blot or western blot analysis.

18.4. Viral Resistance of Transgenic Plants

Genetic identification of a transgene is required for generating transgenic orchids with tolerance or resistance. The resistance test for selection of tolerant or resistant transgenic line is shown in Fig. 18.3. Transgenic orchids are mechanically inoculated with the challenging virus from which the transgene is derived. The challenging nature

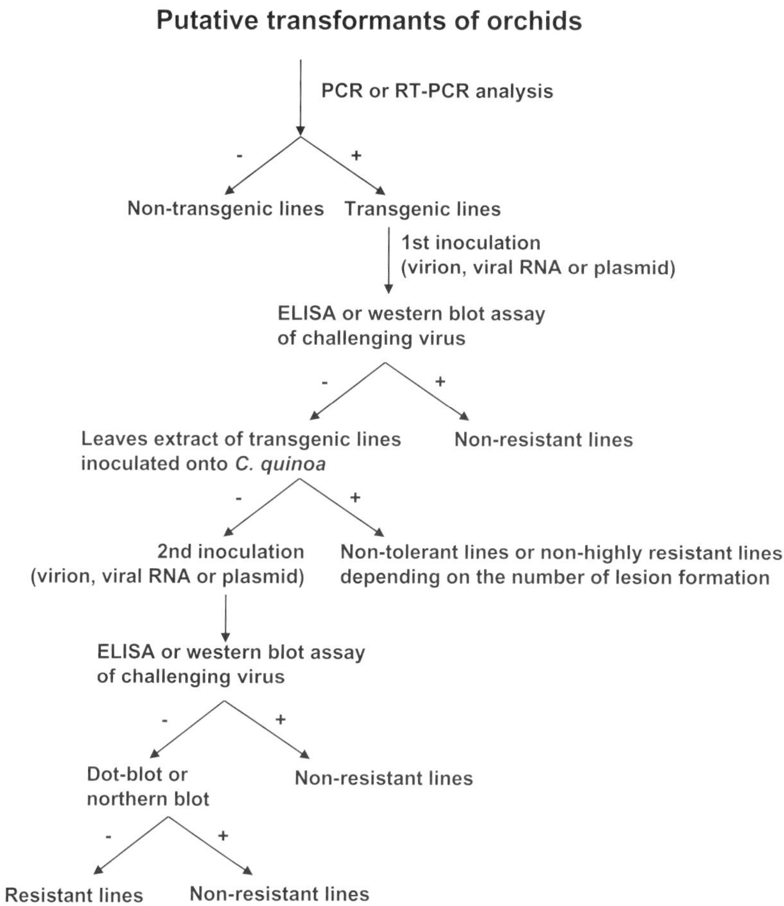

Fig. 18.3. The molecular analysis and resistance test of putative transformants of orchids.

can be the virions, viral RNA or infectious full-length cDNA of the virus. About one month post-inoculation, the resistance of inoculated transgenic lines is evaluated by the formation of symptoms and the titer of challenging virus by ELISA or western blot analysis. Any lines showing susceptibility to the challenged virus are discarded. The remaining putative resistant lines are re-confirmed by infectivity assay. The leaf extracts of these lines are used to inoculate the indicator plants (*Chenopodium quinoa*) and to check the lesion formation. Any lines showing substantial production of local lesions are considered non-immune lines. To doubly confirm the immunity of the resistant lines, all tests are repeated twice. Finally, the transgenic lines that do not show any symptoms or virus accumulation are selected as resistant lines for mass propagation.

18.5. Conclusions

Transgenic resistance seems to be the most effective approach for control of RNA viruses, especially for crops whose genetic resistance traits are scarce in the current plant cultivars. After significant progress was made in the genetic engineering of orchids, the initial concept of PDR was successfully applied with the use of CP transgenes conferring CymMV resistance. The expression of the CymMV CP gene in transgenic orchids induced protection, which may be attributed to the co-existence of both protein- and RNA-mediated interference. For transgenic *Dendrobium*, the level of protection against CymMV was well associated with the expression level of the CP gene of CymMV, although the molecular mechanisms of CP-mediated resistance are not yet fully understood and vary in different viruses (Chang *et al.*, 2005). In addition, the RNA-mediated mechanism involved in conferring resistance to CymMV was reported in transgenic *Phalaenopsis* (Liao *et al.*, 2004) and in our recently developed *Oncidium* Ramsey and Sweet Sugar. Plants generating low or even undetectable levels of CP but substantial levels of small interfering RNA molecules derived from the transgene exhibited highly resistance. These finding indicate that the different levels of resistance in transgenic lines may result from multiple mechanisms.

In our study, each transgenic-resistant *Dendrobium*, *Oncidium* Ramsey or Sweet Sugar lines challenged twice with CymMV virions were still expressing virus-free by western, dot-blot, northern hybridization analysis or infectivity assay. We have observed the resistance of transgenic lines over the years. Although more research is needed to assess the sustainability and stability of the engineered protection, the concept of PDR indeed opens a new avenue for the development of virus-resistant orchids. However, the progress in antiviral resistance in transgenic orchids is lagging, although major breakthroughs have been achieved technically and conceptually in terms of the stable transformation and design of transgenes. No follow-up studies have been reported in the past five years. This absence could be due to the natural slow growth characters of orchids and lengthy selection and regeneration process. For stable transformation, two to three years might be required for the development of transgenic orchids and another one to two years to obtain massive seedlings for resistant tests. Nevertheless, transgenic orchids exhibiting high levels of resistance to ORSV will be obtained in the near future. As well, we expect to obtain transgenic resistance against the two commonly occurring viruses, CymMV and ORSV.

18.6. Future Prospects

18.6.1. *Application of CP-mediated protection for resistance against CymMV and ORSV in other orchids*

Use of the expression of the CP gene has been successful in conferring resistance to CymMV in transgenic *Phalaenopsis, Dendrobium* and *Oncidium.* Because of the widespread infectivity of CymMV and ORSV in cultivated orchids, the high CP sequence homology among the isolates of CymMV and ORSV producing diverse resistance in orchids is likely via an isolate sequence of CP as a transgene. Thus, the technology may also be applicable to other orchids such as *Cymbidium,Cattleya, Phalaenopsis, Oncidium* and *Dendrobium.*

18.6.2. *Integration of virus-derived fragments in transgenic orchids for virus resistance*

Virus-derived genes such as CP, MP and RdRp have been successfully engineered into different crops for virus resistance (Fitchen and Beachy 1993; Sudarshana *et al.*, 2006). The understanding of RNA-mediated resistance derived from post-transcriptional gene silencing has advanced the use of the fragments of complete or truncated viral genes, or even untranslatable regions, as potential candidates for transgenes. Thus, with the emerging viruses whose complete sequence is not fully determined, we can proceed with the transgenic approaches for virus resistance.

In addition, increasing evidence shows that inverted-repeat transgenes offer the best chance to yield resistance than the single-stranded sense or antisense approach. Thus, to overcome the drawback of RNA-mediated resistance, which requires sequence homology of more than 90 percent between the transgene and the challenging virus, several fragments of viral sequences from different isolates or viruses can be fused in a single inverted-repeat construct. This strategy has resulted in multiple virus-resistant plants (Bucher *et al.*, 2006). Because mixed infection of viruses commonly occurs in orchids and the transgenic orchid lines require a long time to generate, genetic engineering of multiple sequences from different viruses would be the ideal strategy to develop transgenic orchid resistance against virus infection.

References

Ajjikuttira PA, Lim-Ho CL, Woon MH *et al.* (2002) Genetic variability in the coat protein genes of two orchid viruses: *Cymbidium mosaic virus* and *Odontoglossum ringspot virus*. *Arch Virol* **147**: 1943–1954.

Anderson JM, Palukaitis P, Zaitlin M. (1992) A defective replicase gene induces resistance to *Cucumber mosaic virus* in transgenic tobacco plants. *Proc Natl Acad Sci USA* **89**: 8759–8763.

Baulcombe DC. (1996) Mechanisms of pathogen-derived resistance to viruses in transgenic plants. *Plant Cell* **8**: 1833–1844.

Beachy RN. (1993) Introduction: Transgenic resistance to plant viruses. *Semin Virol* **4**: 327–328.

Bucher E, Lohuis D, Poppel PMJAv *et al.* (2006) Multiple virus resistance at a high frequency using a single transgene construct. *J Gen Virol* **87**: 697–370.

Chan YL, Lin KH, Sanjaya *et al.* (2005) Gene stacking in *Phalaenopsis* orchid enhances dual tolerance to pathogen attack. *Transgenic Res* **14**: 279–288.

Chang C, Chen YC, Hsu YH *et al.* (2005) Transgenic resistance to *Cymbidium mosaic virus* in *Dendrobium* expressing the viral capsid protein gene. *Transgenic Res* **14**: 41–46.

Chen I, Kawai H, Oku T *et al.* (2006) Introduction of *Odontoglossum ringspot virus* coat protein gene into *Cymbidium niveo-marginatum* mediated by *Agrobacterium tumefaciens* to produce transgenic plants. *J Japan Soc Hort Sci* **75**: 249–255.

Chng CG, Wong SM, Mahtani PH *et al.* (1996) The complete sequence of a Singapore isolate of *Odontoglossum ringspot virus* and comparison with other tobamoviruses. *Gene* **171**: 155–161.

Fitchen JH, Beachy RN. (1993) Genetically engineered protection against viruses in transgenic plants. *Annu Rev Microbiol* **47**: 739–763.

Goldbach R, Bucher E, Prin SM. (2003) Resistance mechanisms to plant viruses: An overview. *Virus Res* **92**: 207–212.

Gottula J, Fuchs M. (2009) Toward a Quarter Century of Pathogen-Derived Resistance and Practical Approaches to Plant virus Disease Control. *Adv Virus Res* **75**: 161–183.

Hadley G, Arditti M, Arditti J. (1987) Orchid diseases- A Compedium. In J. Arditti (ed.), Orchid biology:Reviews and perspectives, IV. Cornell Univ. Press. Ithaca. N.Y., pp. 261–325.

Hideki K, Takanori M, Yukio S, Tetsuo T. (2006). Orchid fleck virus is a rhabdovirus with an unusual bipartite genome. *J Gen Virol* **87**: 2413–2421.

Hu JS. (1990) *Tomato spotted wilt virus on Oncidium* Orchids in Hawaii. *Plant Dis* **76**: 426.

Jensen DD, Gold HA. (1951) A virus ring spot of *Odontoglossum* orchid: Symptoms, transmission and electron microscopy. *Phytopathology* **41**: 648–653.

Liao LJ, Pan IC, Chan YL *et al.* (2004) Transgene silencing in *Phalaenopsis* expressing the coat protein of *Cymbidium mosaic virus* is a manifestation of RNA-mediated resistance. *Mol Breeding* **13**: 229–242.

Lin SS, Henriques R, Wu HW *et al.* (2007) Strategies and mechanisms of plant virus resistance. *Plant Biotechnol Rep* **1**: 125–134.

Lomonosoff GP. (1995) Pathogen-derived resistance to plant viruses. *Annu Rev Phytopathol* **33**: 323–343.

Miller ED, Hemenway C. (1998) History of coat protein-mediated protection. *Methods Mol Biol* **81**: 25–38.

Navalinskienë M, Raugalas J, Samuitienë M. (2005) Viral diseases of flower plants 16. Identification of viruses affecting orchids (*Cymbidium* Sw.). *Biologija* **2**: 29–34.

Obsuwan K, Hieber DA, Mudalige-Jayawickrama RG, Kuehnle AR. (2009). A dysfunctional CymMV movement protein gene confers resistance to CymMV in *Dendrobium* orchid. *Acta Hortic* **836**: 149–154.

Powell-Abel P, Nelson RS, De B *et al.* (1986) Delay of disease development in transgenic plants that express the *Tobacco mosaic virus* coat protein gene. *Science* **232**: 738–743.

Retheesh ST, Bhat AI. (2010) Simultaneous elimination of Cucumber mosaic virus and Cymbidium mosaic virus infecting Vanilla planifolia through meristem culture. *Crop Prot* **29**: 1214–1217.

Sanford JC, Johnson SA. (1985) The concept of pathogen-derived resistance. *J Theor Biol* **113**: 395–405.

Stewart J, Griffiths M. (1995) *Manual of Orchids*. Timber Press, Portland.

Sudarshana MR, Roy G, Falk BW. (2006) Methods for engineering resistance to plant viruses. *Methods Mol Biol* **354**: 183–195.

Vaughan SP, Grisoni M, Kumagai MH, Kuehnle AR. (2008) Characterization of Hawaiian isolates of *Cymbidium mosaic virus* (CymMV) co-infecting *Dendrobium* orchid. *Arch Virol* **153**: 1185–1189.

Wilson TMA. (1993) Strategies to protect crop plants against viruses: Pathogen-derived resistance blossoms. *Proc Natl Acad Sci USA* **90**: 3134–3141.

Wong SM, Chng CG, Lee YH, Lim TM. (1996) An appraisal of the banded and paracrystalline cytoplasmic inclusions induced in *Cymbidium mosaic potexvirus*- and *Odontoglossum ringspot tobamovirus* infected orchid cells using confocal laser scanning microscropy. *Arch Virol* **141**: 231–242.

Wong SM, Mahtani PH, Lee KC *et al.* (1997) *Cymbidium mosaic potexvirus* RNA: Complete nucleotide sequence and phylogenetic analysis. *Arch Virol* **142**: 383–391.

Yamane K, Oyama K, Iuchi E *et al.* (2008) RT-PCR detection of *Odontoglossum ringspot virus, Cymbidium mosaic virus* and topoviruses and association of infections with leaf-yellowing symptoms in *Phalaenopsis. J Phytopathol* **156**: 268–273.

Zettler FW, Ko NJ, Wisler GC *et al.* (1990) Viruses of orchids and their control. *Plant Dis* **74**: 621–625.

Index